THE ENCYCLOPEDIA OF

S·P·A·C·E

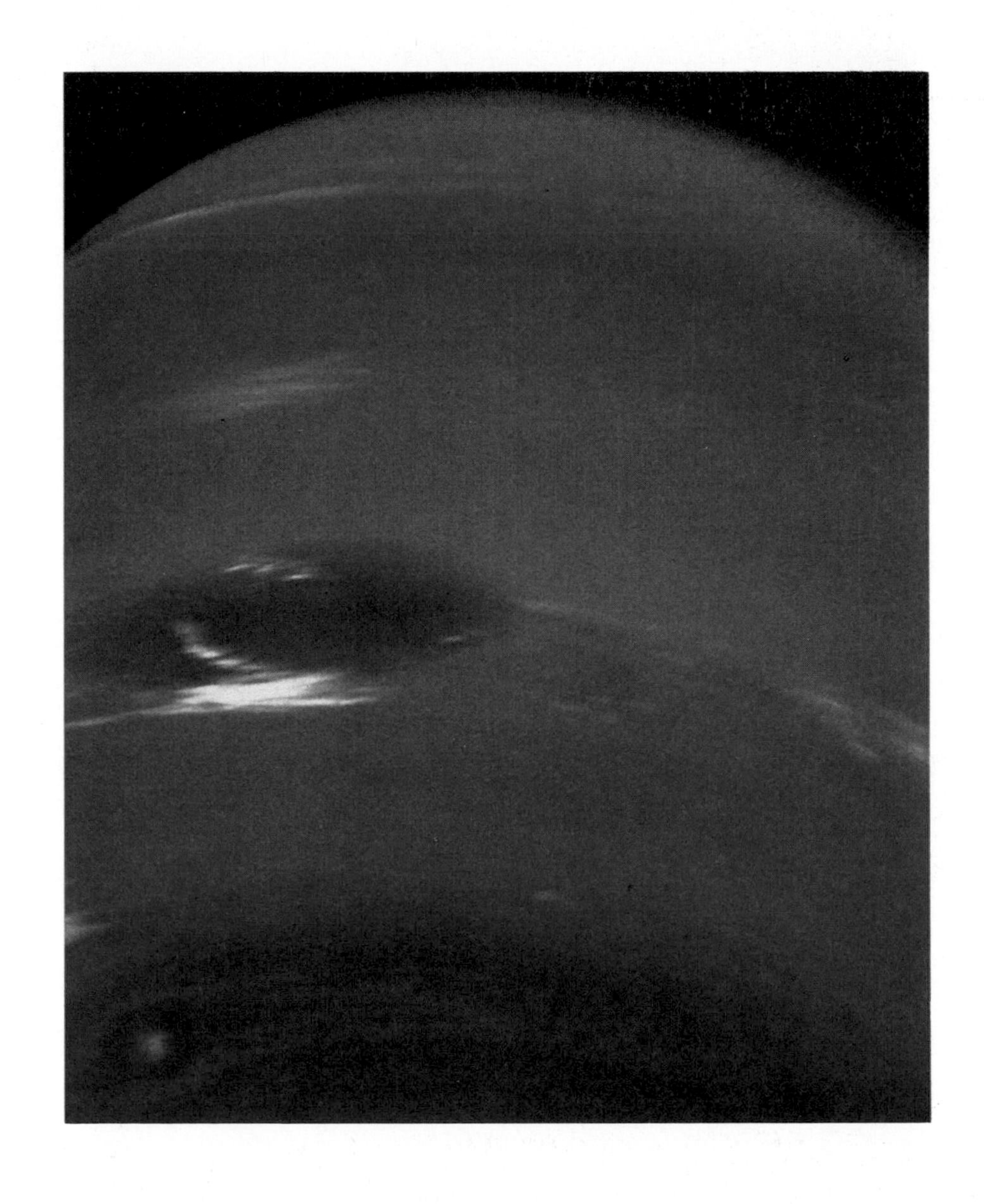

THE ENCYCLOPEDIA OF

S·P·A·C·E

NICHOLAS BOOTH

MALLARD
PRESS

DEDICATION

For Anna
with thanks and love for laughs
and good times (1984–88)

Author's acknowledgments
Though this book bears my name, there are a number of others that out of deep gratitude I would like to thank:

Shirley Campard of Aerospatiale; Daniel Metzle of CNES, Paris; Beatrice Lacoste of ESA Headquarters; Heidi Graf of ESA, ESTEC; Jurrie van der Woude of JPL, Pasadena; Kelly Humphries, Mike Gentry, Lisa Vasquez and Barbara Mason of NASA Johnson Space Center, Houston; Emery Wilson, Hughes Aircraft, California; Rosa Woods and Blanche La Guma of Novosti; Alla Levchenko of TASS; Neil Pattie of British Aerospace; Michelle Smith of the Ministry of Defence.

I would also like to thank Frank Miles and Peter Beer for contributing so readily; Debbie Perego for help with typing – particularly the tables; Angelique Vernet for being around, translating and keeping me sane; Andy Bartyram and Andi Spicer for advice for a particularly recalcitrant word processor; Marianna Klar, Alev Hussein, Meera Chawla and Petra Kaufmann for offering advice which was badly needed.

Photographic acknowledgments
All photographs supplied by NASA with the exception of the following:
Aeronautic Ford: 170; Aerospatiale: 94, 110 bottom; Nicholas Booth: 31, 50, 116, 149, 156; European Space Agency: 38, 41, 53, 55 top, 59, 60, 63, 82, 85, 86, 88, 91, 131, 138, 139, 166, 169, 174, 180, 182, 187, 188; General Electric: 98; Hale Observatories: 81; Marconi Space Systems: 75 bottom; North American Rockwell: 97; Northrop: 124; Novosti Press Agency: 9, 14, 15, 16, 17, 19, 20, 21, 24, 25, 45, 46, 62 top and bottom, 64, 65, 66, 67, 68, 69, 70, 110 top, 111, 112, 120, 121, 132, 133, 134, 135, 141, 142, 147 top, 158 top and bottom, 168 bottom, 173, 175, 190, 191, 196; Perkin Elmer: 184; Picture Report/ESA: 55 bottom; RCA: 77; Rutherford Appleton Laboratory: 83, 84; Science Photo Library/NASA: 51, 52, 80, 194; Tass: 71; Thiokol Corporation: 130; UK Ministry of Defence: 75 top; US Department of Defense: 72, 78

Front cover: The Apollo 9 Lunar Module above the Earth (NASA)
Back cover: The launch of the Space Shuttle Atlantis, 4 May 1989, for mission STS-30 (NASA)
Half title page: One of the many remarkable images of Neptune sent back by Voyager 2 (NASA)
Title page: A human satellite, the Manned Maneuvering Unit (NASA)

First published in the United States of America in 1990 by the Mallard Press
An imprint of BDD Promotional Book Company, Inc.
666 Fifth Avenue
New York, N.Y. 10103

ISBN 0-792-45072-8

Printed in Spain by Gráficas Estella, S.A. Navarra.

CONTENTS

INTRODUCTION: VOYAGES OF DISCOVERY

The date is 24 August 1989. Nearly 5 billion km (3.1 billion miles) from the Sun an unmanned robot is making its closest approach to the planet Neptune. Known as Voyager 2, it was launched from the Earth in 1977 and has flown by all the outer planets (except Pluto) in rapid succession. Along with its identical twin, both Voyager spacecraft have revolutionized our perception of the outer Solar System. The giant planets, their moons and their rings are far stranger than anyone had ever imagined. Neptune is the latest world to come under scrutiny from Voyager 2 – its last destination within the Solar System. The spacecraft will continue on its journey out of the Solar System towards the stars, returning information until it passes out of range early in the next century.

In many ways, the successful completion of Voyager 2's planetary exploration marks the end of the first wave of spacecraft sent to investigate the Solar System. All the planets known to the ancients have been investigated from close range: probes have landed on the surface of Mars and of Venus. Even the dusty heart of Halley's Comet has not escaped direct scrutiny. And, of course, a dozen human beings have walked on the surface of the Moon.

Voyager 2's Neptune encounter has also seen the beginning of the next era of space exploration. Already in 1989, a Soviet spacecraft called Phobos 2 has investigated Mars and its largest moon from close range, sadly with limited success. NASA's Magellan spacecraft has been despatched to Venus, the first planet-bound probe to be launched from the Space Shuttle. Within a few weeks, the Galileo orbiter will be heading towards Jupiter, and early in 1990 the Hubble Space Telescope will finally be deployed above the Earth. Above the turbulence and haze of our atmosphere, Hubble promises to bring about a revolution in astronomy. So Voyager 2 and 1989 mark a convenient break point between the discoveries to date and the promise of further revelations in the future. What better time to review the last three decades of space exploration?

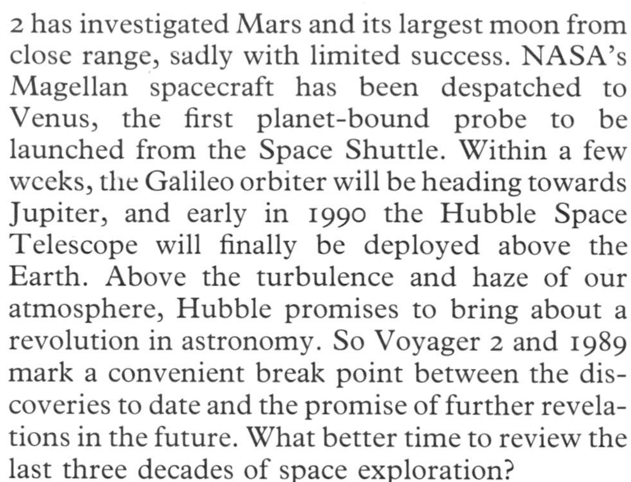

With this encyclopedia, I have tried to do exactly that. It is by no means an easy task, considering that over 3,000 satellites have been launched since Sputnik 1 ushered in the Space Age in October 1957. As far as possible, I have attempted to detail the main achievements in space and tried to look to the future. The first section of the book consists of chapters devoted to major developments, such as the race to the Moon in the Sixties and the events that led to the untimely demise of the Space Shuttle *Challenger*. Each chapter looks at how those developments occurred as well as telling the story behind them. The final contributions to this section look at how space may develop in the near future. Predicting the future is probably more difficult than summarizing the past, but it is clear that the need to reduce launch costs and the construction of space stations will be of paramount importance.

The second half of the book, comprising the encyclopedia proper, looks at the more technical details of each spacecraft and each mission. Trying to decide which spacecraft to include and which to leave out was another exercise fraught with danger. I hope that the final choice is largely representative of the more important spacecraft which have played, and will continue to play, an important role in the development of space.

Nicholas Booth
May 1989

Below: On 24 August 1989, the first phase of Man's exploration of the Solar System came to an end. The Voyager 2 spacecraft, portrayed here looking back at Neptune and its largest moon Triton, had successfully ended its mission to the outer planets. Only Pluto remains unseen by a direct encounter with unmanned spacecraft.

Opposite: Voyager 2's odyssey began in August 1977 with its launch from Cape Canaveral. Within ten hours, the powerful Titan-Centaur launcher had boosted the probe beyond the orbit of the Moon. Voyager 2 brings the tally of probes which have left the Solar System to a total of four. NASA hopes the Voyagers can be tracked until at least 2010.

DEVELOPMENTS IN SPACE

THE CRADLE OF HUMANITY: THE PIONEERS

For centuries, perhaps since our earliest ancestors walked on the planet, human beings have dreamed of travelling into space. The romantic lure of leaving the Earth was as strong to the ancients as it was to the pioneers whose experiments in interplanetary travel turned the dreams into reality earlier this century. One of the earliest writings on the subject comes from 2nd century B.C. Greece, where Lucian of Samoa described a journey to the Moon in a tale with the curious title of 'True History'. He imagined a whirlwind whisking a ship out of the oceans and transporting it to the Moon, where the sailors encountered a race called the Hippogypi who travelled across the lunar surface on three-headed vultures. Lucian's powerful imagination later led him to describe a journey travelling to the stars via a fleet of swans.

In more recent times, the dreamers continued to wax lyrical. By far the most fanciful was the idea of the equally fanciful Cyrano de Bergerac in 1649 that an intrepid explorer attach himself to flasks of dew. The theory was, of course, that when the Sun rose, he would head towards the heavens. Mention should also be made of Hans Pfall, a creation of Edgar Allan Poe, who used a balloon to escape his earthly debts and become the first interplanetary tax exile by living on the Moon.

ROCKETS

The scientific reality of space travel drew closer when, as legend has it, an apple landed on Isaac Newton's lap after falling from a tree. In that same instance, Newton realized that the Moon remained in orbit because of the gravitational influence of the Earth, and that theoretically speaking, every object should attract every other in the Universe. In 1686–7, at the bidding of his friend Edmond Halley (of Comet fame), Newton published his results in a book named *Principia Mathematica*, perhaps the most influential scientific work of all time. Newton presented his life's studies on motion, including the three laws for which every student of physics remembers him. The third was the most important for possible travel in space: 'Every action has an equal and opposite reaction' – the law on which rocket power is based.

Rockets have been around for centuries and in the past have been primarily used for military purposes. Their invention is usually credited to the ancient Chinese, who also invented the gunpowder which powered them. It is known that the Mongols used rockets against the Arabs in the 13th century in the siege of Baghdad. And in the 18th century, the Indians used them against the British at the battles of Seringapatam, where they were so effective that Admiral Congreve copied them and used them to besiege Boulogne in 1807. The phrase 'the rocket's red glare' in *The Star Spangled Banner* comes from the use of Congreve rockets in the War of Independence. Despite their success, improvements in artillery during the 19th century led to the rocket's military demise.

TRAVEL TO THE STARS

In terms of leaving the Earth, it was soon realized that solid rockets would not provide enough thrust to travel into space. This fact did not stop Jules Verne, in his classic 1865 work *From the Earth To The Moon*, from launching three men and two dogs to the Moon by means of a vast cannon. The fact that the forces on take-off would squash the crew and that air resistance would burn the vehicle escaped him. But such unpleasant consequences did not escape the mathematical genius of Konstantin Tsiolkovskii, a scholarly Russian schoolmaster, acknowledged as the father of astronautics.

Born in 1857, Tsiolkovskii was a bright child and became an accomplished scholar. Aged ten years, he suffered scarlet fever and nearly lost his hearing. This made him a voracious reader, and, largely self-taught, he became a teacher in the Moscow suburb of Kaluga. It was during the long summer holidays that he devoted his thoughts to space travel, publishing a work called *Free Space* in 1883. In *Free Space* he realized that Newton's third law was the reason why rockets work: as hot gases escape out of the back of the rocket, the energy they provide push against the body of the rocket forcing it to move forwards. In 1898, he showed that a rocket's final velocity would depend on the velociti of the exhaust gases.

In 1903, the same year that the Wright brothers first took to the skies, Tsiolkovskii produced his most influential work *Exploring Space With Devices*. It was a simple calculation that to enter orbit around the Earth, a spacecraft would have to attain a velocity of 7 km/sec (4.4 miles/sec). The only way he could foresee such power being provided was by the use of liquid fuels, such as hydrogen and oxygen (which were later used in the Saturn V booster which sent men to the Moon). Tsiolkovskii also realized the advantages of liquid propellants: they provided more power, could be controlled better and could be re-ignited in the vacuum of space. He later suggested that a multi-stage rocket (or 'rocket train') would be the

Konstantin Tsiolkovskii, the acknowledged father of astronautics, seen in typical paternal pose with his grandson Alexei in 1930. By the time of his death five years later, Soviet rocketry was well advanced, thanks to his remarkable scientific legacy.

Opposite: The Moon became the target for the first part of the Space Age. This view taken by the Apollo 11 crew perhaps best symbolizes mankind's awareness of the uniqueness and frailty of our own world.

only way in which a rocket could efficiently reach orbit. As a rocket heads upwards, it uses more and more fuel: by having the fuel tanks in different stages, as they became empty they could be discarded. The booster would therefore not have to carry any 'dead weight' to hinder its performance.

Tsiolkovskii's other achievements include discussions of weightlessness, artificial satellites, solar panels and space stations as well as travel to the stars by nuclear rocket. In his later life, the Revolutionary Government which ousted the Czars hailed him as a genius. His death in 1935 prevented him from seeing his most-quoted prophecy come true: 'The Earth is the cradle of the mind, but one cannot live in a cradle forever'.

'A SEVERE STRAIN ON CREDULITY'

In the 1920s, interest in rocketry mushroomed: when Moscow organized an exhibition on 'interplanetary machines' in 1927, exhibits came from Germany, the U.S.A. and Britain among other countries. A new generation of engineers, following on from Tsiolkovskii's ideas, came to the fore, describing space stations and space shuttles while they continued to work on the rudiments of rocketry.

The world's first liquid-fuelled rocket was launched by Robert Goddard in Massachusetts in 1926. A shy, secretive man, adverse publicity led to his being nicknamed 'Moony'.

Whereas Tsiolkovskii gave Russian rocketry the veneer of respectability, his Western counterparts were regarded with ill-concealed suspicion by both scientists and the general public alike. No better example is shown than the case of Robert Goddard, a professor of physics in Worcester, Massachusetts. He became fascinated by space travel and patented over 200 ideas concerning propulsion. In 1919, in an attempt to secure funds to put his ideas into practice, Goddard submitted a paper to the prestigious Smithsonian Institution entitled *A Method of Reaching Extreme Altitudes*. In it he described the basics of rocketry and liquid-fuelled propulsion, and at the very end included a section on how a rocket could eventually be fired to the Moon with gunpowder to create a flash on impact.

This last speculation excited Goddard's local newspaper *The Boston Herald* so much that on 12 January 1920 it breathlessly declared: 'New Rocket Devised by Prof. Goddard May Hit Face of the Moon'. The publicity produced many offers from locals to travel on his as yet unbuilt rocket. It also made him appear as something of a crank: the upmarket *New York Times* picked up the story and huffily dismissed it as 'a severe strain on credulity'.

Such criticism was almost inevitable, but nevertheless was extremely wounding to Goddard who shied away from publicity as a result. But it did lead to some funding and on 16 March 1926 he launched the world's first liquid-propelled rocket. Fuelled by liquid oxygen and gasoline, Goddard's prototype reached the giddy height of 56 m (184 ft) in $2\frac{1}{2}$ seconds. Four more test flights took place, and in 1929 his work came to the attention of the aviator Lindbergh, whose interest secured funds from the Guggenheim Foundation. Goddard moved to a ranch in New Mexico, where he built ever more complex vehicles, some of which reached the height of 2 km (1.25 miles). The Second World War interrupted his work, and he died in 1945. Despite the earlier ridicule, his pioneering work was recognized by the U.S. government which belatedly granted his widow and the Guggenheim Foundation $1 million in 1960.

GERMANY BEFORE THE WAR

During the late 1920s, a group of German rocket enthusiasts were regularly meeting in a Berlin alehouse to discuss their ideas and how best to put them into practice. They soon named themselves the Society For Space Travel, known by its initials, VfR. Their number included Hermann Oberth, an Austrian by birth, who in 1923 had produced a booklet on space travel entitled *The*

Wernher von Braun, seen at the time of the first Apollo Moonlanding, shows the structure of one of his most advanced rocket engines. His work on the mammoth Saturn V booster gave NASA the lead in the race to the Moon. His untimely death in 1977 left his greatest wish unfulfilled – that of journeying into space himself.

Rocket Into Interplanetary Space. He had achieved greater notoriety in 1928, when he was consultant to Fritz Lang for the film *Frau Im Mond* ('Woman In The Moon'), for which a rocket had been built for publicity purposes. A working prototype had followed which unfortunately exploded on the launchpad. In 1930 they were joined by an 18-year-old engineering student called Wernher von Braun, whose expertise helped put together the first primitive liquid-fuelled rockets. Von Braun was something of an *enfant terrible*: at Berlin university he had experimented with centrifuges in which mice were subjected to the forces that would be encountered on lift-off. On one unfortunate occasion the centrifuge door had flung open, splattering mousey remnants over his lodging's wall, and he was summarily turned out onto the street by his landlady.

The VfR's experiments quickly came to the attention of the German Army, forbidden under the terms of the Armistice after the First World War to build vast arsenals of weaponry. Nothing was mentioned about rockets in the agreement, thus giving the Army the impetus to build them. In May 1931, the VfR launched a liquid-fuelled rocket which reached a height of 60 m (200 ft). The following year, a demonstration of another primitive rocket was arranged at a munitions dump in Kummersdorf, near Berlin. Under the watchful eye of Captain Walter Dornberger, who was in charge of solid-fuel rocket development in the Army, von Braun and his VfR colleagues launched a test payload which crashed before its parachute opened.

Nevertheless, this inauspicious start was more than enough to convince Dornberger that a liquid-fuelled rocket could become a viable weapon of war. Von Braun was soon employed by the Army under Dornberger's supervision to develop such rockets, and later his VfR colleagues were drafted in, where they could experiment to their hearts' content on liquid-fuelled rocketry. By 1937 they had moved to Peenemunde on the Baltic Coast, and began work on a 14 m (46 ft) high rocket fuelled by alcohol and liquid oxygen. Known as the A4, it was successfully fired on 3 October 1942 when it reached a height of 85 km (53 miles), splashing down 190 km (120 miles) downrange. Dornberger turned to von Braun and said, 'Today the spaceship is born!' It was no exaggeration.

Before the rocket could be introduced as a weapon of war, the irrational whim of Adolf Hitler was soon felt by the Peenemunde group. Though he had enthusiastically backed the rockets to begin with, Hitler dreamed early in 1943 that they wouldn't work. Not for the last time would rocketry fall to the whim of a politician, and funding was temporarily withdrawn. Von Braun was even arrested by the Gestapo for the crime of wanting to use the rockets to explore space rather than for war. The situation became so bad that at one point the Peenemunde group had to fill in forms to ask for vital supplies. Even pencils had to be requisitioned, using the description: 'narrow-gauge hexagonal dowlings with lead interiors'.

By 1944, the tide of the war was changing so much that Hitler cast aside his nightmare and ordered the A4 into production as the world's first strategic missile. Renamed the V2, or 'Revenge Weapon', it was launched by mobile batteries which gave a greater flexibility. Records show that from September 1944 onwards, no less than 4,320 V2s were launched as retaliation against the Allied bombing of German cities. Though some of the first V2s were fired against Paris and Antwerp, most were aimed towards London. No less than 1,115 were fired towards this prime target, and Civilian Defence records show that over 2,500 Londoners were killed as a result, with nearly 6,000 suffering serious injuries. Though von Braun's work was always motivated by 'aiming for the stars', there were many who suffered because his creation was sometimes used to aim a bit lower.

THE PAPERCLIP AFFAIR

For von Braun and the others, the closing in of Allied forces and the imminent demise of Hitler's Thousand Year Reich was more of a pressing concern. On 2 May 1945 most of the technicians headed westwards in a convoy of vehicles to surrender to the U.S. Seventh Army. Capture by the Americans seemed vastly preferable to that by the Russians, particularly after they had destroyed most of the facilities at Peenemunde to avoid it coming into Soviet hands. When the Russian army reached the base a few days later, it found a handful of technicians and very little else of use. Among the surviving documents were plans for a two stage rocket based on the V2, known as the A9/10 which would have enabled warheads to be fired on New York had it been built.

In the chaos of post-war Germany, yet another peculiar episode in the development of space travel occurred. Though the Allies were supposed to be sharing all the information from the German experiments, U.S. Army Intelligence wanted to ensure that the most important technicians went to the States. When personnel files of the Peenemunde pioneers were examined by the Americans, paperclips were placed on those of the more desirable scientists. By September 1945, the first group of Germans (including Dornberger and von Braun) were spirited away to New Mexico. Dornberger had also informed his captors of the whereabouts of much of the remaining V2 spares that had been manufactured for the German war effort. Around a hundred or so complete V2s were shipped to New Mexico and fired by von Braun under the supervision of the U.S. Army. When the Russians asked for their share, tractor parts were instead crated up and sent across to Moscow.

Rocketry blossomed in the United States in the post-war period when the developing Cold War saw the necessity for improved missiles which could conceivably carry nuclear warheads to hit targets in the Soviet Union. Starting in early 1946, 64 V2s were launched with a variety of scientific instruments from the White Sands Proving Ground in New Mexico. An uprated version, using a WAC-Corporal upper stage, built by the Jet Propulsion Laboratory in Pasadena, California, was launched to a record height of 400 km (250 miles) in February 1949. The following July, another WAC Corporal was launched from a sand spit in Florida known as Cape Canaveral.

By the start of the 1950s, the three armed services had decided to build their own missile launchers. Von Braun's Army Group moved from New Mexico to Huntsville, Alabama, where the Army Ballistic Missile Agency (ABMA) was formed. Its first task was to build a rocket with a range of 800 km (500 miles) known as an Intermediate-Range Ballistic Missile (IRBM). This first missile was christened the Redstone, and a more powerful version called the Jupiter was later proposed.

There were so many ex-Peenemunde personnel within the ABMA that its location soon became known as 'Hunsville' in Army parlance. Army-funded research into rocketry also continued at the Jet Propulsion Laboratory in Pasadena. The USAF developed the Thor IRBM with a range of 2,400 km (1,500 miles) before its Atlas Inter-Continental Ballistic Missile (ICBM) – capable of reaching 8,000 km (5,000 miles) – would come

Opposite: After moving to the United States by the sleight of hand of Army Intelligence, V-2 rockets were modified to carry an additional stage and investigate the upper atmosphere. Tests were conducted from a little-known sandspit in Florida called Cape Canaveral.

into service in the mid-1950s. The Navy had adopted the technology that went into the V2, and advanced it with a pivoting mount for the rocket engine, powered by liquid oxygen and kerosene, and known as the Viking launcher.

EARTH SATELLITES

In July 1955, the world's scientific community announced the International Geophysical Year (IGY), which oddly enough was to last from 1 July 1957 to 31 December 1958. Its aim was to study the nature of the Earth's physical characteristics, including the upper atmosphere which had recently been studied by a succession of high-altitude rockets (such as the Navy's Viking) known as sounding rockets. Both the Soviet Union and the United States announced that they would launch satellites to orbit the Earth for the purposes of scientific investigation.

The general public was taken aback by such announcements. In the mid-1950s, space travel still seemed to be in the realm of pure fantasy. Though von Braun had authored an article in 1952 in *Colliers* magazine putting forward the notion of a wheel-shaped space station, and Arthur C. Clarke had proposed artificial satellites for the purposes of telecommunications in 1945, they still seemed beyond the realm of possibility. The scorn heaped on Goddard in the 1920s was mild by comparison. The *Daily Mirror* declared in an editorial that: 'Our candid opinion is that all talk of going to the Moon, and all talk of signals from the Moon, is sheer balderdash – in fact, just moonshine.' Such scepticism was not restricted to hardened, cynical journalists. Forest Moulton, a distinguished American astronomer declared that 'there is not the slightest possibility of such a journey' on being asked about travelling to the Moon. As late as 1956, the then Astronomer Royal, Richard Woolley, suggested space travel was 'utter bilge' on being questioned by reporters, a remark later put down to jet lag after a journey from Australia.

But with the potential reality of space travel becoming more apparent, a number of agencies had investigated the possibilities. Strange as it may seem, there seemed very little use for artificial satellites other than to reflect national pride. The U.S. Navy had been first off the mark with studies of an artificial satellite as early as November 1945 for its Bureau of Aeronautics. The $8 million price tag for the project seemed a trifle too much, so the newly-created aerospace think-tank based in California known as RAND (Research And Development) was asked to look into it as well. They agreed that such a project was viable and published a report in 1946 entitled 'Preliminary Design For An Experimental World-Circling Spaceship' in which they envisaged a scientific satellite launched by a four stage rocket. Such projects remained little more than curiosities in official files: the business of building missiles was of greater concern.

It was von Braun's group at Huntsville and the interest of a U.S. Navy's Aeronautics Branch chief engineer called Alexander Satin, who came up with the first definite plans for a small scientific satellite. Starting with a series of meetings in 1954, they evolved 'Project Orbiter' which they planned would be launched by the Redstone. By this time, another group at the Naval Research Laboratory was planning for its own design of satellite. Known as Vanguard, it would use the Viking as a first stage, a second stage based on an Aerobee rocket (which had successfully been used for high-altitude sounding flights) and a third stage developed from scratch.

There was intense discussion within the Pentagon as each of the three services felt it was their divine right to launch the first satellite. Inter-service squabbling over whose modified ICBM would be chosen was halted when President Eisenhower declared that the U.S. Navy's Vanguard would be America's scientific contribution

By the winter of 1933/4, scientists from the GIRD laboratory in Moscow had developed advanced missile prototypes. Seen at the extreme left of the group, with obligatory cap, is Sergei Korolev who ensured that Russia entered the Space Age with a number of impressive firsts.

to the IGY. It was a decision that was to be bitterly regretted, because the honour of launching the world's first artificial satellite was to fall to the Russians and to a man who, officially at any rate, did not exist.

SERGEI KOROLEV

Thanks to the legacy of Tsiolkovskii, Russian rocketry had got off to a good start. In 1927 the Gas Dynamics Laboratory was set up in Leningrad which later succeeded in building a number of powerful liquid-fuelled engines. Five years later, another organization known as GIRD came into being in Moscow (its Russian initials meant 'Group For The Study Of Reaction Propulsion') which later launched the first Soviet liquid-fuelled rocket. In 1933, they were combined under the direction of the Red Army and Marshal Tukhachevsky, who directed them to build winged rockets and rocket powered gliders.

The fledgling Russian effort was led by a young engineer, Ukrainian by birth, who had started out his professional career as an aircraft designer. But after meeting Tsiolkovskii in the late 1920s, Sergei Pavlovich Korolev turned his considerable talents to rocketry. Any number of ambitious projects were started by him, including a rocket-powered glider which he test flew in the passenger seat. In 1934 he authored a book called *Rocket Flight In The Stratosphere* hinting at the direction in which he would have liked to take his research. Like von Braun in Germany, Korolev was happy to undertake military-funded research in the hope that he could one day fulfil his dream of space travel.

Korolev's work was progressing well when, once again, political interference came into play. But Josef Stalin's interference far exceeded anything that Hitler dreamed. After ruthlessly assuming power after the death of Lenin, his rulership of the Soviet Union had become even more authoritarian than the Tsars'. Like most dictators, he became very paranoid about conspiracies and treason against his rule. In June 1937, Marshal Tukhachevsky was arrested and executed by a firing squad as were nearly half the officers in the Red Army. It is estimated that 6 million Russians disappeared during this period of 'The Terror'.

The work that Marshal Tukhachevsky was overseeing immediately came under suspicion, as did those associated with the research. Accordingly, Korolev was sent to Siberia to work in the Kolyma goldmines, a place from which very few people returned. Though he only remained there for a year, his health suffered and he was never to fully recover from the hardships. In 1938 he was sent back to Moscow under house arrest where he designed military aircraft. During the Second World War, Korolev was transferred to Omsk, still under guard in case he tried to join the Nazis.

The flared appearance of Sergei Korolev's A-1 booster acted as a standard for Soviet rockets. His design, which clustered rocket engines together, ensured that sufficient thrust could be achieved to reach orbital velocity. Known as the 'Semyorka' ('old number seven'), it is seen displayed here at the Paris Air Show in 1968.

As a result of 'The Terror', Soviet rocketry's early leads were diminished, a fact that was brought home when Peenemunde was discovered. Sadly, the atmosphere of hostility and recrimination continued after the war. When Korolev saw a V2 launched from the Baltic under British guidance, he was not allowed inside the test site as an 'official' Soviet observer. Because his name had been removed from all files, he did not exist. He had to suffer the ignominy of watching the proceedings disguised as a captain from outside a perimeter wall.

As was happening in the United States, the Cold War gave greater urgency to missile building. The Soviets tested a number of captured V2s starting in October 1947 near an isolated part of the river Volga, fairly near to a town called Kapustin Yar (translated, meaning 'Cabbage Patch'). Under Korolev's direction, about a dozen were tested and a Soviet version called the Pobeda ('Victory') was built under his supervision. During the late 1940s it became both the prototype

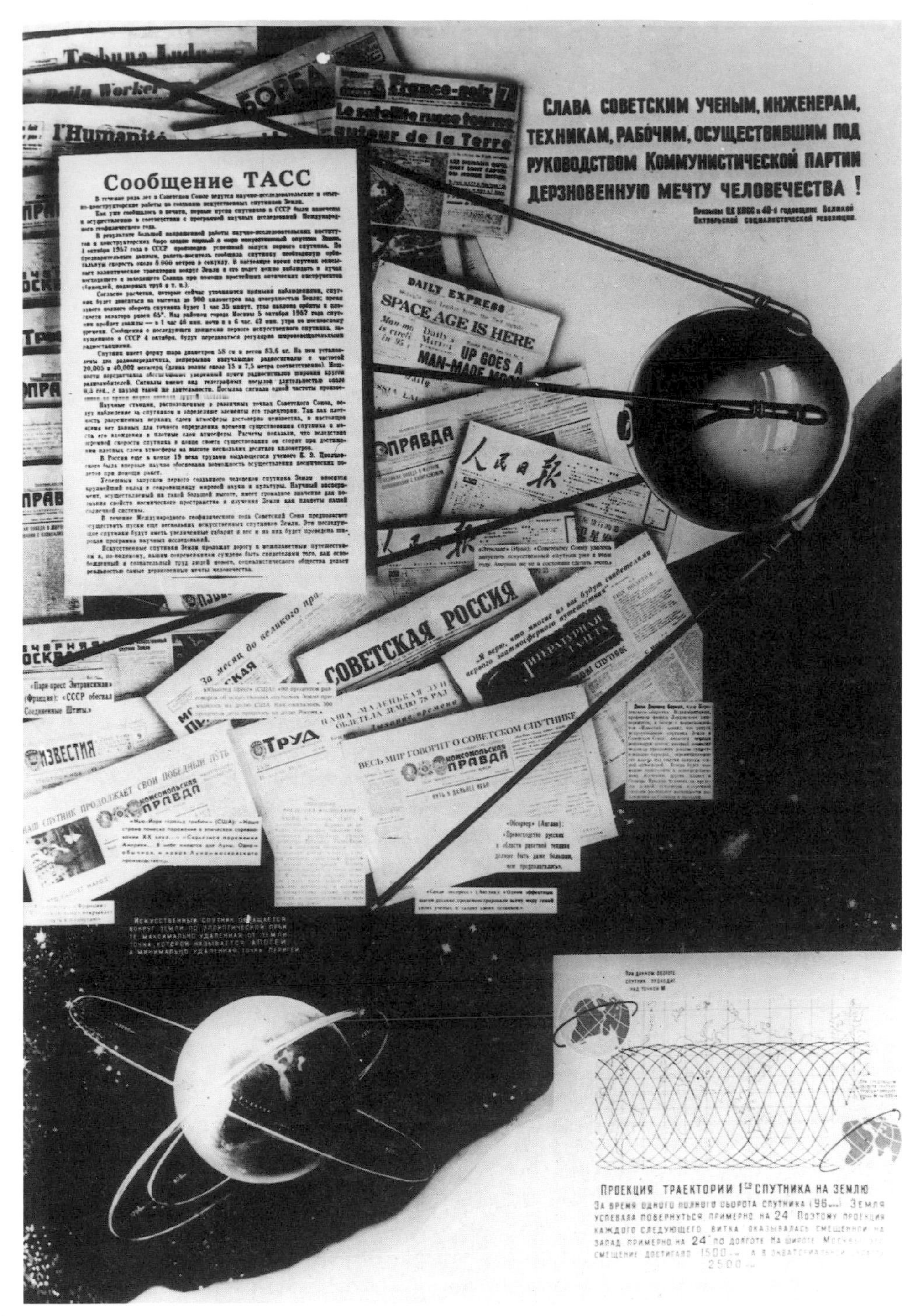

On 5 October 1957, the Western world woke up to the disturbing news that Russia had launched the world's first artificial satellite the night before. Seen here, against a backdrop of newspaper headlines, is the ball-shaped satellite. Its launch caused panic in the United States, even though one Senator vainly referred to it as a 'silly bauble'.

Soviet ICBM and was used for 'sounding' experiments. Captured German rocket scientists were consulted in their fields of expertise, but they did not contribute to the construction of Soviet rockets.

In the early 1950s, Korolev was given the go-ahead to build an ICBM capable of reaching the United States. Korolev's design genius came into its own. Designated the A-1, the missile was a giant 28 m (92 ft) in length, and nearly 3 m (10 ft) in diameter. Its central core housed the fuel tanks and four main engines which all burned simultaneously. Four strap-on boosters were added, each composed of four engines themselves. This clustering of engines ensured that the rocket could be mass produced quickly and would develop enough thrust to launch a nuclear warhead across the seas.

It was also powerful enough to be able to launch 1,600 kg (3,527 lb) into orbit. Throughout the early 1950s, a number of Soviet scientists were suggesting that a satellite launch would soon be possible. In 1955, the Soviet Academy of Sciences set up a committee to coordinate development of an earth satellite project. As a result of this, the day after Eisenhower announced Vanguard, they announced their intention to launch a Russian contender. Soviet scientists had a powerful ally in the form of Nikita Khruschev, a politician already coming to the fore in the aftermath of Stalin's death in 1953. In his memoirs, Khruschev recalled that his first view of a prototype A-1 in 1955 was rather like sheep seeing a gate for the first time.

In total secrecy, it was decided to develop a new purpose-built launch site in the flat desert land near the Aral Sea. In May 1955 work began on the new 'Cosmodrome', consisting of a variety of launchpads and facilities, which was not completed until March 1957. The site was near the small town of Tyuratam, conveniently located by a major railway connecting Moscow to Tashkent. The Soviets always referred to the site as the Baikonur Cosmodrome in an effort to confuse the West, Baikonur being a railhead hundreds of miles to the North. After a number of tests in mid-1957, Korolev's missile successfully flew over 6,000 kilometres (3,750 miles) on 3 August, after which it was ready for a far greater task.

FELLOW TRAVELLER

Wintry darkness had already set in at the Baikonur Cosmodrome on Friday 4 October 1957 when the Space Age dawned. A fully-fuelled A-1 booster, known by Korolev and his team as the 'Semyorka' or 'Number 7', stood motionless, cradled by giant gantry arms on the launchpad. The decision to launch the world's first satellite had been made months before by the Politburo but now it was finally about to happen. The satellite was essentially a simple radio transmitter contained inside a polished metal sphere with four whip aerials trailing behind it. Though called Preliminary Satellite in all the design papers, it was soon christened with the more poetic name of Sputnik, meaning 'Fellow Traveller'.

A series of frustrating delays were caused by one technical problem after another and the clock in the underground bunker containing Korolev and his team was rapidly reaching midnight. Just before Korolev gave the go-ahead for the final countdown, a lone bugler stood on the freezing launch pad to salute the task that was about to be attempted. Within minutes, the gantries of the launch pad fell back and with a mighty roar the 20 separate engines of the rocket fired into life. A vast flame lit up the surrounding darkness as the world's first satellite headed into orbit. An agonizing $1\frac{1}{2}$ hours passed before the satellite passed overhead signalling that it was successfully in orbit. The unmistakable 'beep-beep-beep' of Sputnik's radio transmission could be heard, evidence that Korolev and his team had fulfilled their life's ambition.

The radio transmissions reverberated around the world. Khruschev and the Russian people hailed it as a great technical achievement showing the superiority of the Soviet system. No finer testimony to Korolev's guiding genius is that the basic design of the Semyorka is still in use today. His idea of clustering the rocket engines has been vindicated in the rockets that have been built since then. Sadly, Korolev did not bask in any of the glory associated with making history, as Khruschev was far too canny to allow anyone other than himself to take credit for the Sputniks. Korolev's identity was hidden under the title of the mysterious 'Chief Designer' which would be invoked whenever a Soviet success in space took place. Details about Korolev and his life were not released until after his death in 1966.

In the United States, Sputnik resulted in feelings that bordered on mass hysteria. The satellite's launch signalled one clear message to the American people, already in the grip of anti-Soviet paranoia. Sputnik 1 could easily have been a nuclear warhead, underlined by the fact that it weighed nearly ten times more than the proposed Vanguard satellite. American ICBMs would not be ready to be deployed tactically until the early 1960s, and the launch of Vanguard was still two months away. People began to talk openly of the 'missile gap' and President Eisenhower's political opponents began to criticize the complacency of his administration. When nuclear physicist Edward Teller referred to Sputnik as a greater defeat than Pearl Harbor he was not exaggerating the mood of the country.

'KAPUTNIK!'

Yet further humiliation at the hands of the Communist menace was to take place. Another Sputnik was launched on 3 November, a staggering five times heavier than the first, and it carried a dog as its passenger. The first animal to orbit the Earth was a mongrel bitch called Laika who had been used in earlier rocket tests. Contained in a pressurized cabin, she seemed quite content and happy as the first living creature to venture beyond the atmosphere. She showed no signs of ill-health before passing away a week later after her oxygen supplies had run out. As if on cue, almost immediately a delegation of dog lovers gathered outside the Soviet Embassy in London to protest.

America's entry into the space age was scheduled to take place a month later. December 6th dawned and the historic moment was televised live across the United States. A tall, thin vanguard rocket was seen to rise a few feet off the ground and explode in a ferocious fireball. The radio transmitter inside the satellite continued to beep minutes afterwards, until a technician was able to put it out of its misery. The next day newspaper headlines like 'Kaputnik!' and 'Flopnik!' were

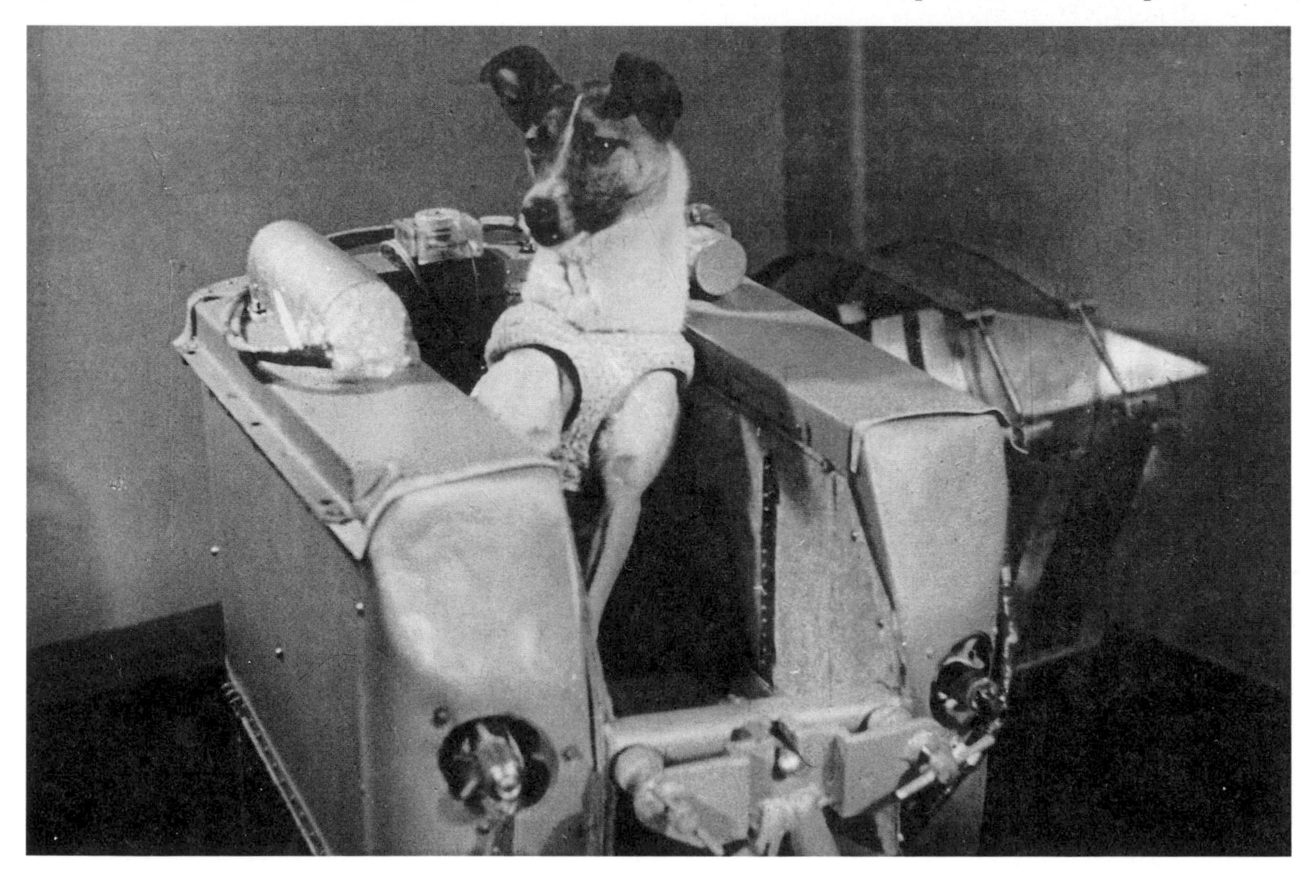

From Sputnik to 'Muttnik'! The first living creature to fly aloft was Laika, a mongrel, seen here in the canister in which she orbited in Sputnik 2. Her flight proved beyond doubt that humans could follow in her footsteps.

The 'Mercury Seven' – seven of the finest American test pilots deemed to have the 'right stuff' for orbital flight. They are, from left to right: (top) Shepard, Grissom and Cooper; (bottom) Schirra, Slayton, Glenn and Carpenter. Only Shepard landed on the Moon, the result of his self-assignment as Chief Astronaut!

typical. To describe the U.S. Navy as embarrassed and Khruschev as amused would be, to say the least, understatements.

Sensing that enough was enough, the Department of Defense gave the go-ahead to von Braun's team in Huntsville to assemble a booster and get it right. They were given 90 days to save America's national prestige, and quickly resurrected the idea of Project Orbiter. Scientists at the Jet Propulsion Laboratory in California hastily assembled a satellite called Explorer and a fourth rocket stage later called Juno to kick it into orbit. Von Braun's Jupiter, Redstone's big brother, was prepared for launch and moved to Cape Canaveral. On 31 January 1958, Explorer 1 successfully reached orbit and almost immediately returned information to show that the Earth was surrounded by radiation belts.

THE BIRTH OF NASA

The squabbling among the armed services was blamed for the fiasco of the delays and ignominious failures. It was something that came home to President Eisenhower, who already was beginning to tire of the power of the 'military-industrial' complex. He decided that America's space programme should be entrusted to civilians. Military space systems such as reconnaissance satellites would be left to the Department of Defense. After lengthy discussions, Eisenhower and Senate Majority leader Lyndon Johnson agreed to propose a piece of legislation numbered S3609 and entitled 'The National Aeronautics and Space Act of 1958'. It was passed very quickly and became law on 1 October 1958. Employees of an agency called the National Advisory Committee on Aeronautics which had conducted federal aviation research for decades soon found their pay cheques bore the name of the National Aeronautics and Space Administration or NASA.

Under Keith Glennan, who had been 'poached' from the Case Institute of Technology, the new agency's task was to bring together the disparate elements of America's space effort under one administrative umbrella. Though there were still problems with duplication of effort and arguments with the military over policy decisions, NASA soon amassed an impressive scientific and engineering expertise. Within a week, Glennan announced that NASA's first major task was to launch human beings into space as part of a project called Mercury. Over 500 eager test pilots, the cream of America's armed services, were invited to apply to become astronauts, and within a few months the number was whittled down to seven. Feted and idolized wherever they went, the 'Mercury Seven' soon became synonymous with the aspirations of the agency which would launch them into space.

Under Sergei Korolev, the Soviet space programme headed towards the same goal of placing a man in orbit. Along the way, they scored some further astonishing firsts. In January 1959, an unmanned probe called Lunik 1 attempted to reach the Moon, and although it missed the Moon, it became the first man-made object to enter orbit around the Sun. On being questioned about the achievement by a reporter, one Pentagon official was so disgruntled that he declared that if he threw a baseball in the air that too would be in orbit around the Sun. In September Lunik 2 hit the Moon carrying a pennant of the Hammer and Sickle to stake its prize. And on the second anniversary of Sputnik 1, Lunik 3 passed the far side of the Moon and returned the first pictures of the hemisphere of our neighbour which cannot be seen by astronomers. The naming of features with titles like Mare Moscovrae (The Sea of Moscow) left the world in no doubt whose achievement it was.

It was the Moon which was to become the focus for the next chapter in space exploration – one that even the most prescient would have been hard pushed to predict.

THIS NEW OCEAN: THE MOON RACE

Perhaps generations hence, historians will record the 1960s as the highwater mark of the 20th century. It was a time of unparalleled social and cultural upheaval marked by the Beatles, 'Love-Ins' and the mini-skirt, yet tempered by a dark underbelly of student unrest, war in Vietnam and famine in Africa. But in the future those events may pale into insignificance when compared with the fact that in that turbulent decade's closing months, human beings walked on the surface of another planet for the very first time.

THE LEADING PLAYERS

How the superpowers came to race to the Moon may be understood in terms of the political Cold War between them. Throughout the 1950s, relations between the U.S. and U.S.S.R. worsened, reaching their nadir in the early years of the 1960s. It was an ideological struggle between capitalism and Communism personified by its two leading players, Nikita Khruschev and John F. Kennedy. Khruschev, the canny son of a peasant farmer, characterized the Western perception of the Russian menace. He was a bluff, bear-like man, often given to hilarity but also notoriously short-tempered, venting uncontrollable bursts of rage with mock theatricals like banging his shoe against the bargaining table. Khruschev realized very quickly the power of space spectaculars, allowing much of the glory and kudos to reflect on himself. Kennedy, the bright new congressman from an illustrious Irish dynasty, represented the post-war aspirations of the American people. In all things, Kennedy was determined to show American pre-eminence; space was no exception. Much of the developments in space in the 1960s and beyond can be understood in terms of the ideas that the two leaders set into motion.

1960 was an election year in the United States. Kennedy was campaigning on a platform that the Soviet lead in space was putting the United States on the wrong side of a 'missile gap'. In October, when Khruschev visited New York and the United Nations, he brought with him models of a Mars probe which his technicians under the mysterious 'Chief Designer' were preparing for launch. The two launches ended in disaster, both failing to leave Earth's orbit, a fact which prompted Khruschev to vent much anger on his return. Fearful technicians under Marshal Nedelin readied a third booster and a back-up probe against the clock before the opportunity to launch to Mars closed. When the countdown reached zero, the rocket remained stationary and

This unusual shot of Yuri Gagarin was not released until 1986, when *glasnost* began to permeate the Soviet space programme. The brave cosmonaut is seen here playing with his children, dressed as a Crusader. The Soviet people remain intensely proud of their space heroes.

the technicians were ordered out onto the pad to inspect the booster, which was against all the rules. Without warning, the booster ignited, and the launch site was engulfed in a ferocious explosion which killed many key personnel. It was a blow from which the Soviet space programme never fully recovered, and was to lose them valuable time in the race to the Moon.

FIRST MAN IN SPACE

Unaware of this disaster, the world watched in awe as further Soviet animal launches prepared the way for manned flights. When Kennedy came to power in early 1961, the Soviets were unequivocally ahead in space, a fact that cast a spell over the nascent American effort. In January 1961, a test launch of the Mercury capsule ended in disaster when the Atlas booster exploded. In the Oval Office, it was more than cause for concern: Kennedy appointed his deputy, Lyndon B. Johnson, to look into the fast developing space race. In March he reported that the nation would back a major space effort. Almost immediately, Kennedy enquired of Wernher von Braun: 'Do we have a chance of beating [the Soviets] by putting a laboratory in space, or by a trip around the Moon and back with a man? Is there any space program which promises dramatic results which we could win?'

But before von Braun could reply, Radio Moscow astonished the world: 'The first cosmic ship named Vostok, with a man on board, was orbited around the Earth from the Soviet Union. He is an airman, Major Yuri Gagarin.' Within 90 minutes, the brave flier was back on Earth after completing one orbit in the Vostok capsule. There were many people in the West who simply didn't believe the news, despite the fact that radio transmissions from Gagarin in orbit were picked up on short wave radios.

An unusual and informal portrait of the first Soviet cosmonauts who flew the Vostok and Voskhod missions. From left to right: (top) Gagarin, Bykovsky, Yegorov, Belyayev, Popovich, Komarov; (bottom) Feoktistov, Tereshkova, Leonov, Nikolayev and Titov. Yegorov and Feoktistov are the notable civilians, distinguished space medicine and spacecraft engineering experts.

Khruschev was jubilant, particularly when Red Square was filled with ecstatic crowds to celebrate the return of the handsome young pilot. It was a blow for the 'Mercury Seven' – America's first astronauts-to-be – who were still only preparing for their first flight. Alan Shepard's suborbital lob just under a month later could hardly compare with Gagarin's accomplishment, a fact which caused the Russian to comment sourly: 'We sent some dogs up and down just like Alan Shepard.' Yet this first manned Mercury launch, under the glare of live TV and radio coverage, more than made up for Gagarin's flight in the eyes of the American people. In Kennedy's mind, perhaps galvanized by looking for something to distract press attention from the CIA-backed Bay of Pigs invasion of Cuba which had failed days before, a new goal would be set: the Moon.

Nearly a year before, a NASA engineer called John Houbolt, more for amusement than anything, had shown that a trip to the Moon was possible by rendezvous in lunar orbit. Von Braun unearthed plans of his follow-on project from the Jupiter, known as the Saturn V, that was powerful enough to send over 140,000 kg (300,000 lb) towards the Moon. These developments were exactly what Kennedy needed to hear. On 25 May he addressed a joint session of Congress: 'Now is the time to take greater strides in space, time for a great new American enterprise, time for this nation to take a clearly leading role in space achievement, which in many ways holds the key to our future on Earth. I believe this nation should commit itself to achieving the goal, before this decade is out, of landing a man on the Moon and returning him safely to the Earth. No single space project in this period will be more impressive to mankind, or more important for the long-range exploration of space; and none will be so difficult or expensive to accomplish.'

MERCURY AND VOSTOK

Despite this bold rhetoric, the United States still hadn't launched a human being into orbit. NASA deemed another sub-orbital test necessary, and Virgil 'Gus' Grissom was chosen for the task. Things, it seemed, were fated to get worse before they got better: after a perfect flight, on landing in the Atlantic, the hatch of the Mercury capsule blew off and it sank. Grissom was rescued from the sea after swallowing much water and with a large shark exploring his rapidly-discarded helmet.

In August, Gherman Titov spent more than a day in orbit, becoming the first human to wake up in orbit, proof – if needed – that the Soviets were still ahead. During his 17 orbits, he became the first victim of space sickness, a condition which was to affect many other spacefliers, caused by the

After their emergency landing in the forests near to Perm in Siberia, the Voskhod 2 cosmonauts were eventually located by rescue teams. Pavel Belyayev is seen talking to mission controllers with a radio handset, while Alexei Leonov (in a white communications cap) looks on at left.

effect of weightlessness on the balance system in the inner ear.

It wasn't until the following February that John Glenn took his place in the pantheon of all-American heroes as the first U.S. astronaut to orbit the Earth. Again his flight was covered live by all the TV networks, who were not aware of the drama surrounding his return. A sensor suggested to mission control that the capsule's heatshield – its protection from the fiery heat of re-entry – had become detached. Rather than jettison the booster pack attached to the heatshield, they were kept in place just to make sure. It thankfully transpired that the sensor was wrong, but Glenn's re-entry was a true baptism of fire, with flaming chunks of his booster pack shooting past the window.

Whereas NASA had a definitive goal by Presidential decree, the Soviet plans for going to the Moon were at the mercy of Kruschev's whims. As space was such excellent propaganda, he demanded that two Vostoks be launched so they would be in orbit at the same time. So, in August 1962, cosmonauts Andrian Nikolayev and Pavel Popovich were launched within two days. At one point their orbits passed within a few miles of each other. Western newspapers were in awe of the achievement, but as NASA officials pointed out in vain, the spacecraft were automatically controlled and could not change orbit. Khruschev duly appeared with the families of the two on their return, overlooking Red Square in Moscow to face the exultant crowds.

In the meantime, NASA attempted to catch up with more ambitious flights in the Mercury series. The second orbital flight saw Scott Carpenter waste so much attitude control fuel in order to better view the Earth's surface, that he landed 400 km (250 miles) off course and was ordered never to fly again. In October 1962, Wally Schirra flew for six orbits, with so few problems that his was referred to as 'the textbook flight'. Bouyant with the success of Schirra's flight, Gordon Cooper drew the curtain on Project Mercury in May 1963. He spent nearly 34 hours in orbit, where he managed to overcome a host of minor mishaps caused by his control systems short-circuiting. He successfully returned to Earth under manual control.

Two months later two more Vostoks – the last in the series – repeated the previous mission of the heavenly duo. But far from being a repeat, one of the two was a woman. Valentina Tereshkova had been picked from the obscurity of a paper mill for the simple reason that she was an accomplished parachutist, her favourite hobby. She was trained for the one flight, after which she became a roving ambassador extolling the virtues of Soviet communism and womanhood, but due to her lacklustre speeches she was described as humorous as a speak-your-weight machine.

THE VOSKHOD FOLLIES

For NASA, the next step in space were two-men missions called Gemini where the theories of orbital rendezvous could be put into practice. It was no exaggeration to say, as NASA often did, that Gemini was the bridge to the Moon. The assassination of John F. Kennedy in November 1963 added a greater poignancy and urgency to his dream of a Moonlanding by the turn of the decade. The launching facilities at Cape Canaveral were named in his honour, as was the Cape itself.

On hearing of Gemini, Khruschev demanded that there be a three-man flight before the first Gemini. Though a three-man Soyuz capsule was in the pipeline, the only option for Chief Designer Korolev was to adopt a Vostok capsule. Only by removing all safety margins, including the removal of emergency egress equipment, could a three man crew be accommodated – and even then, without spacesuits. It is a wonder how cosmonauts Vladimir Komarov, Konstantin Feoktistov and Boris Yegorov survived, but survive they did – unlike the political career of their mentor. Khruschev was deposed during their flight in Voskhod 1 in October 1964, and Leonid Brezhnev took his place.

However, another legacy from Khruschev's reckless propagandizing in space remained to be fulfilled. It also nearly caused the first in-flight fatality. Five days before the first manned Gemini mission, Voskhod 2 was launched on 18 March 1965, this time with a two-man crew equipped with the luxury of spacesuits for a very simple reason. NASA had announced that Gemini would see the first 'extra vehicular activities' (EVAs) – or spacewalks. Commander Alexei Leonov would pip the Americans at the post with a ten minute spacewalk. He tumbled and floated at the end of an umbilical chord nearly 4.5 m (15 ft) long, enjoying the view and relishing the experience. But returning was far from easy, as he found out when he tried to enter the hastily-adapted airlock back to the capsule and failed. His spacesuit had 'ballooned' in the vacuum, and he could not get his legs into the airlock. Only by lowering the pressure in his suit, narrowly avoiding deep-sea diver's 'bends' because of the enforced decompression, was a fatality avoided. Worse was to come before re-entry: the onboard computer

Ed White became the first American to walk in space. He spent 21 minutes outside the Gemini 4 spacecraft in June 1965, attached to the capsule by a 7.6-m (25-ft) umbilical line. Had there been problems, the line would have been cut in accordance with mission rules!

failed to fire the retro-rocket motors, and the pilot, Pavel Belyayev, had to fire them under manual control on the next orbit. As a result, they landed a thousand miles off course in a forest in the Ural mountains. They had to wait over a day before rescue teams could get to them, also rescuing them from the attentions of a curious pack of wolves.

It was Russia's last manned spaceflight for nearly two years, a fact that did not escape the notice of eager Western observers. The final legacy of Khruschev's regime was to produce a workload that more or less hounded Sergei Korolev into his grave. Since the early 1960s, he had been working on a new series of manned spacecraft called Soyuz (meaning 'Union') which would be used to fly cosmonauts to the Moon. His original plans envisaged different variants of the basic Soyuz which would be launched separately to the Moon. But Khruschev's continual demands for 'spectaculars' altered his plans time and time again. By 1965, development of Soyuz was so far behind that the Chief Designer worked ever longer hours to get the project on schedule. The period in the labour camps had seriously damaged his health, and the increased work began to take its toll. In January 1966, he was admitted to hospital where in view of his importance the Soviet Minister of Health took charge of removing tumours from his colon. During the operation a further tumour was discovered and, instead of attempting to remove it at a later date, the Minister continued beyond the accepted limits of anaesthetics. Korolev's heart literally gave in, and despite attempts to revive him, he was dead within minutes. On 16 January 1966 he was buried in the Kremlin, and for the first time his identity and singular contribution to Soviet space exploration was acknowledged.

BRIDGING THE GAP

During the two years that no cosmonauts flew in 1965 and 1966, Gemini managed to make up the ground to match the Soviets' early lead and notch up a plethora of achievements. Gemini 3 was the first manned test, with Virgil Grissom and John Young as pilots. Their new craft, nicknamed by Grissom 'the unsinkable Molly Brown' after a popular musical of the time, and with reference to the problems at the end of his previous mission, worked like a dream. Using a handheld computer, they were able to change orbit many times, and Young even offered his commander a corned beef sandwich which he had sneaked on board. Ed White became the first American to walk in space on the Gemini 4 mission, where he literally had to be ordered back into the cabin, so enjoyable did he find the experience.

Gemini 5 showed that astronauts could survive for 14 days in space, the duration of a flight to the Moon and back. Geminis 6 and 7 were jockeyed to within a metre of each other, after having changed their orbits as would need happen in lunar orbit before a landing could be attempted. Gemini 8 successfully docked with a target, and even survived the near fatal jamming of one of its thrusters, causing both to spin round once every second. Only by the quick thinking of its commander, a former test pilot called Neil Armstrong, was the situation rectified. A further three flights, all testing docking manoeuvring and space walking techniques, saw Gemini draw to a successful close in December 1966. NASA was ready to go to the Moon.

Safe! David Scott (left) and Neil Armstrong (right) are rescued by frogmen who attached a flotation collar to their Gemini 8 capsule. A jammed thruster had caused the craft to rotate once a second. Armstrong's quick reactions saved the mission, a calmness which impressed his peers.

DEATH AND DISASTER

1967 was to have been the make or break year. The first Apollo flight was scheduled for February, and it is now known that the Soviets were planning to do the first dockings and transfers of crews with their Soyuz capsule. In late January, the first Apollo capsule, known by its serial number of Apollo 204, was at Cape Kennedy with its three crewmen practising for their flight. Commander Gus Grissom was fully expected to become the first man on the Moon. Ed White, the first American to walk in space, would also figure prominently in the early Moonlandings. Both veterans were joined by Roger Chaffee, a newcomer or 'rookie'.

But things started to go wrong almost from the

very start in a series of practice tests out on the launchpad. An engine nozzle shattered when the capsule was accidentally dropped, static crackled in the communications links, and repeatedly failing computers all led to a far from harmonious set of dress rehearsals. Apollo 204, its commander decided, was jinxed, and after experiencing his Mercury capsule sinking, Grissom felt he knew a thing or two about jinxes. Fearing his wrath, engineers attempted to ready the spacecraft for a flight at the end of February. But after a long day of rehearsals on 27 January disaster struck: an electrical short from an exposed wire in the spacecraft caused a spark to ignite some velcro strip in the cabin. In the pure oxygen atmosphere, the flame exploded to flashpoint, and within 30 seconds the crew were asphyxiated. It took the emergency team nearly 30 minutes to open the hatch instead of the expected 90 seconds.

The 204 fire led to a total overhaul of the Apollo programme, including the redesign and fireproofing of many components. In an effort to meet Kennedy's deadline of the end of the decade, many safety margins had been flaunted with makeshift and shoddy engineering. The 18-month delay before men took an Apollo capsule into space allowed NASA to correct many of the problems with the complex technology, despite its devastating effect on morale.

It certainly looked like the way ahead for landing men on the Moon was clearly open for the Soviets, particularly when rumours of a new spacecraft and link-up were picked up by Western newspapermen in Moscow. On 23 April Soyuz 1 was launched with Vladimir Komarov on board, with the launch of another craft expected. But nothing happened: Komarov, Radio Moscow reported, made 18 orbits of the Earth, and was in fine health. Then suddenly, martial music was played and on television a black-bordered portrait of the cosmonaut was shown. Komarov was dead: his brand new vehicle had tumbled out of the sky, with his parachute system entangling as he headed earthwards.

As the first in-flight fatality, it was particularly catastrophic for the Soviet space programme, especially after the death of Korolev the year before. (Worse was to come in 1968 when Yuri Gagarin was killed in an aeroplane crash.) Various rumours about the Soyuz 1 crash circulated, particularly after no body was ever displayed publicly. Years later, a photograph surfaced of Komarov training with cosmonauts Khrunov and Yeliseyev wearing spacesuits, suggesting a link-up with a second Soyuz was in the offing. But whatever may have been planned, Komarov was buried in the wall of the Kremlin with full military honours at a state funeral. Like the Americans before them, Russian engineers went away to improve their spacecraft and lick their technological wounds.

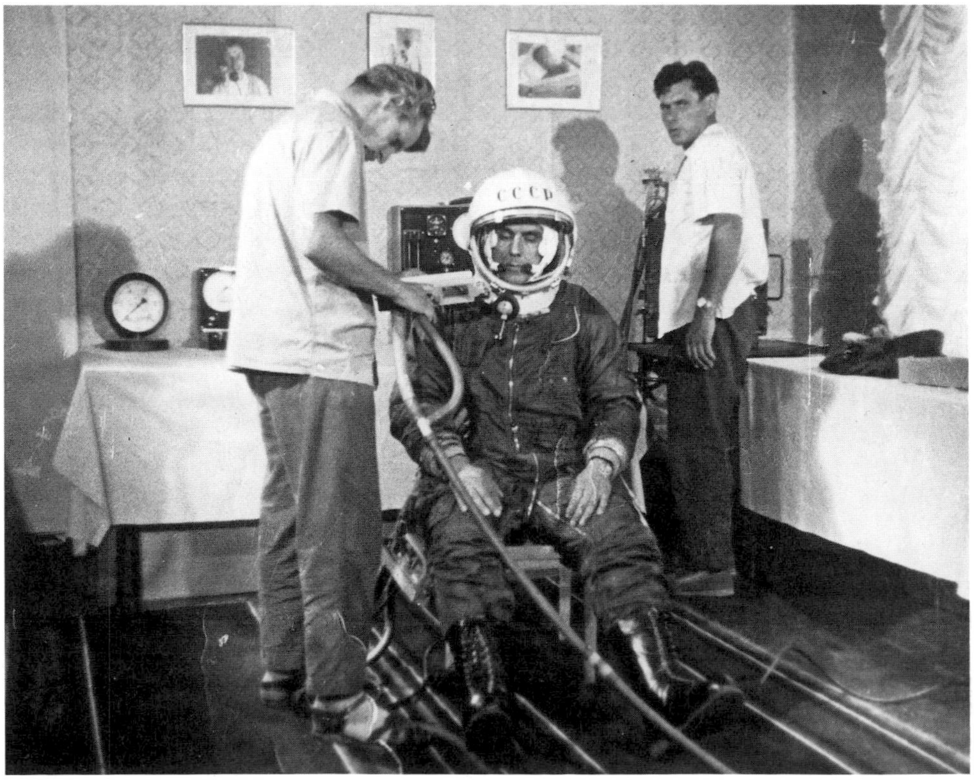

Vladimir Komarov became the first Soviet casualty of space flight in April 1967. He is seen here training for an earlier Vostok flight as a back-up. In the early 1960s, a heart disorder had led to his removal from flight operations. Only by obtaining sworn statements from Moscow specialists which said that the condition was not serious was he able to regain flight status.

NECK AND NECK

At the end of 1967, those wounds had more or less healed. Two Cosmos craft (186 and 188) were launched at the end of October and performed the first totally automatic rendezvous between two spacecraft. Western observers believed that such a mission was planned for Komarov and that the Cosmos vehicles were Soyuz capsules flown unmanned. On 9 November 1967, Wernher von Braun's mighty Saturn V was tested for the first time with an unmanned Apollo capsule as its payload. The massive rocket worked and the capsule re-entered the Earth's atmosphere at speeds which it would encounter on returning from the Moon. In January 1968, Apollo 5 was launched atop a Saturn V and tested the engines which would be used on the Lunar Module in which astronauts would descend to and take off from the Moon. And finally, in April, the ultimate test of the Saturn V/Apollo combination was made. Though excess vibrations caused the whole booster to shake up and down (known to the engineers as the 'pogo' effect) it was declared that they could be eliminated and the Saturn V was ready for manned flight; further test flights were deemed unnecessary.

Erring on the side of caution, the first manned Apollo flight was launched on the smaller Saturn 1B launcher and scheduled for mid-October 1968. Apollo 7 was to be a ten-day mission to test all the Apollo flight systems – and to NASA's relief, they worked perfectly. In fact, most problems were caused by the crew led by Wally Schirra who was the only astronaut to fly the first three generations of American manned space-

In late September 1968, Zond 5 became the first sea recovery of a Soviet spacecraft. Brought back to Bombay, it was swiftly despatched to the U.S.S.R. Most Western authorities believe it was an unmanned dress rehearsal for a flight around the Moon.

craft. All three crew members developed headcolds and soon became tense and irritable when Mission Control changed the sequence of their workload. Ironically, using a portable camera, the crew gave the first television broadcasts from orbit and later won an Emmy for their enthusiastic, not to say 'hammy', performances.

An Emmy was not in the offing for Georgi Beregovoi who successfully tested the redesigned Soyuz four days after the Apollo 7 crew returned to Earth. The official news releases were at pains to point out that he showed restraint in his broadcasts at all times, presumably to ridicule the vaudevillian antics of Schirra and company. After reaching orbit on 26 October, it was announced that an unmanned Soyuz 2 had been launched the day before. Soyuz 3 with Beregovoi aboard approached within 200 m (650 ft) of the automatic craft, but no docking was made. It is not known if a docking was planned, but the successful completion of the flight of Soyuz 3 was enough to expunge the memory of Komarov's death.

As the end of 1968 beckoned, both superpowers were neck and neck in the Moon race. The next stage was to see which would first launch human beings into orbit around our celestial neighbour.

THE MOON AT CHRISTMAS

When NASA's new administrator, Tom Paine, was sworn in by President Johnson in October 1968, he was quoted as saying that although he believed that America would be the first to land on the Moon, he would not be surprised if the first to orbit it were the Russians. Congress had long since become accustomed to the scaremongering of NASA Administrators. Paine's predecessor, Jim Webb, had revealed the existence of a Soviet superbooster in Congressional hearings the year before. Seasoned Washington officials believed it was a ploy to get more money for NASA in the aftermath of the Apollo 204 fire. They soon referred to this superbooster as 'Webb's Giant' in the kind of voice that suggested it was little more than a figment of his imagination. Webb insisted that the CIA and the military intelligence community had hard evidence from spy satellites showing that it was even bigger than the Saturn V, but to little avail. Yet a year later, Tom Paine's words were not taken as idle speculation for an unmanned Russian Zond spacecraft had already circumnavigated the Moon and returned home safely.

The Zond series had been initiated in 1964 as 'automatic interplanetary stations', essentially testing the technology required to send probes to the planets. But suddenly in September 1968, Zond 5 was launched on a flight path that was nothing more than a dress rehearsal for a manned flight around the Moon. Though it was launched on 15 September and orbited the Moon on the 19th, Moscow added to the mystery and denied that Zond 5 was anywhere near the Moon. Then a day later, Radio Moscow sheepishly announced Zond 5 had been 'in the vicinity of the Moon' and was carrying tortoises, insects, plants and seeds. Radio signals picked up by the Jodrell Bank telescope in Cheshire heard a tape-recording of a voice. On 22 September, Zond 5 entered the Earth's atmosphere and for the first time a Russian spacecraft was recovered from the sea. A Soviet ship positioned in the Indian Ocean picked up the capsule with its strange occupants and docked at Bombay, from where a transport plane whisked it back to the Soviet Union. Photographs taken of the Zond capsule as it bobbed in the sea and was transferred to the docks revealed it to be remarkably similar to the Soyuz descent capsule. Zond 5 was an unmanned Soyuz capsule which had been modified to carry a single cosmonaut to the Moon and back. Calculations by Western observers showed that using the newly-introduced Proton booster, the capsule could carry just enough supplies and fuel to last the seven day journey.

Events accelerated: less than a week after Beregovoi's return, Zond 6 repeated Zond 5's flight path with one remarkable difference: as it hurtled through the Earth's atmosphere during re-entry, it had turned its heat shield at a right angle to the direction of motion. The resultant aerodynamic lift had made it skip across the atmosphere like a stone across water and successfully landed in Kazakhstan. It was clear that a Zond with a cosmonaut as passenger would be launched at the next opportunity to fly to the Moon. From Baikonur, this date was 8 December.

For NASA the feeling of being pipped at the post was particularly galling. Apollo 7 had been so successful that the series of three further Earth orbit tests had been cancelled. Apollo 6 had been intended to test the Lunar Module in Earth orbit, but there were problems and it was not quite ready. So the original plan for Apollo 9 was brought forward. Apollo 8 would be launched on the Saturn V to the Moon and would then enter orbit before returning home. Because of Cape Canaveral's different location, the earliest Frank Borman, Jim Lovell and Bill Anders could be launched was 21 December.

On 6 December, Pavel Belyayev arrived at Baikonur ready to enter the Zond capsule for a flight to the Moon. The cosmonaut who had been the pilot on Voskhod 2 and had made the manual re-entry after the automatic landing computers failed seemed a logical choice. Photographs of him taken at this time show him sitting inside a Soyuz capsule in which an Earth globe in the instrument panel was noticeable by its absence. It is rumoured he was strapped inside the Soyuz/Zond craft waiting for the countdown to begin. But when the launch window opened on 8 December, nothing was launched from Baikonur. To this day, the Russians have not announced the reason why and Belyayev took his secret to the grave as he died two years later after an appendix operation went fatally wrong.

So the honour of becoming the first men to reach the Moon was left to the crew of Apollo 8. After being launched by the Saturn V on 21 December, the crew spent a routine three days en route for the Moon. The engine of their Apollo Service Module would have to successfully fire to allow them to orbit the Moon and fire again to let them return home. After passing out of radio contact behind the Moon, Bill Anders' jaunty 'Please be informed there is a Santa Claus' told an expectant world the engines had worked. They made ten orbits of the Moon, and returned the first detailed TV pictures ever seen of its surface and of the Earth as a globe.

To many people, memories of Christmas 1968 are indelibly linked to TV pictures of astronauts in a cramped cabin with the legend 'LIVE FROM THE MOON' underneath. On Christmas morning, Frank Borman read out the first sentences of Genesis from the Bible. Though he was a devout Christian and lay preacher, the idea to do this came from an official of the U.S. Information Service. Across the United States, however, irate atheists and agnostics telephoned the TV networks to complain. The service module engine fired successfully, and within two days the crew of Apollo 8 were home.

Seeing the success of the first manned lunar mission, the Politburo hastily met on 1 January

Wishful thinking? Twenty years after the first Moonlandings, enough evidence has surfaced to suggest that the Soviets were developing the required technology. This fanciful painting entitled 'Ocean of Storms' shows a cosmonaut close to the mortal remains of a lunar lander. Perhaps *glasnost* will open Soviet history books to the West about this aspect of its space history.

1969 to discuss space goals. It was pretty clear that NASA had the technological advantage, especially as the Russian superbooster had yet to fly. So it was decided that the main thrust of the Soviet programme would go towards building space stations in low Earth orbit, though the elements of the Moon programme would continue to be tested. On 14 January, Soyuz 4 was launched with Vladimir Shatalov as its lone occupant: a day later he was joined by Soyuz 5 and its crew of three. On the 16th, both craft docked, and two of the Soyuz 5 cosmonauts (Khrunov and Yeliseyev) transferred to the other craft and returned with Shatalov back to Earth. Boris Volynov was left to return a day later. They were reunited in Moscow on 24 January as they headed in an official motorcade for the Kremlin and were nearly the victims of an assassination attempt. A disgruntled Army officer fired at their car thinking it contained Leonid Brezhnev and killed the driver. They survived, slightly bruised, and probably wished they could head off again for the comparative safety of Earth orbit!

THE FINAL LAP

For Apollo, the self-imposed deadline of the end of the decade was fast approaching. In March, Apollo 9 finally tested the Lunar Module in low Earth orbit, the first time a manned spacecraft had been flown which could not return to Earth (it had no heatshield). The LM behaved flawlessly and in May another LM was taken into orbit around the Moon by the crew of Apollo 10. With John Young in the Command Module, Tom Stafford and Gene Cernan (who had flown Gemini 9 together) took the LM named 'Snoopy' to within 9 km (5.6 miles) of the lunar surface. After discarding the landing engine stage, the astronauts headed up towards Young in the ascent stage when the vehicle began to 'buck like a bronco'. It was later found that a switch had been left in an 'on' position which should have been switched 'off'. The relatively primitive technology of the time meant that manually-operated toggle switches had to be used to selectively control the amount of data entering the computers and control systems. The Apollo 10 crew returned safely home, but problems with switches would surface on the next flight, mankind's first attempt to land on the Moon.

The crew for Apollo 11 had been announced by NASA in January 1969. Its commander was to be Neil Armstrong who had been born just a few miles from where the Wright Brothers had first taken to the skies. Edwin 'Buzz' Aldrin would land with him (he was quick to point out to journalists that his mother's maiden name was Moon). As an acknowledged expert in the fine art of celestial navigation and rendezvous, his experience would be invaluable. The command module pilot was Michael Collins, who would have flown to the Moon on Apollo 8 had he not required surgery. The flight was scheduled to begin on 16 July.

Tranquillity Base. Buzz Aldrin is seen next to the Apollo 11 Lunar Module *Eagle* in the Sea of Tranquillity. An experiment to measure the solar wind is seen to his right. Perhaps more than anything his description of 'Magnificent Desolation' summarized the scene.

In the meantime, the Russians attempted to launch Webb's Giant. Though the information is still classified, enough evidence has surfaced over the years to show that the Soviets had developed a superbooster, sometimes referred to as the 'G-vehicle'. Six weeks before Apollo 11 was launched, the booster exploded as it was being fuelled on the launchpad at Baikonur. Military spy satellite pictures are said to show extensive devastation to the site. Two more failed tests in 1970 and 1971 led to the booster's abandonment. So as a final last-ditch publicity attempt, the Soviets launched Luna 15 three days before Apollo 11. It was to be an unmanned sample return mission which could return lunar soil before the crew of Apollo 11 could. The gambit failed: Luna 15 crashed into the Sea of Storms.

THE EAGLE HAS LANDED

After a flawless journey, Apollo 11 reached the Moon on 19 July. The next day, the Lunar Module 'Eagle' separated from Command Module 'Columbia', in which Collins would maintain a lonely vigil. If anything went wrong with the attempted landing, Collins was powerless; the official mission guidebooks said that whatever happened, Collins would have to return on his own. After the 13th lunar orbit, Eagle's engines fired to lower it towards the chosen landing site in the Sea of Tranquillity.

It is estimated that half a billion people worldwide watched Armstrong and Aldrin become the first human beings to land on the surface of another planet. TV pictures from the crew and mission control were relayed by communications satellites to every corner of the globe. What most of the viewers and TV commentators didn't realize at the time was how close the crew came to disaster.

A vital component of 'Eagle's' equipment was the radar system which automatically passed information to the onboard computer which would then calculate the rate of descent. It was vital information for the astronauts who were flying the spacecraft. As they approached the Moon's surface, the computer started to malfunction because it was being fed too much information too quickly. Buzz Aldrin repeatedly reported a '1201 alarm', jargon which meant that the computer was overloading. Houston acknowledged the problem and an engineer called Steve Bales, whose job was to monitor the computer's behaviour, had to look into it. Mission rules were strict: in the event of a computer failure, the landing would have to be aborted. He quickly realized that the problem was caused by a switch that should have been switched off.

So Armstrong assumed manual control and 'Eagle' touched down with only 20 seconds of hovering fuel left. Wires dangling from the Module's footpads signalled to the engines to switch off, leading Armstrong to announce 'Houston, Tranquillity Base here. The Eagle has landed.' Mission control – only too aware of the 1201 drama – was relieved. The astronaut in direct contact with the crew (known as CapCom – Capsule Communicator) was Charlie Duke, whose laconic Carolinan tones were heard to reply: 'Roger, Tranquillity. We copy you on the ground. We've got a bunch of guys about to turn blue. We're breathing again.'

Rather than take a short sleep, Armstrong and Aldrin decided to get out onto the lunar surface as quickly as possible. Though donning suits in the cramped environment of the Lunar lander proved difficult, Armstrong was ready to make history as he slowly descended the ladder outside the vehicle. After stepping off the ladder he announced: 'That's one small step for man, one giant leap for mankind.' He was to reveal later that the phrase had been suggested by his wife.

He was joined 20 minutes later by Aldrin, and together they spent a total of $2\frac{1}{2}$ hours on the surface of the Moon. They found the best way to walk in the one-sixth gravity was to hop like a kangaroo and in the meantime collected nearly 22 kg (48 lb) of lunar soil. Before accepting a phone

Three weeks in the Lunar Receiving Laboratory, a trailer-sized quarantine facility, awaited the triumphant crew. After eight days away from Earth, they enjoyed the luxury of a shower. A few days after leaving quarantine, the crew were awarded presidential awards by Richard Nixon at a dinner attended by all 50 state governors.

call from President Nixon the two astronauts erected an American flag which had to be attached to a wire frame in order to make it 'fly' in the lunar vacuum. Considering the history they were making, they made a curious omission: no photographs were ever taken of Armstrong on the Moon.

Back inside the Lunar Module, they inhaled surface dust which had stuck to their spacesuits and sneezed violently for many minutes afterwards. They reported it smelled like gunpowder. After blasting off from the lunar surface, the two astronauts successfully docked with Collins in 'Columbia' and returned home on 24 July. Plans for their recovery by the U.S.S. *John F. Kennedy* were shelved by the White House as they were to be greeted by President Nixon. They were recovered by divers from the U.S.S. *Hornet* instead and because of fears that the lunar soil might contain harmful bacteria or contamination, the returning heroes had to suffer the indignity of quarantine in a modified trailer for the next three weeks. Clad inside biological isolation garments, the crew entered the quarantine facility marked 'Hornet Plus Three' which was later transported to Houston. Results showed that there was nothing harmful in the soil and when it was suitably illuminated and watered, plants could actually be grown in it.

After their quarantine, the Apollo 11 astronauts were feted with seemingly endless ticker-tape parades and they paid numerous goodwill visits abroad. Life was never quite the same for any of them again. For months afterwards they were followed by reporters and photographers no matter where they went. Such attention took its toll on Buzz Aldrin, who had already suffered, by his admission, the ignominy of coming second for the first time in his life. He found it most difficult to readjust to life back on Earth and later suffered from severe bouts of depression and alcoholism.

'HOUSTON, WE'VE HAD A PROBLEM!'

Before the decade ended, two more Americans were to land on the Moon meeting Kennedy's deadline with ease. Pete Conrad and Al Bean took the Apollo 12 Lunar Module 'Intrepid' to a successful landing in the Ocean of Storms. With Dick Gordon they had had an eventful launch through a rainy squall on 14 November when their Saturn V was hit by lightning. The resulting surge in electricity had triggered most of the spacecraft's fuses, and they were bathed in an eerie green glow from the circuit-breakers inside the Command Module. They came very close to an abort, but things were back to normal once they reached Earth orbit and headed towards the Moon. As Conrad stepped off 'Intrepid's' ladder he commented, 'It might have been a small step for Neil, but it's a hell of a leap for me!' They landed very close to Surveyor 3, which had been there for 31 months by the time they arrived. Bean and Conrad removed parts of the lander and returned with them to Earth where technicians were surprised to find signs of contamination on the Surveyor craft. It was later revealed the offending virus had probably come from a technician who had sneezed before the unmanned lander had been launched. Apollo 12's stay on the Moon was spoiled for TV viewers as Bean pointed the TV camera too near the Sun and its sensitive optics burned out.

President Nixon met the Apollo 13 astronauts in Honolulu after their successful splashdown in the Pacific. Jack Swigert (left) eventually became a senator but died of cancer before taking his seat. Fred Haise (middle) stayed with NASA to test the Space Shuttle. Jim Lovell (right) left NASA soon after, aware that he would not get another Apollo flight.

A new decade dawned with the war in Vietnam, student unrest and escalating inflation of more pressing concern than astronauts walking on the Moon. Having met Kennedy's deadline, Apollo became the victim of its own success. When Apollo 13 took off on 11 April 1970, nobody took much notice, and what little interest remained was focussed on the superstitious aspects of it being number 13. Two days before lift-off, command module pilot Ken Mattingly had been replaced by his back-up, Jack Swigert, because of a suspected exposure to measles. After a launch at 13.13 local time, one of the five second-stage engines shut down early. Longer firings of the other engines made up for the deficiency, and the crew settled down to their routine. TV viewers, long accustomed to Apollo astronauts heading for the Moon, switched off in droves.

Who needed real life drama when the American public could escape to the cinema to see *Marooned*, a movie about stranded astronauts starring Gregory Peck? But once again, a switch was to cause problems, this time connected to a heat regulator in one of the Service Module's liquid oxygen tanks. Testing of the electrical systems before launch caused a surge of current to pass through the heater, and a switch became welded

Swansong! Apollo 17's launch signalled the end of 'mankind's greatest adventure'. The mighty Saturn V was launched moonwards for the last time. A total of 381 kg (840 lb) of soil was returned by all 12 astronauts who landed on the Moon.

into the 'on' position. When it was filled with liquid oxygen, it became little more than a time bomb waiting to go off. It did so two days later, after the crew had overseen an automatic mixing of the propellants.

'Houston, we've had a problem' was Swigert's way of telling Mission Control that something had gone wrong. They had heard a tremendous bang from a ferocious explosion. Very soon they realized that the oxygen tank had been ripped apart, as gas could be seen venting into space. Because the fuels were used to generate power, the crew also noticed that their electricity reserves were dwindling. With only an hour's worth of power left at most, a Moonlanding was out of the question. Houston advised the crew to enter the Lunar Module 'Aquarius' and use it as a lifeboat.

An anxious world watched as the stricken crew used 'Aquarius's' life-support system to keep them alive. Only 38 hours-worth of power remained for the crew, much too short a time for them to return back to Earth. Only by powering everything down, which meant living in darkness and cold, would they survive. The Commander, Apollo 8 veteran Jim Lovell, was amazed at the calmness of his crewmates in taking pictures of the lunar surface as they sped by. Subcontractors and technicians worked out a rescue procedure, aided by the fact that Swigert had actually written the emergency procedures manual. A new set of flight instructions was passed up to the crew, who used the LM's descent engine to speed behind the Moon and back to Earth. The tired crew made mistakes, and there remained the possibility that there would not be enough oxygen or power for the three crew to survive. They began to talk about choosing straws to decide whose life would be sacrificed to allow the other two to survive.

Thankfully it was a decision that did not have to be taken. Three and a half hours before re-entry, the crippled Service Module was jettisoned and the crew saw the extent of the damage caused by the explosion. They quickly transferred to the Command Module and reluctantly jettisoned 'Aquarius'. Fears that the heatshield had been damaged during the explosion proved groundless and they landed safely in the Pacific. But it had been a near miss.

GOODBYE TO THE MOON

The near fatality of the Apollo 13 mission showed the Apollo programme in a new light. Far from being a wonderful, visionary endeavour the American public began to see it as a waste of time and resources. The programme ground to an untimely demise. When Apollo 14 landed on the Moon in January 1971, Commander Al Shepard produced two golfballs which he teed off, much to Mission Control's amusement, across the horizon. But many people worldwide didn't share the amusement: many were incensed by the gesture. Was it fair that billions of dollars were spent on sending men to play golf on the Moon while there were millions of people starving on Earth? Worse was to come with Apollo 15 in July of that year. On returning to terra firma, it was revealed that astronauts Jim Irwin and Dave Scott had illicitly franked some stamps on the lunar surface. Though their intention was to sell them privately and use the money to set up trust funds for their children, they were repeatedly sold and re-sold. Some of the stamps changed hands for many hundreds of dollars apiece and the crew were severely reprimanded and destined never to fly again.

By 1972, there wasn't much left *to* fly, as the

end was nigh for Apollo. Though NASA wanted to return to the Moon through to Apollo 20, the last three flights in the series were cancelled by Congress. There was enough trouble getting NASA's next project, the Space Shuttle, through the legislative process without the added vexation of increasingly unpopular Moon flights. So in December 1972, Apollo 17 became the grand finale with its spectacular night launch of the mighty Saturn V booster. Night turned to day at Cape Kennedy as they headed for the Moon with geologist Jack Schmitt part of the crew, the only trained scientist to land on the Moon. In all, the Apollo 17 crew spent a total of 75 hours in the Taurus-Littrow valley. The dubious honour of being the last man to bid farewell to the Moon fell to the Commander, Gene Cernan. With the words, 'Let's get this mother outta here!', he fired the engines of the Lunar Module 'Challenger' and the final Apollo crew began their journey home.

CONCLUSION

Twenty years after the first Moonlanding may be too short a time period to gain a true insight into Apollo and its ultimate significance. There are those who believe that for the $25 billion spent over a decade, very little was achieved except for a collection of Moonrocks and a few spin-offs like non-stick frying pans. Many (this author included) believe that Apollo's lasting achievement was to make human beings realize the fragility of the environment in which we live. Along the way, the electronics revolution of the early 1970s was also triggered by Apollo, a revolution which has changed our way of life.

Whichever way one views Apollo, it pointed the way for future developments in space. After the end of the Moonlandings, the only manned projects which NASA had on the cards were the 'Apollo Applications Program' and the 'Apollo–Soyuz Test Project' (ASTP), a joint project with the Russians. By converting the upper stage of the Saturn V into a space station, NASA astronauts learned to live in space for many weeks at a time. The project was later given the less prosaic description of Skylab. ASTP was agreed in the atmosphere of detente between the superpowers in the early 1970s, and the linking took place in July 1975. Despite talk of further joint missions, nothing was to happen. The next time Americans were to fly in space after ASTP was April 1981 aboard the Space Shuttle *Columbia*. The Shuttle was developed in an era of economic cutbacks which would later haunt its designers in the tragedy of the *Challenger* explosion.

For the Russians, after losing the race to the Moon, they devoted their energies to building a series of Salyut space stations in Earth orbit. Throughout the 1970s, Soviet cosmonauts learned to live in space and overcome hazards and dangers worthy of any adventure story. Official Soviet announcements assumed a policy of amnesia, continually denying they ever wanted to send cosmonauts to the Moon.

It is left to future generations to decide who really were the losers and winners.

Jack Schmitt, the only trained geologist to land on the Moon, is seen standing by a massive boulder in the Taurus-Littrow valley. By the time of Apollo 17, more ambitious landing sites were attempted. To the right of the boulder is the Lunar Rover used to ferry astronauts across the surface.

TRIUMPH AND TRAGEDY: THE SPACE SHUTTLE STORY

In the late 1950s, the sight – and more memorably the sound – of supersonic rocketplanes above the Mojave Desert in California became a familiar occurrence. A series of remarkable supersonic test planes – rated X-1 to X-15 – regularly plied the most rarefied regions of the blue beyond, pushing back the frontiers of aeronautics. The test pilots who flew the remarkable flying machines were based at the Edwards Air Force Base, an airstrip built around dried saltbeds in the desert scrubland. Many of them felt that, given the political backing and funding, supersonic planes could be modified to allow them to be flown beyond the confines of the upper atmosphere and to orbit the Earth. Such hopes, in the frenzied climate of beating the Russians to orbit human beings around the Earth, remained unfulfilled. Nevertheless, the experience of flying the rocketplanes trained some remarkable pilots: one was Neil Armstrong, the first man to walk on the Moon. Another was Joe Engle, who, 20 years later, would fly into space in a vehicle that was the first true spaceplane – the Space Shuttle.

Even in the 1950s, the idea of spaceplanes wasn't new. The theoretical implications of repeated flight into orbit by the same winged vehicle had been discussed at length by various aeronautical pioneers starting with the German Eugene Sänger in the early years of the Second World War. In the late 1950s, spurred on by the success of the X-series of rocketplanes, the USAF proposed a hypersonic glider which would undergo Dynamic Ascent & Soaring Flight and hence was known as Dyna Soar for short. But the need to get astronauts into the dark black yonder in bell-shaped Mercury capsules led to its cancel-

In February 1984, astronauts Bruce McCandless and Bob Stewart became the first human satellites when they tested the Manned Maneuvering Unit. Exhilarated, they called each other 'Flash Gordon' and 'Buck Rogers'.

Many of the astronauts who eventually piloted the Space Shuttle were test pilots who flew the X-15 rocketplanes in the late 1950s. Dropped from 14,000 m (45,000 ft) from a B52 bomber, the X-15 flew to the very edge of our atmosphere.

lation on the grounds of duplication in 1963. Yet many within NASA and the USAF felt that the idea of hypersonic gliders to be flown in space was an idea whose time would come. To prepare for the future, a series of small gliding lifting bodies were tested.

A SHUTTLE IS BORN

By 1969, however, the need for economic access to Earth orbit became paramount as the early enthusiasm for space endeavours began to wane. Though the newly-elected President, Richard Nixon, was about to bask in the glory of Apollo 11's landing on the Moon, he came to power in an atmosphere tinged with economic recession and civil unrest. With Apollo reaching its climax, and only a converted Saturn V booster upper stage as a makeshift space station definitely on the cards, NASA had no new goals. So Nixon formed a Task Group to look into future plans for NASA, headed by its Administrator, Tom Paine. In September they reported their findings: NASA's next goal should be a manned landing on Mars by the mid-1980s. To achieve this goal, stations would have to be built in orbit and then launched to the red planet, requiring a reusable spaceplane to refuel and supply the astronauts before their departure.

In this, its earliest genesis, the Space Shuttle was envisaged as a dual spaceplane: two crewed vehicles would lift off with one part returning back to base, after powering the smaller into orbit on its back. The total bill of the Mars venture would come to $24 billion, a figure which disturbed the President beyond all reckoning in the tightening financial realities promised by the 1970s. Economic recession and the escalating war in Vietnam meant that only the spaceplane part of the Task Force's plans remained intact.

When NASA asked for $10 million to develop the Shuttle it was politely refused by Congress, who sanctioned only $5 million to begin with. For the first time in its manned programme, the agency was subject to penny-pinching, and the Shuttle's design soon changed. A totally reusable spaceplane sitting atop a giant fuel tank (which would be jettisoned) emerged from drawing boards. The only way such a vehicle could get into space was with additional thrust provided by powerful strap-on boosters. Financial restraints meant that two solid-fuel boosters, using the same fuel as the Minuteman missile, appeared either side of the fuel tank. Such a remedy was not popular; though they had been used on unmanned missions, 'solids' were less controllable than liquid-fuelled rockets. Once ignited, like fireworks, they could not be switched off. Furthermore, emergency 'solids' for launch pad aborts and jet engines for the orbiter to fly through the atmosphere in the event of emergencies were dropped on the grounds of cost. These decisions would lead to catastrophe 15 years later when *Challenger* was lost above the coast of Florida.

DESIGN CHANGES

Throughout 1971 Congress debated the value of the Shuttle, with NASA emphasizing its economic benefits. It was self-evident that reusability would be the key to economical access to space, but the claims for its cost effectiveness were highly controversial. With no space station to which cargo could be freighted, the Shuttle

The original design for the Space Transportation System (STS) envisaged two piloted vehicles. This NASA artwork from 1970 shows the basic outlines of the vehicles. This option, which would not have required solid rocket boosters, would probably have been safer, though more expensive. Because of this latter reason, it was cancelled.

assumed the role of a satellite launcher. Early NASA estimates suggested that if the Shuttle flew only 30 times per year it would pay for itself handsomely. A decade later problems of flying the highly complex Shuttle meant that it was barely managing a third that number. This indicates how exaggerated those claims were. To complicate matters even further, NASA enlisted the support of the U.S. military to help make the Shuttle seem more attractive on the grounds of national security. Reluctantly the USAF agreed, but imposed design changes on the proposed spaceliner: it would be a snub-nosed, delta-winged glider with a wide payload bay. Wide payload bays allowed the deployment of large satellites – and implicitly, the removal of unfriendly ones in time of conflict. Its aerodynamic design would allow it plenty of room to manoeuvre through the atmosphere to land at friendly airbases in emergencies.

Congress approved the Shuttle project in late 1971, noting and agreeing that its motivating force was the drive towards making spaceflight routine. Inaugurating the programme in January 1972, President Nixon announced: 'The United States should proceed at once with the development of an entirely new type of space transportation system design to help transform the space frontier of the 1970s into familiar territory, easily accessible for human endeavour in the 1980s and 1990s. It will revolutionize transportation into near space by routinizing it. It will take the astronomical costs out of astronautics.'

TEETHING TROUBLES

Unfortunately, the technology required to make the Shuttle fly was not easy to develop. Though its delta-winged design meant that it would generate more lift than more conventional (yet more stable) designs, it meant that the Shuttle orbiter would require very complex flight control systems. This necessitated hydraulic power systems to control its aerodynamic surfaces which taxed its builders to the cutting edge of technology. During its ascent through the atmosphere, its main engines would have to be moved or gimballed to provide directional thrust to augment that from the solid rocket boosters. The main engines had to be reliable enough to be used for at least 50 flights, and weight restrictions meant that they would have to be very efficient. Not surprisingly, the engines took far longer to develop than had been originally envisaged and very quickly were over budget. Another problem was the orbiter's revolutionary heat shield, comprised of individual silica tiles. They had a tendency to break with the slightest pressure, and there were difficulties keeping them glued to the orbiter's fuselage.

Despite the delays which these problems

caused, the first Shuttle to be built, known by its designation of OV-101 was rolled out on 17 September 1976. NASA had wanted to name the orbiter *Constitution*, but a letter-writing campaign by fans of the TV series 'Star Trek' led to the name *Enterprise* being adopted. (Appropriately enough, a number of the actors from the TV series attended the ceremony.) *Enterprise* was designed only for atmospheric tests as it had no engines or heat shield. After being dropped from the back of a Boeing 747 high above the California desert, the aerodynamic performance of the vehicle would be tested as it came in to land at Edwards Air Force Base. Twenty years after the first rocketplane tests, a new generation of vehicle proved its worth in the autumn of 1977. Joe Engle was one of the four pilots who tested the airworthiness of the orbiter, and along with his colleagues, was surprised at how quickly the 75-ton vehicle responded to their manoeuvres.

NASA had originally hoped that the Shuttle would fly into space in 1978. But troubles with the main engines and the heat tiles meant that the second Shuttle to be built, *Columbia*, arrived at Cape Canaveral in March 1979 on the back of a 747. Its surfaces looked distinctly like a patchwork quilt, as hundreds of tiles had been dislodged during the flight. Further work led to greater overruns in both cost and time. There was a point in 1979 when the Carter administration met to discuss the project's viability with a view to cancelling it. But with extra money from Congress the development continued apace. Sadly, NASA's other projects suffered at the hands of the project which many were starting to call a 'white elephant'.

'WHAT A WAY TO COME TO CALIFORNIA!'

By the beginning of 1981 the Shuttle was finally ready to fly. The crew had been chosen the previous year and had spent long hours training for the flight. John Young, who had flown four times previously (twice to the Moon), was its commander. The 51-year-old's experience would be invaluable on the first test, though age had taken its toll on his eyesight; his helmet had been specially designed for him to wear spectacles. His crewmate, Bob Crippen, was chosen at Young's express wish because of his knowledge of the orbiter's complex computer operations, even though he'd never flown in space before. Without computers, it would not be possible to fly the Shuttle.

For the first time in NASA's history, a manned vehicle was to be flown without prior testing. Partly because it required humans to fly it, but also to save costs, *Columbia* would take to the skies without being man-rated: only ejector seats with parachutes would be available should anything go

In the autumn of 1977, *Enterprise* was tested for its 'handling' above the Californian desert. A specially adapted 747, known as the Shuttle Carrier Aircraft, was used to ferry the shuttle. The odd-shaped cover at the shuttle's rear was used to minimize aerodynamic buffeting.

The crew for the first flight of the Space Shuttle *Columbia* in 1981 had John Young (standing) as its commander. He had already flown into space four times beforehand. His pilot, Bob Crippen, was a rookie who had originally been a military astronaut.

wrong. Launch was scheduled for 10 April 1981. Press interest reached new heights, as did public interest, testified by the crowds that gathered around the Cape. But only minutes before the launch was due, *Columbia*'s onboard computers started to play up. One of the five main computers was 'talking' to the others some 1/4,000th of a second out of sequence. This minute time lag was enough to scrub the launch. So two days later, NASA tried again and succeeded: the launch of STS-1 (Space Transportation System) went ahead flawlessly. Many space enthusiasts were quick to point out that the new era of the reusable spaceplane was happening 20 years to the day that Yuri Gagarin entered the history books.

In orbit, *Columbia* behaved perfectly: the crew reported that the bird was behaving like a dream. Its payload bay doors were successfully opened, allowing excess heat from the vehicle to dissipate into space. There was a moment of high drama when TV cameras revealed heat tiles missing from around the Shuttle's orbital manoeuvring engine pods. Fears that other tiles in more sensitive areas had been lost were assuaged by the fact (not admitted at the time) that U.S. Air Force Long-Range Cameras were used to scan the dark underbelly of Columbia. Two days later, after closing its payload bay doors, the crew faced their most hazardous ordeal – re-entry. *Columbia* would drop from orbit at 25 times the speed of sound, and begin its long descent to Edwards Air Force Base. Despite all the computer simulations, would the vehicle survive?

Very soon the waiting world got its answer as long-range TV cameras spotted *Columbia*. 'What a way to come to California!' hollered Crippen as they descended over the coast and headed towards the Mojave desert. Exactly as planned, the Shuttle landed on Runway 23 of the dried saltbeds of the Edwards Air Force Base. Mission Control's greeting of 'Welcome home *Columbia*, beautiful!' was echoed with tumultuous applause the world over. The crew emerged later, grinning from ear to ear, in the sure knowledge that they had ushered in a new era of space exploration.

PROBLEMS AND PROGRESS

NASA had originally decreed that six test flights would be needed before the Space Transportation System was declared fully operational. In the event only four were required. In November 1981, Joe Engle finally entered space when he and Dick Truly piloted *Columbia* on its second mission. Problems with the onboard power cells meant that the mission had to be curtailed from five days to two. In March 1982, the sight of the Shuttle's main fuel tank – left an unpainted orange colour – set the pattern for later flights. Instead of painting the tank white, an extra 270 kg (595 lb) could be carried on board. Because of flooding at Edwards, the third mission came to an end when *Columbia* landed at Northrop strip in

New Mexico, a back-up landing site. Gypsum salts there got into the main engines so that parts of them had to be replaced.

STS-4 saw for the first time restricted TV coverage of Ken Mattingly and Hank Hartsfield's activities because they were flying classified experiments for the Department of Defense (DOD). This was a taste of things to come, as the DOD was a strong customer of the Shuttle, a fact which many observers regretted. Their fears were intensified after *Columbia* landed at Edwards and the crew were greeted by President and Mrs. Reagan. Instead of announcing his approval of a space station for NASA (for which the Shuttle had been originally designed), Reagan merely affirmed that space would continue to play an important role in national security.

Columbia's first operational flight took place in November 1982, when four crew members were accommodated by removing the ejector seats, wearing only blue overalls with helmets for emergency oxygen supplies. They successfully deployed two telecommunications satellites, and delighted in showing TV viewers a card which announced: 'Ace Moving Company. Fast And Courteous Service. We Deliver.'

Such a boast could not be made by the crew of the next flight in April 1983, which saw the third orbiter *Challenger* enter into service. An expected increase in telecommunications as the Shuttle became operational led NASA to develop a trio of communications satellites called TDRSS (Tracking Data Relay Satellite System). *Challenger*'s task was to deploy the first of them in orbit. However, the Inertial Upper Stage which was supposed to boost the satellite into geostationary orbit failed to fire properly, and TDRS-1 was deployed in a totally useless orbit. The lack of data communication ability resulting from this accident was to have serious repercussions on later flights.

SEARCH AND RESCUE

Lost communications satellites were to become a recurrent feature of the Shuttle programme. It was hardly the astronauts' fault: after leaving the safety of their hands and the Shuttle's remote manipulator arm the satellites were entirely at the mercy of the booster motors. Two further communications satellites were lost in February 1984 when another rocket motor, the Payload Assist Module, misfired in both cases. The following April, a Syncom satellite failed to operate when its radio antenna failed to open after being 'sprung' from the payload bay by a giant spring device.

In both these cases, however, the satellites were rescued in later Shuttle missions. Yet many observers questioned the wisdom of going to the trouble of rescuing satellites: to them it was like shooting yourself in the foot and then taking the credit for applying a bandage. And when the two satellites lost in February 1984 were returned to Earth the following November, nobody could be persuaded to buy the 'second-hand, as new' satellites for two years.

Despite the problems there were many successes. Dr. Sally Ride became America's first woman in space on the seventh Shuttle flight in June 1983. She was at pains to point out she was first and foremost an astronaut, with her gender playing a very little part in her choice. This view was somewhat compromised by Mission Control who had announced: 'Lift-off of STS-7 and America's first woman astronaut.' After the flight she declared that it was 'the most fun' she had ever had, an experience she was to repeat in October 1984 with Kathy Sullivan, who was to become the first American woman to perform a spacewalk.

On the tenth flight in February 1984, Bruce McCandless became the first human satellite. He had joined the astronaut corps in 1966 and served as a back-up to many of the crews who flew on Apollo and Skylab missions. Since the mid-1970s, he had been part of the development team for a remarkable device known as the Manned Maneuvering Unit or MMU. The MMU was essentially a spacesuit with a back-pack plus thrusters in which an astronaut could fly untethered to the Shuttle. It would allow astronauts greater flexibility to rescue stranded satellites. He was to test the device out on his first foray

Trouble in store? On *Columbia's* maiden voyage, delicate heat tiles were seen to be missing when the crew examined their maneuvering engine 'pods'. Though these tiles were not particularly critical, fears that others would be missing from the under belly reached epidemic proportions among the press reporters on the ground. Luckily they were unfounded.

into space on 3 February and declared: 'That may have been a small step for Neil, but it was a heck of a big leap for me!' He journeyed as far as 50 metres (160 ft) away from *Challenger* and had the time of his life.

But the MMU wasn't designed for fun. On the next mission, astronauts George Nelson and James van Hoften rescued the Solar Maximum satellite, unthinkable without the MMUs. The satellite had been launched in 1980 to investigate the Sun and the effect its radiation had on the Earth's environment, but after nine months it had failed because a handful of fuses had blown. Despite problems on the first attempt, Nelson and van Hoften succeeded in 'capturing' the satellite and repairing it in the payload bay. They had given Solar Max a new lease of life.

PAYING PASSENGERS

Another unqualified success for the Shuttle programme was Spacelab, a research laboratory built for NASA by the European Space Agency (ESA). Spacelab consisted of a pressurized module with external instruments mounted on a pallet which could be fitted inside the Shuttle payload bay. The pressurized module and the pallets could be flown together or separately, but on Spacelab 1 they were flown together. The mission was flown in November 1983 on STS-9 under the command of John Young, who was making his sixth flight into space. One of the crew of six, the largest number of astronauts yet flown on one mission, was the first ESA payload specialist Dr. Ulf Merbold. He was a West German with research interests in crystal lattice defects which could best be investigated in the weightlessness of Earth orbit. Along with NASA's Byron Lichtenberg and Owen Garriott, they operated 20 instruments inside the pressurized module and 16 on the pallet with great enthusiasm. Despite the wide range of instruments (covering everything from X-ray telescopes to a special sled inside the module which moved back and forth to test astronauts' inner ear responses) NASA declared the flight to be a total success.

The second Spacelab mission was flown aboard *Challenger* in April 1985, and known as Spacelab 3 (problems with Spacelab 2 resulted in its delay to a later flight). Jet Propulsion Laboratory physicist Taylor Wang experimented with a device that levitated liquid drops with sound waves in an attempt to manufacture pure materials in space. Another experiment saw the growth of a mercuric oxide crystal to 3–4 times bigger than expected. Its manufacture in zero gravity meant that it contained no defects and was purer and larger than any Earth-grown crystals. The test sample was worth many millions of dollars.

A number of animals, including two squirrel monkeys and 24 rats were also flown onboard Spacelab 3, contained within cages that cost a staggering $10 million to build. Four monkeys were to be flown originally, but then they were found to have a virulent strain of Herpes which the crew could catch. Unfortunately when the cages were opened in zero gravity, food and droppings floated out as far as *Challenger*'s cockpit much to the crew's horror. They were then forced to don surgical masks and to use vacuum cleaners to remove the mess.

The Spacelab 2 experiment package flew on *Challenger* three months later. The pressurized module wasn't flown, and an array of astronomical telescopes sensitive to different wavelengths were carried on pallets in the payload bay. In November 1985, Spacelab D-1 was flown, the D standing for Deutschland as the mission was paid for by the West German government. Reinhold Furrer and Ernst Messerschmid, who performed a variety of materials processing experiments, became West Germany's first national astronauts. ESA payload specialist Wubbo Ockels, Dutch by

The sight of a Shuttle launch became all too familiar to T.V. audiences around the world. Any thoughts of the routineness of the Shuttle program were belied by problems behind the scenes. Here *Challenger's* last successful launch in November 1985 is shown.

birth, performed a variety of medical experiments in the Spacelab module.

The Spacelab payload specialists were among the more popular of the 'paying passengers' on Shuttle flights. By mid-1985, political considerations had come into play when Senator Jake Garn became an observer on the 16th Shuttle flight in April 1985. As chairman of the Senate appropriations committee responsible for NASA's budget, he had pestered the agency for four years to be allowed to fly. Not wanting to bite the hand that feeds, NASA had given him the barest minimum of training and the opportunity to fly. It set a precedent, for on the mission before *Challenger*'s explosion Congressman Bill Nelson – who held a similar position to Garn in Congress – followed in his footsteps to fly on the Shuttle.

Of even greater worry for the astronaut crews was that NASA offered payload specialist positions to engineers and representatives of the aerospace corporations whose satellites or experiments were being flown. McDonnell Douglas engineer Charlie Walker flew no less than three times to use his electrophoresis experiment to develop new methods of drug manufacture. In due course, engineers from RCA and the Mexican Morelos satellite group would take their turn in the Shuttle cockpit, where they did very little but watch the mission specialists release their satellites. In June 1985, a Saudi Prince by the name of Sultan Salman Al-Saud flew on the *Discovery* mission which released the Arabsat 1B communications satellite. As a TV advertising executive – and a nephew of King Fahd of Saudi Arabia – his mission was regarded with horror by veteran crews. Many felt that NASA had become a travel agent for flying VIPs in an attempt to minimize the upset to their parent companies if anything went wrong. Said one astronaut: 'There has to be a difference between some senator or Arab prince and the qualified scientists who fly aboard the Shuttle.'

FURTHER DELAYS

By mid-1985 NASA's original notion of one flight per week looked a long way off: flights were averaging one per month, if that. NASA was finding that in flying as complicated a vehicle as the Shuttle there were headaches aplenty. Silica heat tiles had to be replaced and parts of engines had to be cannibalized from other orbiters. Starting with the first Shuttle flight of 1984, NASA had introduced a new numbering system designed to help in the smooth flow of Shuttle flights. So the tenth Shuttle mission became '41-B' where the '4' indicated that the money for the flight had come from NASA's 1984 budget allocation, and the '1' indicated a launch from Cape Kennedy. The letter indicated the order of flight during that financial year. Because military Shuttle flights would start out from the Vandenburg Air Force Base, it was designated site '2' – starting with flight 62-A in July 1986, which would see Secretary of the Air Force Edward 'Pete' Aldridge as an observer/passenger.

Search and Rescue. In August 1985, astronauts 'Ox' van Hoften and Bill Fisher rescued a Syncom satellite that had been incorrectly deployed the previous April. Though human beings were versatile enough to handle the delicate operations involved, many questioned the Shuttle's worthiness.

To many outside observers, the new numbering system seemed too labyrinthine: flight 51-B was slotted between flights 51-D and 51-G. The doomed flight of *Challenger* in early 1986 was designated 51-L, slotted in after the first three flights with the '61' prefix. It was all very confusing, particularly when flights were delayed or cancelled altogether like mission 51-E, after which the crew were split up and re-assigned. The bumping of flights and passengers meant that in some cases the computer software tailored for each flight was not available for use in the flight simulator until a week before lift-off.

Worse still were the near misses. In August 1984, Hank Hartsfield's crew stayed put in *Discovery* on its maiden flight after a faulty ignition valve caused a small fire to break out, only four seconds before the solid rocket boosters were to be ignited. Water jets managed to extinguish the flame, but the crew remained patiently aboard the potential bomb for 40 minutes while propellants from the main tank were drained. Something

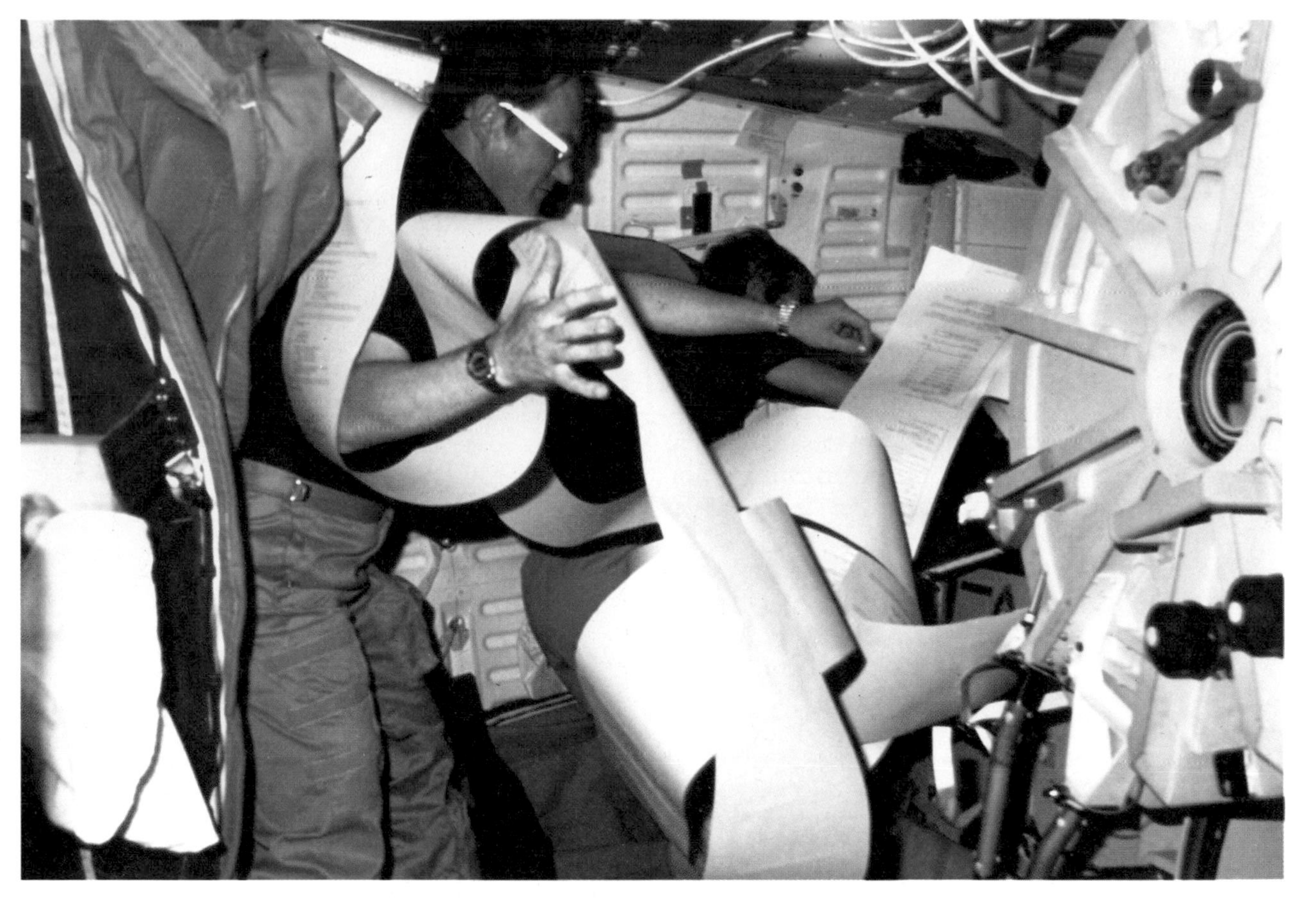

Ad Astra Per Bureaucracia – Commander Karol Bobko wrestles with 'overnight' teletype commands from mission control. Merely keeping a Shuttle in orbit required very detailed marshalling of resources, especially when there were problems. On this flight, in April 1985, a Syncom satellite was lost. Note the main airlock to the right of the picture.

similar happened on the flight which launched Spacelab 2 into orbit, 51-F, in July 1985. On 12 July, a coolant valve within the main engines failed to close and though no fire occurred, they were within three seconds of solid ignition. After the launch was successfully achieved on 29 July, the central main engine shut down because it was overheating. By readjusting the fuel to make the other two engines work more efficiently, *Challenger* ended up in an orbit 70 km (44 miles) lower than planned. 'It's been an adventuresome time,' commented the flight commander, Gordon Fullerton, on safely reaching orbit.

Another headache was the fact that only a handful of landings had been made at the Kennedy Space Center. NASA had made much of the fact that alligators had to be cleared before a Shuttle could land on the 4,500 metre (2.8 miles) long concrete runway. But there was nothing NASA could do about poor weather at the Cape which continually meant that landings had to be made at Edwards AFB in California. Some flights were 'waived off' with only minutes to spare before a Kennedy landing was attempted. By the end of 1984, only two of the ten operational Shuttle flights had landed at Cape Kennedy. The Edwards landings meant that the Shuttle orbiter had to be mated to a 747 to be transported back to Florida, adding further delays to the launch schedule.

In April 1985, strong crosswinds at the Cape were encountered by *Discovery* as it came in to land at the end of mission 51-D. Commander Karol Bobko had to apply differential braking pressures on the individual landing wheels to keep the Shuttle moving in a straight line and in so doing one of the tyres burst. From that point onwards, no more Kennedy landings were allowed.

THE YEAR FOR SPACE SCIENCE

Though 1986 dawned in an atmosphere pervaded by worries about maintaining the strict flight schedule, the year promised to see the Shuttle reach maturity with the launch of a series of probes to the planets. NASA had confidently announced 1986 to be 'The Year For Space Science'. In May, two probes would be launched within a week of each other. One called Ulysses, a joint project with the European Space Agency, would head up and over the poles of the Sun, to investigate regions of our star that had never been seen before. The other was Galileo, designed to reach Jupiter, drop a probe into its multi-hued atmosphere, and spend nearly two years orbiting the world, looking at its four largest, and inexplicably curious moons. September would see the launch of the Hubble Space Telescope, the largest optical instrument ever sent into space. It promised a quantum leap in the understanding of the Universe with its giant 239 cm (94 in) mirror harnessed above the turbulent interference of our atmosphere. Though NASA didn't send a probe to Halley's Comet on its return to the inner Solar System in early March, mission 61-E on *Columbia* would be launched at that time to allow astronomer astronauts to use a telescopic instrument called Astro to make extensive observations of the comet's nucleus.

The May launches were critical: their timing could not be altered, for interplanetary mechanics could not be changed. This meant that *Columbia*'s 61-E mission in March and *Challenger*'s January flight could not slip, because of the refurbishment time required to ready each Shuttle vehicle for its next flight. But already, 12 days into the new year *Columbia*'s first flight of 1986 had been delayed, originally planned for launch in December 1985. After five scrubbed launch dates (including one delay which stopped the mission in its tracks only 20 seconds before launch) *Columbia* finally reached orbit on 12 January, with Bill Nelson, the Floridan congressman aboard. Though they successfully deployed a Syncom satellite, an experiment called the Comet Halley Active Monitoring Program (CHAMP) failed because of a faulty battery. It was hardly an auspicious start for 'the year for space science'. To try to ensure that the Astro mission in March would lift off on schedule, NASA wanted to bring *Columbia* home a day early. In the event, bad weather meant that there was a delay and the crew had to come down in California, adding further delays.

The scheduling headaches were becoming so acute that NASA wanted *Challenger* on mission 51-L launched as soon as possible. The 25th flight of the Shuttle series, *Challenger* would be heavier than before as it was carrying a replacement TDRS satellite for the one lost in 1983. But most attention focussed on Christa McAuliffe, a 37-year-old teacher who had been chosen out of 10,690 educators to take part in the Teacher In Space Project. The project had been started at the behest of the White House to appease criticism of the Reagan administration's cut-backs in education, but it also was to emphasize the routine nature of the spaceflight. Later in the year, a journalist would fly in space as eventually would the singer John Denver.

Challenger's Commander was Dick Scobee, making his second flight in space. He had already announced to friends that he would retire from NASA after returning to Earth. Hughes aircraft engineer Greg Jarvis was finally to fly as a payload specialist, after being repeatedly 'bumped' from earlier flights. Had it not been for Congressman Nelson's late addition to the crew of the previous mission, Greg Jarvis would have flown with them. Apart from Mike Smith, the pilot who was making his first flight, the other members of the crew were making their second journeys into space.

But merely launching *Challenger* into space was proving difficult. Launch attempts over the weekend of 24–25 January were cancelled because of predictions of bad weather – but in the event, the weather was fine. Another launch attempt on Monday, 27 January was doomed when the crew access hatch to the orbiter had failed to close on the launchpad. It appeared that the handle on the outside was loose, and time was wasted while a battery-operated drill could be found. When a drill was eventually located, the batteries were found to be dead. These farcical problems ensured that by the time the hatch was fastened properly, the opportunity to launch was well and truly over. A further worry was the fact that ice had repeatedly formed on the launch pads overnight. The next day one of the launch technicians was heard to remark that *Challenger* looked like something out of *Dr. Zhivago*.

'OBVIOUSLY A MAJOR MALFUNCTION'

Finally, mission 51-L was launched shortly after 11.38 local time on Tuesday, 28 January. The vehicle soared gracefully into the air, a sight long since familiar to television viewers. Not that there

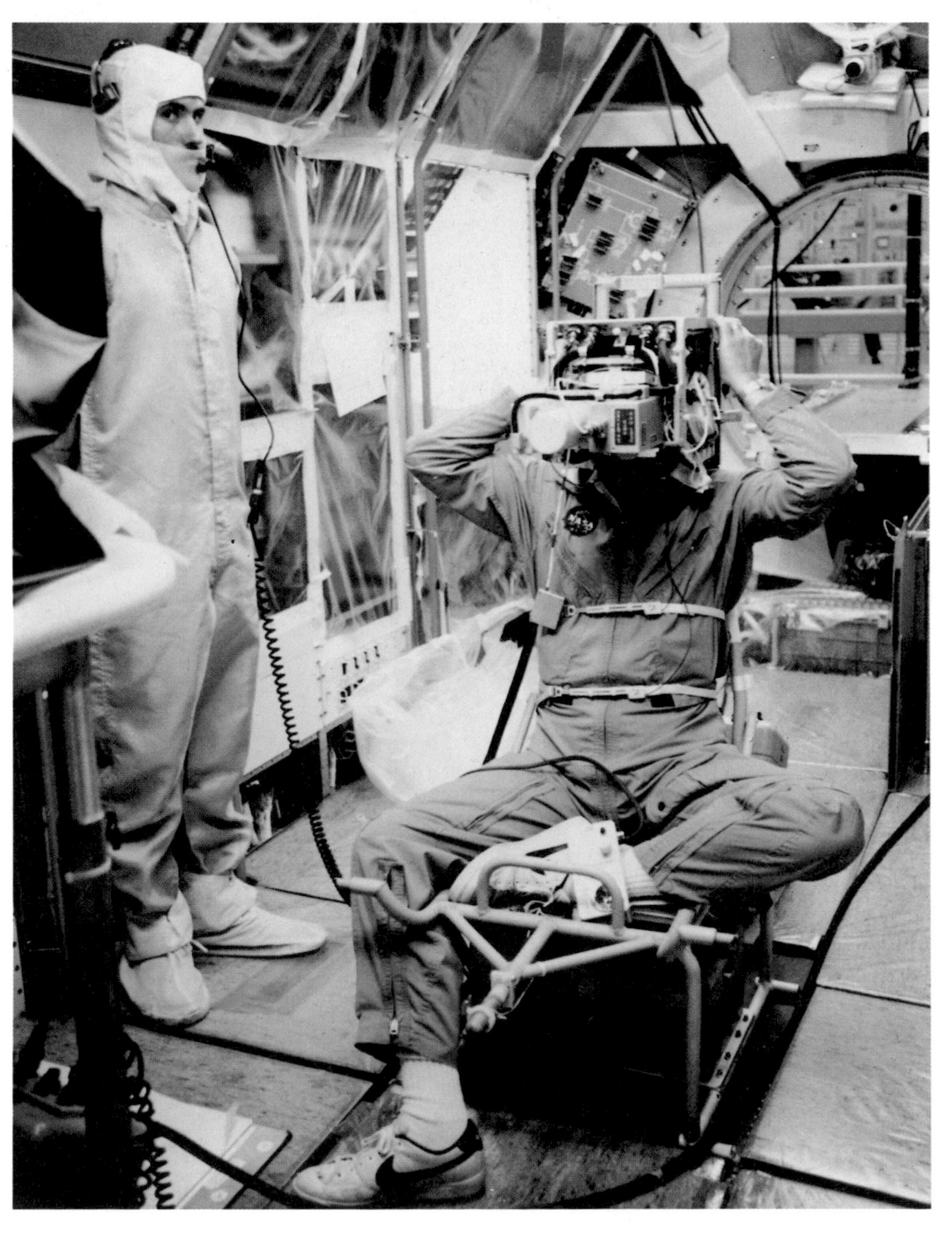

Perhaps the most successful aspect of the Shuttle programme was Spacelab, the European laboratory carried in its hold. Here, Dr. Bob Parker tests a 'helmet' and rotating chair which investigated the human balance mechanisms. Further Spacelab flights involving European and Japanese personnel will occur in 1991.

were very many of them outside of NASA's centres, because most of the TV networks had lost interest in the Shuttle, and were not covering the event, teacher or no teacher aboard. Only a cable network was televising the launch 'live'.

A minute into the flight, 10,600 m (35,000 ft) above the Atlantic, the Shuttle astronauts prepared for the most hazardous part of their ascent – Max Q. This was the point when the aerodynamic forces on the vehicle were at their greatest. Pilot Mike Smith had noted 'there's a lot of wind around here today', as *Challenger* continued to accelerate upwards. Mission Control gave the go-ahead for the Shuttle main engines to be operated at 104 per cent of their rated performance, which was acknowledged by Dick Scobee as 'Go at throttle up!'

An ominous puff of smoke from *Challenger's* right-hand 'solid' is seen at the bottom of this picture, just above the curve of the external fuel tank. Later analysis showed that this was caused by an 'O'-ring vapourizing.

A few seconds later, Mike Smith realized something was wrong. During lift-off his main task was to monitor the engines' performance. The pressure indicators in the main engines showed that they were losing power – and fuel. Automatically, he looked out of his window and saw the frightening sight of one of the Solid Rocket Boosters smashing into the External Tank, the cause of the pressure drop. He mumbled, 'Oh-oh', and six-tenths of a second later, *Challenger* exploded. The orbiter was destroyed in the explosion, though the crew cabin seems to have stayed intact. Rapid decompression rendered most of the crew unconscious. Four of the air packs provided in case of emergencies were activated, but to little avail. The force of the cabin landing in the Atlantic was more than enough to kill the crew.

The Range Safety Officer at the Cape (whose job it was to make sure that civilian lives are not threatened by launch mishaps) automatically armed explosives in the solid rocket boosters and blew them apart. Debris continued to fall out of the sky for 20 minutes afterwards. Out on the launch grandstands, onlookers – including the families of the crew – stood aghast at the ominous white cloud in the sky. At first they believed that the orbiter had survived and would be making an emergency return. But such hopes were dashed by the voice of Mission Control.

Thirteen seconds after launch, monitoring of the Shuttle flight automatically passed to the Johnson Space Center in Houston. Mission commentator Steve Nesbitt was monitoring the launch well away from TV monitors, with only the engineering and scientific data relays available to him – a bizarre decision forced by the whim of NASA officials in Washington. The data on his screen suddenly stopped. Measuring his words very carefully, Nesbitt said: 'Flight controllers looking very carefully at the situation. Obviously a major malfunction. We have no downlink. We have a report from the flight dynamics officer that the vehicle has exploded.'

AFTERMATH

The results of the explosion were immediate. All Shuttle flights were suspended, and a Commission was set up by the President to investigate the cause of the explosion. Ronald Reagan's annual State of the Union message the following day became a moving tribute to the fallen crew. All over the United States, flags were flown at half mast, such was the emotion felt at the human loss. At a service held at the Johnson Space Center the following Friday, the President confirmed that manned space flight would continue when the

The hidden fear of 25 years of manned space exploration exploded into reality above Cape Canaveral on 28 January 1986. Seconds after the explosion which destroyed *Challenger* a swollen cloud of vapour became visible. Debris from the explosion is already visible falling towards the Earth.

cause of the accident was found. He added that this would have been the wish of the fallen crew.

The Presidential Commission made its report in June. Chaired by William Rogers, a former Secretary of State, it included astronauts Neil Armstrong and Sally Ride, as well as Nobel laureate Richard Feynman amongst its 13 members. After examining all the evidence, including debris salvaged from the ocean floor, they announced their findings. The cause of the accident was that a casing joint in one of the solid rocket boosters (known as an 'O'-ring) had failed. The 'O'-ring was essentially a synthetic rubber ring contained inside a socket whose function was to ensure that the separate sections of the booster (joined together before launch) held together. TV pictures taken two seconds after *Challenger* lifted off from the pad spotted a puff of black smoke from a position where an 'O'-ring was located in the right-hand booster. Within a minute, a flame appeared at the same location and eventually engulfed the whole vehicle. Booster wreckage from the ocean floor was located and analyses corroborated the 'O'-ring hypothesis.

The Commission also unearthed a veritable can of worms, with top NASA management revealed as making questionable decisions about the Shuttle programme. The agency that had successfully landed Americans on the Moon was revealed in a new light of incompetence and failure. Problems with 'O'-rings had been noted, but remained hidden in administrative files, such was the pressure to maintain the flight schedule. The new NASA management decreed that the Shuttle would no longer carry commercial payloads, only scientific and military ones. The dreams of a commercial airline were shattered.

The Commission made a number of recommendations to NASA to improve the Shuttle vehicle and its boosters. New safety features such as an emergency egress system and re-designed booster joints were incorporated into the Shuttle. It was to take 20 months before the agency was ready to launch the next Shuttle. On 29 September 1988 – only two days before NASA's 30th anniversary – *Discovery* and its crew of seven successfully returned to flight. The symbolism of a phoenix rising from the ashes seemed entirely appropriate, as the crew successfully deployed a TDRS satellite. But more symbolic, perhaps, were the words of Mission Control as *Discovery* successfully touched down at the Edwards Air Force Base. 'Welcome home, *Discovery*. A successful end to a new beginning.'

THE GOLDEN AGE OF PLANETARY EXPLORATION

For planetary astronomers, the start of the Space Age could not have been better timed. As the 1950s drew to a close, 8ations with even the best astronomical equipment made it difficult to advance any further in their studies of the distant planets which shimmered in their telescope eyepieces. It was clear that the only way forward was to visit the worlds for themselves. The dawn of the Space Age in 1957 saw that oft-expressed wish turn rapidly into reality.

The scientific basis for exploring the planets was dramatically demonstrated by the results of the first U.S. Satellite, Explorer 1. Its simple geiger counter almost immediately discovered the Earth's radiation belts, soon to be named in honour of James van Allen, the scientist who built the instrument. This discovery led to the development of satellites in Earth orbit which investigated the way in which particles and plasmas from the Sun interacted with our planet's magnetic field and how our weather is affected as a result. It was clear that by visiting the other planets, spacecraft equipped with scientific experiments could tell us a great deal about our planetary neighbours.

In January 1959, the Russians launched Lunik 1 towards the Moon – only their fourth probe after Sputnik 1. Although it missed the Moon completely, it showed that Soviet space scientists had been planning the exploration of the planets for a long time. We can only guess at the rationale behind their work, but in the case of the United States, history records that the Jet Propulsion Laboratory (JPL) was asked by NASA in 1959 to

Unmanned exploration of the Moon neatly complemented that by the Apollo missions. For the first time, astronomers were able to see the rugged far side of our celestial neighbour.

come up with a logical plan to explore the planets. Starting with the Moon, JPL outlined that Venus and Mars should be next, followed by the outer planets. It was altogether a very ambitious plan, with the first human landings on Mars planned for 1971!

In reality, the race to land men on the Moon in the 1960s saw the financial floodgates open, and naturally enough, planetary exploration rode on the coat-tails of this expenditure. Sadly for NASA, when funds dried up in the 1970s, planetary science suffered. With cost overruns in the development of the Space Shuttle, planetary science was soon relegated to a position akin to Cinderella compared to her ugly sisters.

MAPPING THE MOON

In the first few years of the Space Age, the Moon became the focus for the fledgling planetary exploration programmes of both East and West. The Soviets scored an early success with Lunik 2 which became the first probe to hit the surface of the Moon in September 1959. It was followed in early October by Lunik 3 which successfully flew behind the Moon, returning the first views of its 'hidden' hemisphere that can never be seen from the Earth. The Soviets scored a number of other lunar 'firsts' including the first successful soft-landing, achieved on 3 February 1966 by Luna 9, after five previous failures. The small, ball-shaped lander returned the first pictures of the lunar surface, which were intercepted by the Jodrell Bank radio telescopes and shown to the world before the official versions were released, much to the Soviets' ire. In April, Luna 10 became the first probe to enter lunar orbit.

NASA's first attempts to reach the Moon in 1959 and 1960 were dogged by repeated failure. It was clear that new designs of spacecraft were needed, especially if the same technology would be adapted to send probes to the planets. To support the Apollo programme, NASA decided on a three-point plan: a series of Ranger spacecraft would take detailed photographs as they headed towards the surface and their destruction; Surveyor spacecraft would make the first soft landings; and Lunar Orbiters would map the whole of the Moon from which Apollo landing sites would be chosen.

The first Rangers suffered heartbreaking failures in 1962 and 1963. This was mainly because their launch vehicles were not very powerful, and did not allow the luxury of back-up systems to be carried. However, on 31 July 1964, Ranger 7 successfully reached the Moon after, appropriately enough, 13 failed attempts. It returned over 4,000 pictures of the Sea of Clouds before impact, and was followed by two further successful missions in the Ranger series. In May 1966, Surveyor 1 achieved the seemingly impossible: it became the first in a new series of spacecraft to successfully land on the Moon at the first attempt. Only two of the series of seven (Surveyors 2 and 4) failed, and the results of the programme were enough to pave the way for manned landings and scotch bizarre notions that the Moon was covered with vast expanses of dust which would totally engulf landing vehicles. During 1966–7, five Lunar Orbiters mapped the whole of the Moon in order to finalize the choice of Apollo landing sites.

After Surveyor 7's successful touch down on 10 January 1968, NASA's next Moonlanding was destined to be Apollo 11 with Armstrong and Aldrin aboard. With Apollo reaching its climax in early 1969, the Soviets made ever more desperate attempts to steal some of the propaganda thunder. The Zond missions in late 1968 were most likely tests for a manned fly-by, and in desperation, they attempted unmanned sample returns in early 1969. Three days before Apollo 11 was launched, Luna 15 was a last-ditch attempt at returning lunar soil, but it crashed. Undaunted, the Soviets spent the early years of the 1970s perfecting automated techniques on the Moon. They successfully despatched two roving vehicles – or Lunokhods – across the Moon's surface, both of which returned much useful information. The final Luna probes in the early 1970s saw the final generation of orbiters and sample return missions. The last took place in August 1976, when

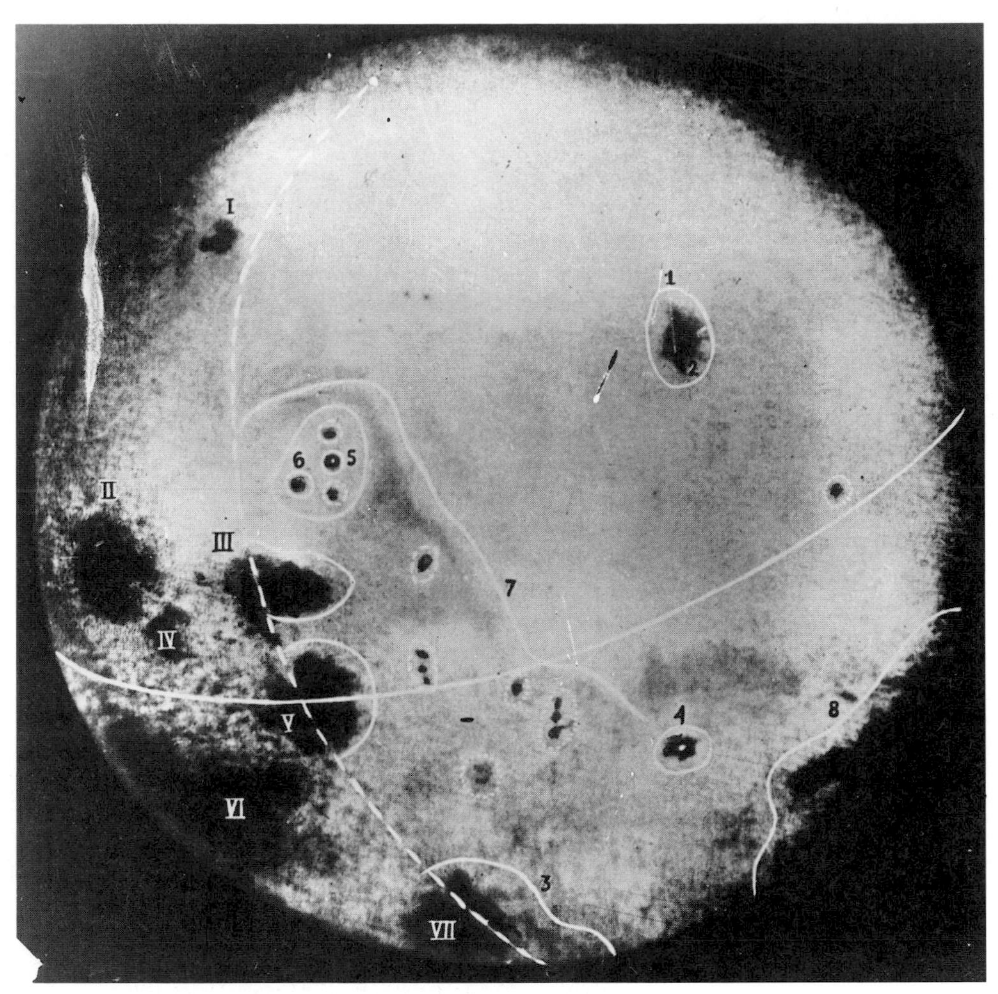

The first views of the Moon's far side were provided by the Luna 3 spacecraft. They were crude by today's standards, but in 1959 they were regarded as near miraculous. New features which were named by Soviet astronomers include: 1) The Sea of Moscow; 4) Tsiolkovskii; and 7) Soviet Mountains. Roman numerals indicate objects on the visible face of the Moon.

In October 1967, Venera 7 became the first successful probe to penetrate the dense Venusian atmosphere. The craft's strengthened parachute lines are seen in this mock-up displayed in Moscow.

Luna 24 returned a small handful of rocks back to Earth.

Since then nothing has ventured back to our celestial neighbour. The great irony is that, despite the highly intensive surveys – both manned and unmanned – many fundamental questions remain about the Moon. A prominent lunar scientist has aptly described the Moon as 'the joker in the planetary pack'. Before the first lunar samples were returned, many scientists expected that the rocks would solve most of the fundamental mysteries when they were analysed. Sadly, this proved to be false: sample analysis has not revealed whether the Moon was once part of the Earth or whether it was captured by the Earth millennia ago. Questions remain: Does the Moon have a molten core? When did volcanism stop taking place? What about the chemical composition of the whole of the Moon? Only the chemistry of about a quarter of the total surface area is known, with virtually nothing known about the chemistry of the far side and the poles. A new generation of polar orbiting probes is expected to answer these many questions in the 1990s.

VENUS AND MARS

Throughout the 1960s, attention focused on the inner Solar System and the Earth's neighbours in space – Venus and Mars. Before the space age very little information was known about them, which led to the development of many bizarre theories concerning both. Venus was known to be heavily covered with clouds rendering its surface to telescopic observers a featureless haze, and carbon dioxide had been detected in the atmosphere. When it was suggested that water was the prominent component of the clouds, the notion of seas of soda water became popular. Others decided that swamps infested with dinosaurs would be found lurking under the cloud tops. In the realms of scientific reality, it had been found that Venus rotated in completely the opposite direction to the other planets and it rotated extremely slowly. The planet's day was longer than its year: that is, it took only 225 Earth days to complete one orbit of the Sun compared to the 243 it took to rotate on its axis. Clearly, Venus was a strange world.

This strange world was to be the target for the first successful probe to reach another planet – NASA's Mariner 2 in December 1962. Coming as it did during the troubled development of the Ranger spacecraft, it acted as a morale boost for the Jet Propulsion Laboratory. Its results were revelatory: the surface of Venus, it seemed, was hotter than the melting point of lead and sweltered beneath a dense carbon dioxide atmosphere. There seemed to be no evidence for a magnetic field. More startling information came from a series of Soviet Venera spacecraft which scored a considerable number of 'firsts'. In October 1967, Venera 4 successfully penetrated the thick atmosphere of Venus and returned radio signals for over an hour. This was quite an achievement considering the atmosphere at the surface had a pressure some 90 times greater than the Earth's. In December 1970, Venera 7 returned data for 23 minutes from the surface before it was claimed by the intensely hostile conditions.

Mars, too, was of paramount interest to the superpowers and they started to explore it in earnest during the mid-1960s. If possible, the pre-Space Age view of the red planet was even more preposterous. Reports by 19th-century astronomers of linear markings on the surface were ascribed to canals being built by intelligent Martians. The American astronomer Percival Lowell became their champion, and in the early part of the century the notion of the canals acting as an irrigation network became popular.

The first attempts to send probes there were made by the Soviets, but they failed miserably. It wasn't until July 1965 that Mariner 4 successfully flew past returning 11 photographs of a small part of the southern hemisphere. They showed a crater-pocked surface rather than canals, and soon Mars became the 'dead' planet in popular conception. Mariner 4 also showed that the atmosphere of Mars was much thinner than had been previously thought and for the most part composed of carbon dioxide. In turn, two more Mariners (6 and 7) flew past Mars in 1969, just a few days after the first manned Moonlandings. They revealed nearly 20 per cent of the surface, and that the south polar caps were made of frozen carbon dioxide.

Every 17 years, Mars comes particularly close to the Earth, a fact which was put to good advantage by space planners. 1971 was just such a vintage year and NASA successfully launched Mariner 9, the first Mars orbiter, after its identical twin crashed into the Atlantic a few weeks before. The Soviets attempted to make the first landings on Mars with Mars 2 and 3, but were thwarted by the development of a fierce dust storm which was raging by the time they arrived there. The probes could not be reprogrammed, and the landers were dropped onto the surface only to be engulfed by the dust. In 1974, they tried again with four probes, only one of which, the orbiter Mars 5, was a success.

ON TO MERCURY

The Soviet Mars spacecraft were handicapped by the unreliability of their technology, so until it could be improved, Russian scientists turned their attentions to Venus, which is much nearer to the Earth. After all, Venera spacecraft had survived the crushing heat and pressure of the inhospitable Venusian atmosphere. In October 1975, Veneras 9 and 10 survived for long enough to return the first pictures of the Venusian surface. They revealed a strange, rocky landscape in which oddly-shaped stones could be seen. The dense cloud layers meant that the Sun could not be seen directly. Soviet scientists described the lighting level as similar to that which is found at midday on a cloudy winter's day in Moscow.

In 1973, NASA's Mariner 10 had flown past Venus en route to Mercury. It returned detailed pictures of the cloud tops, which in the ultraviolet began to show features. The clouds above the equator were moving at 100 m/sec (328 ft/sec) in the opposite direction to the Sun, despite the slow rotation of the planet underneath. Despite the heat at the surface, the temperatures of the cloud tops were found to be −23°C (−9°F).

The technique of 'gravity assist' was used for the first time by Mariner 10. It used the gravitational field of Venus to allow it to head towards Mercury without need for an extra rocket stage. Mankind's first views of Mercury came in March 1974 when the small planet closest to the Sun was revealed as a rocky, airless world heavily scarred by craters from impacts millennia before, superficially like the Moon. Appearing like a vast bullseye on the surface was a huge multi-ringed crater some 1,300 km (812 miles) across and named Caloris Basin. By carefully adjusting Mariner 10's orbit around the Sun, Jet Propulsion Laboratory engineers allowed two more close

Innermost Mercury, the densest of the Earthlike planets, is seen in this mosaic of Mariner 10 photographs. The heavily-cratered surface bears witness to intense bombardment in the first epoch of the Solar System's history.

passes of Mercury by the spacecraft. In all, Mariner 10 mapped just over half of the surface area of Mercury, and its pictures remain the only detailed views we have had to date.

THE NEW MARS

By far the greatest achievement of planetary exploration in the mid-1970s was NASA's missions to Mars. After arriving in orbit in November 1971, Mariner 9 simply waited for the dust (which had affected the Russian Mars spacecraft) to clear and returned the first pictures of the whole of the planet. By sheer bad luck, the earlier Mariners had missed all the most interesting features: massive volcanoes, extensive canyons and dried-up river valleys. After analysing the 8,000 pictures it returned, NASA summed up Mariner 9's results as 'The New Mars'.

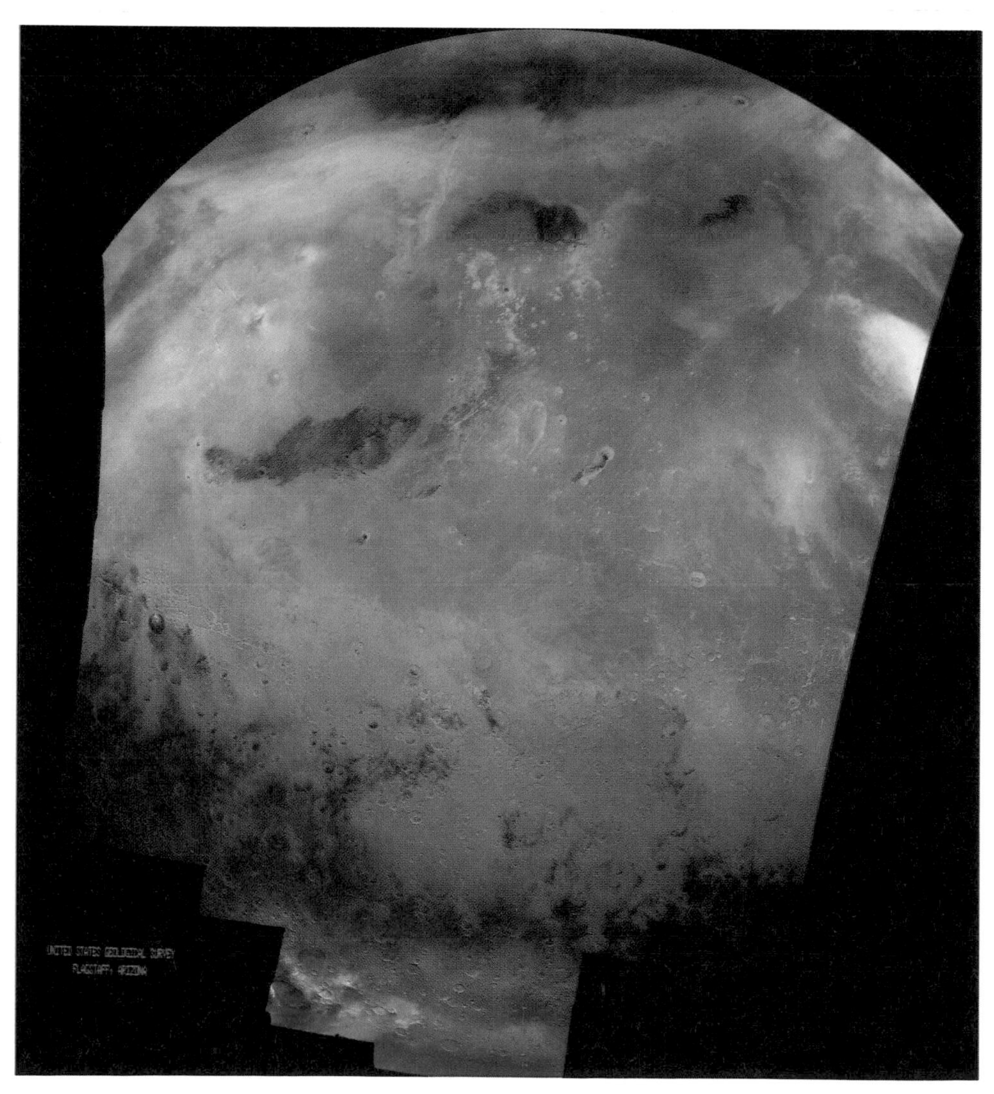

This highly-detailed mosaic of Mars was assembled from over 100 separate images from the Viking orbiters. The plains of the northern hemisphere contrast with the heavily-cratered south, something which remains a mystery. Ice can be seen on the margins of the Antarctic regions.

The next step was even more audacious: to land two probes onto the surface. Named Viking, each consisted of an orbiter slightly bigger than Mariner 9 and a lander equipped with instruments to see if microbial life existed on the Red Planet. Though the data from the Mariners had shown that conditions on Mars were hostile, the possibility remained for Martian microbes. In due course, the Vikings were launched in Autumn 1975 with the hope that Viking 1 could be landed to coincide with America's 200th birthday on 4 July 1976. However, after arriving in orbit in May 1976, its better TV cameras found that the proposed landing sites were far too rough. What had seemed like featureless plains to Mariner 9 were shown to be filled with craters and boulders which could destroy any landing attempts. A hectic search for new landing sites over the next few weeks led to the choosing of a relatively smooth region to the west of a dried up river valley in Chryse Planitia – the Plain of Gold.

On 20 July 1976, the Viking 1 lander was despatched to the surface and successfully touched down, seven years to the day since humans first walked on the surface of the Moon. Within minutes its cameras were returning pictures of a barren, rocky landscape. Later, colour pictures revealed the sky was coloured pink due to dust in the atmosphere. Two months later Viking 2 came down in Utopia, revealing much the same landscape. But most attention focussed on the Viking landers' biological experiments. Each spacecraft could only perform limited tests to search for any metabolic activity in the soil. Sadly, the results were deemed inconclusive but the lack of evidence for organic – that is carbon-containing – material without which life cannot exist narrowed its possibilities. However, that isn't the final word on the subject. Samples were only taken from two sites on the surface, and there remains the possibility that life may exist – or may have existed – elsewhere. The answer to the puzzle about life on Mars remains as tantalizingly elusive as it ever was.

NASA had hoped that the Viking spacecraft would last for 90 days at most. However, both orbiters operated for many months and were only shut down after they ran out of attitude control fuel which kept them stabilized in orbit. But the landers continued to operate until April 1980 (Viking 1) and November 1982 (Viking 2). Their legacy is an intimacy with a planet that is matched only by that extended to our own.

IN THE GIANT'S REALM

By the late 1960s, both the technology and the opportunity to journey to the outer planets became available. Unlike the inner, earthlike planets, those that reside in the outer recesses of the Solar System are giant balls of gas with vast retinues of moons and rings. Even by the standards of the inner planets, the gas giants were more alien. Journeying towards them would not be an easy task: the distances involved were enormous and there was also the cold and the potential hazards of the interplanetary environment to contend with. In a quietly optimistic mood, NASA built two simple spacecraft to test the interplanetary water: Pioneers 10 and 11. Identical, and primitive by today's standards,

The second Viking lander revealed the surface of Mars to be a windswept desertscape. The trench (mid-picture) and canister (left) were the result of sample takings which revealed no presence of life. The horizon appeared tilted because one of the lander's footpads came to rest on a stone!

they were launched in March 1972 and April 1973 respectively. The first reached Jupiter in December 1973, the second a year later. Both negotiated the hazardous asteroid belt – the region of interplanetary debris between the orbits of Mars and Jupiter – and passed within the lethal radiation zones of the giant planet. Their primitive cameras revealed much about the giant cloud-covered world, and the spacecraft certainly lived up to their name. Pioneer 11 was re-directed halfway across the Solar System so that it could fly past Saturn. After an epic journey of over five years, it successfully flew past the ringed giant in September 1979 and returned pictures which hinted that there was far more intricate detail within the rings than had been thought.

By then, two more envoys had been launched from Earth, called Voyager; being more sophisticated than the Pioneers, they would be capable of investigating such intricacies as the rings of Saturn. In the late 1960s, planners had noticed that the outer planets would be so aligned in the mid-1970s that a spacecraft could fly past Jupiter, Saturn, Uranus and Neptune in turn, by successively making use of each planet's gravitational influence to pick up speed and head towards the next planet. By aiming at a precise point near each planet, a game of interplanetary billiards referred to as 'gravity assist' could be played to look at the outer Solar System in detail. Two spacecraft were launched to investigate Jupiter and Saturn, with an option that the second could continue and reach Uranus and Neptune.

Both Voyagers were launched in late 1977 to take advantage of the planetary alignments and Voyager 1 reached Jupiter in March 1979. The greater clarity of the Voyager TV pictures revealed Jupiter as a psychedelic maelstrom of colour which atmospheric scientists were at a loss to understand. A vast cloud system known as the Great Red Spot – three times the size of the Earth – whorled with scarcely believable turbulence. Though meteorologists pronounced that all their theories had been 'shot to hell', there seemed to be small-scale order within the planet-wide chaos. Jupiter's atmosphere seemed to behave in the same way as the Earth's oceans did. But nothing prepared the geologists involved with Voyager for the wealth of detail it revealed about Jupiter's four largest moons, known as the Galilean moons after the Italian scientist Galileo who discovered them with the first astronomical telescope in 1610. The sense of novelty, said the geologists, could not have been greater had Voyager been exploring a new Solar System. Callisto was found to be the most heavily-cratered body ever seen: Ganymede's surface was covered by vast icy furrows: Europa was smooth, covered by icy cracks: and Io appeared for all the world like a bizarre, celestial pizza. A few days after Voyager 1's closest approach to Jupiter, analysis of its pictures of Io revealed it to be volcanically active. No less than eight plumes, spewing sulphurous material high above the moon, were seen.

In July, Voyager 2 reached Jupiter and followed up with observations of Jupiter's atmosphere which showed that the weather had changed considerably. The volcanoes had altered too as one of them had ceased to be active. Following Voyager 1's tentative discovery of a ring, Voyager 2 was able to return direct TV pictures of it.

As well as the cameras, the Voyagers carried a full complement of particle and fields detectors. Though obviously not as immediate as the TV pictures, their results showed Jupiter to be remarkable in other ways. They found that Jupiter's magnetosphere – the region influenced by the planet's magnetic field – was the largest structure in the Solar System. An electrical current of more than a million amps flowed between Jupiter and Io as a result of the magnetic interactions between the two. Certain radio emissions detected by the Voyagers were found to be caused by lightning in Jupiter's atmosphere.

For both craft, the next stop on their interplanetary itinerary was Saturn, where, once again, they totally revolutionized human perception of the 'ringed' planet. From the Earth, only three rings can be seen which appear to be 'solid', although they are not. Theorists in the 19th century showed that they would be made up of many individual rings, the debris from a moon that had fallen foul of the planet's immense gravitational pull, perhaps. In November 1980, Voyager 1 found literally thousands of rings, including ones where there appeared to be gaps as seen from Earth. Some of the rings seemed to defy the laws of nature because they were 'kinked', later ascribed to the influence of small moons within the rings and curious effects of the planet's magnetic interactions.

Saturn's retinue of moons also came under close scrutiny, including Titan, larger than our Moon, with a dense atmosphere of nitrogen. It has been likened to the Earth billions of years ago while life was forming, as it also contains organic materials such as methane and ethane. Voyager's cameras revealed the atmosphere to be a hazy orange colour with no hint of what the surface may be like. So Voyager scientists were left to speculate that the surface might be a bizarre mixture of methane seas and icebergs. A smaller icy moon called Tethys, only 960 km (600 miles) across, was seen to be scarred by a crater nearly a third of its diameter in size, looking similar to the Death Star in the film *Star Wars*.

After its brief encounter with Saturn, Voyager 1 headed upwards and outwards from the Sun. It is still returning data about the stream of particles emanating from the Sun – the solar wind – and its effects on the interplanetary environment. Voyager 2 followed in its wake in August 1981 and returned much more data about Saturn, the rings and the moons. However, the spacecraft malfunctioned at a critical moment. The platform which houses its scanning instruments (including the TV cameras) jammed, so that the instruments could not be targeted properly. Engineers were able to free it a few days later, and relieved scientists wore badges which announced 'Uranus or bust' – an event that was to occur four years later.

Voyager in the realm of the Giant Jupiter, the largest planet, was seen in all its glory by Voyager 1 in March 1979. Its turbulent atmosphere is captured in this view, along with two of the Galilean moons in the foreground. Europa (right) appears yellow, while Io (left) appears above the Great Red Spot.

NEARER TO HOME

While the Voyagers were blazing the trail through the outer Solar System, Venus was undergoing even closer scrutiny. In December 1978, two Soviet and two American probes arrived there within a few weeks of each other. The Soviet Veneras 11 and 12 landed on the surface, and revealed that lightning discharges were taking place in the boiling atmosphere. NASA's Pioneer Venus 1 became its first craft to enter orbit around the planet, while the second Pioneer Venus craft dropped no less than five probes into the atmosphere. The data return from all these spacecraft was tremendous. Winds within the atmosphere were measured directly, as was its chemical composition: sulphuric acid, no less, was a component of the uppermost cloud layers. Because 96 per cent of the atmosphere is made of carbon dioxide, heat is effectively 'trapped' beneath the clouds leading to the hell-like conditions (and known as

This computer-enhanced view of Saturn's rings shows intricate detail within them. Rather than the three rings seen from Earth, the Voyagers took their number into the thousands. The thin, outermost 'F-ring' displays brightness variations caused by kinks within it.

the 'runaway greenhouse effect'). Though the planet is now arid, analysis of the atmosphere by radioactive dating techniques hinted that 4 billion years ago Venus may have had oceans.

In 1982, Veneras 13 and 14 succeeded in landing and returned colour pictures of the Venusian surface as well as taking limited chemical analyses of the surface rocks. The results were consistent with earlier landers, and by this time scientists were far more interested in what the general picture of the surface looked like from orbit. The Pioneer Venus orbiter was the first spacecraft to carry a radar mapping instrument which could penetrate the dense clouds and reveal the structure of the surface. Its results were dramatic: continents, mountains and rift valleys were observed, as well as two vast volcanic regions. It seemed that like the Earth, the surface of Venus was affected by plate tectonics – the relative motions of the uppermost regions of its crust. It was left to Veneras 15 and 16 in 1983 to return even more detailed radar views of the surface which showed that there were craters, hills and ridges. Their success was a fitting finale to the highly successful Venera programme.

Despite their success, the Pioneer Venus spacecraft were NASA's last planetary probes to be launched in the 1970s. The sad reality of cost overruns with the Space Shuttle meant that ambitious new missions had to be cancelled. During the early 1980s, American scientists began to refer to the first two decades of Solar System missions with the wistful phrase of 'the golden age of planetary exploration.' Almost miraculously, a handful of missions that had been mooted in the 1970s survived the cutbacks enforced by the Reagan administration. But fate would intervene in the most tragic of circumstances before they could come to fruition.

As we have seen (see page 40), NASA declared 1986 to be 'the year for space science' with launches set for the Galileo orbiter to Jupiter, the Magellan radar mapper to Venus and the international Ulysses probe to fly above the poles of the Sun. The year began well enough with mankind's first detailed look at the planet Uranus, courtesy of the Voyager 2 spacecraft. It was still working after a journey of 3 billion miles that had taken nearly a decade to complete. The cold, blue-coloured gas giant was full of many surprises. For a start, the axis of its rotation is inclined at 98° to its orbit around the Sun, but Voyager found that the cloud-tops were at the same temperature, −215°C (−355°F). This is particularly odd because the poles receive much more sunlight than the equatorial regions and should be much warmer. Until only a few days before the spacecraft reached the planet, no radio noise was detected from the planet. Then, suddenly, the planet's magnetic field was discovered, inclined at 60° to the axis of rotation. Because of the peculiar tilt of the planet itself, the magnetic field is swept into a strange corkscrew-shape that spirals in synchronization with the planet's rotation. While the Voyager scientists tried to make sense of these peculiar facts, the spacecraft's TV cameras detected two further rings to the ones seen from Earth and ten new moons. The final tally is 11 rings and 15 moons. Most of the latter were icy bodies, as the freezing cold of the outer Solar System has ensured that they have undergone only minimal geological activity. But the joker in the pack was Miranda, only 480 km (300 miles) across, found to be an icy hybrid that exhibited geological features seen on other moons in the outer Solar System – faults, criss-crossed grooves, canyons and cliff-like upthrusts. Miranda seems to be the aggregate of a number of moons that were shattered by the gravitational pull of Uranus.

Despite the puzzles, the Voyager scientists were euphoric; the outer Solar System was far more mysterious than even Jupiter and Saturn had prepared them for. But that euphoria was short-lived: four days after Voyager reached Uranus, *Challenger* was destroyed in the skies above Florida. Along with the lives of the crew went all hopes of a revival of the U.S. planetary programme. It was a reality heightened by the return of Halley's Comet in March of that year. Cutbacks had already decreed that NASA would not send probes to the comet. A telescope to investigate Halley, carried aboard *Challenger*, was also destroyed in the explosion.

A COMET CALLED HALLEY

Nevertheless, when the comet did return, a small flotilla of international spacecraft was sent to investigate it. The comet takes 76 years to orbit the Sun, and its path is well enough known to allow spacecraft trajectories to be determined well in advance. Despite the decline in the U.S. planetary programme, the Halley missions saw regular exchanges of information and planning meetings between East and West. Though they look graceful in the night skies, the appearance of comets is deceptive. The long, graceful 'tail' of a comet is little more than a vast stream of gas and dust thrown off its icy core (known as the 'nucleus') by the buffeting and heating of the solar wind. That such gas and dust travels at speeds of many kilometres a second presented many problems for the designers of spacecraft. The international cooperation ensured that the engineers could crack common design problems and enable the scientists to get the maximum data return from each spacecraft.

The first Halley probes to be launched were the Soviet Union's VeGa 1 and 2 just before Christmas 1984. They were so named after the contrac-

Venus was unveiled by the Pioneer Venus Orbiter from 1979 onwards. By March 1981, when this radar map was compiled, 93 per cent of the surface had been mapped. Clearly seen are the mountains, plains and basins of the underlying terrain. The Pioneer Venus Orbiter is still operating a decade after launch.

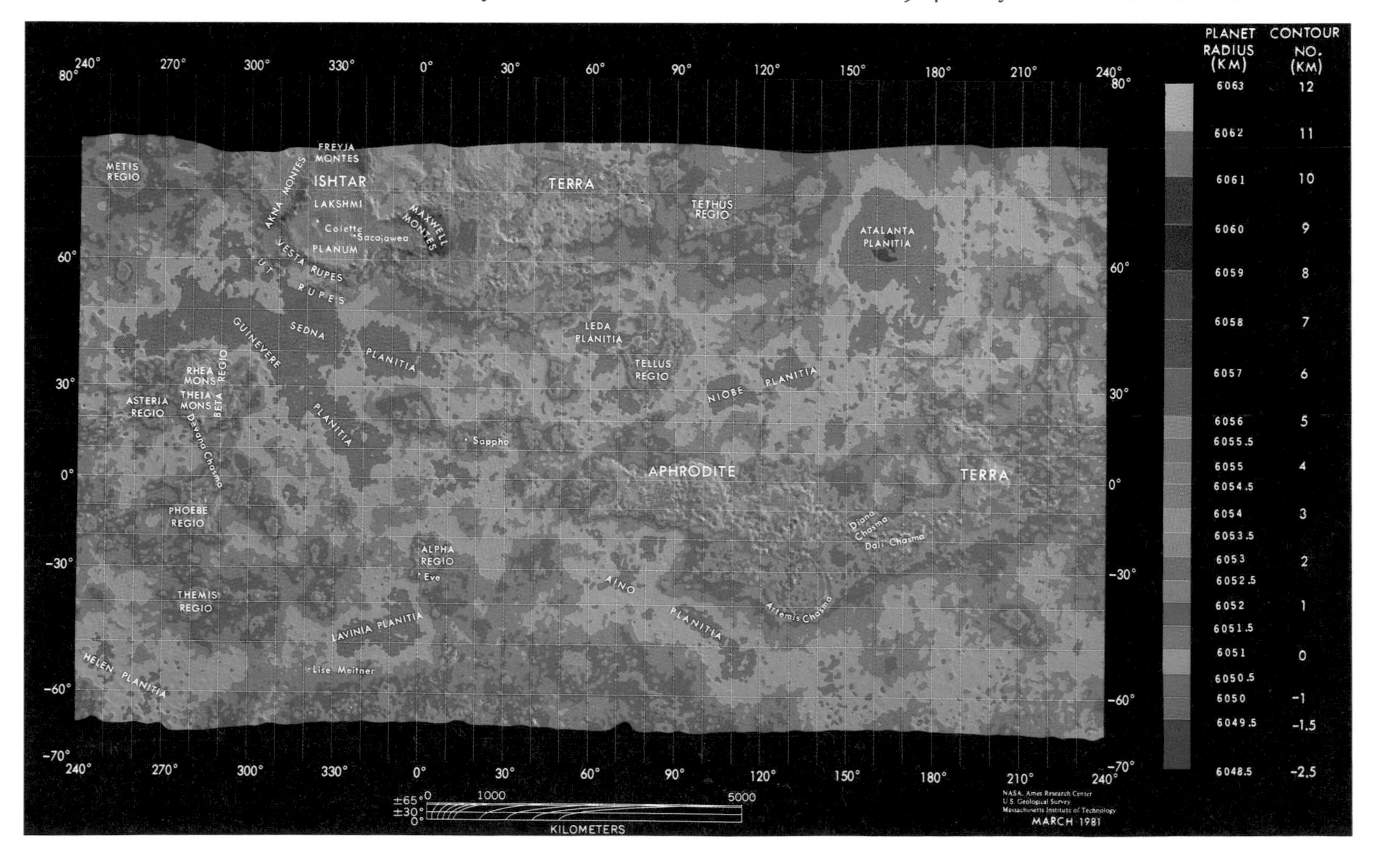

tion of the Russian words for Venus (Venera) and Halley (Gallei), its two targets. In June 1985, the VeGas flew past Venus and dropped balloons into its atmosphere which provided detailed information about currents within the cloud decks. This revolutionary way of exploring our planetary neighbour was another triumph for international cooperation: the idea was proposed by the French and the balloons were also tracked by NASA's Deep Space Network.

By the time the VeGas reached Halley in March 1986, two Japanese probes called Sagikake ('Pioneer') and Suisei ('Comet') had flown ahead of the cometary nucleus to investigate the interactions with the solar wind. The VeGas were targeted to fly within 6,000 km (3,750 miles) of the nucleus and Soviet scientists were confident that they would survive. In September 1985 NASA's interplanetary probe ISEE-3 had survived passage through the tail of a comet called Giacobini-Zinner to which it had been re-directed and was re-named the 'International Cometary Explorer' in honour of its achievement. In due course, the VeGas became the first human artefacts to enter the nucleus of Halley's Comet, returning TV pictures in early March. These were then used by the European Space Agency (ESA) to target its Giotto spacecraft within 600 km (375 miles) of the nucleus.

Giotto was the jewel in the crown of the flotilla and another triumph for international cooperation as ESA's first deep space mission. The spacecraft itself was built in Britain, launched by an Ariane booster from French Guiana and controlled from the European Space Operations Centre in Germany. Heading towards the nucleus at a speed of 60 km/sec (37 miles/sec), Giotto's designers had to develop a special 'bumper' shield to minimize the damage from impacts from the dust particles. Their gambit paid off, as Giotto was able to reach within 1,400 km (875 miles) of the nucleus, before a dust impact started it wobbling just 14 seconds before closest approach. The comet was revealed as nothing more than a dark, icy body in the shape of a peanut some 10 km (6.25 miles) long, a pristine remnant of the Solar System's birth.

THE FUTURE

As a result of the Halley missions, international cooperation pointed the way forward for planetary exploration, a fact echoed by missions which have come to fruition in 1989. Despite heartbreaking failures, Phobos 1 and 2 were the very acme of collaborative ventures. Both craft returned important data which have whetted the appetites of planetary scientists. No less than 11 countries cooperated in their construction and, again, NASA helped with their tracking. The Magellan mission, NASA's first planetary mission in a decade, became the first to be launched from the Space Shuttle, in May 1989. Scientific planning for this Venus radar mapper has extensively involved cooperation from Soviet scientists with their experience gained from the Venera 15 and 16 missions.

Giotto, Europe's first Solar System spacecraft, returned hundreds of pictures of the nucleus of Halley's Comet. This enhanced view was synthesized from seven images taken between 26,000 and 2,800 km (16,156 and 1,740 miles) as Giotto sped towards the comet. The bright plumes of gas jets from the dark nucleus are seen on the Sunward side.

After the hiatus caused by the *Challenger* accident, the U.S. space programme seems to be back on track. In October, the much-delayed Galileo orbiter to Jupiter will finally be on its way. By Easter 1990, the Hubble Space Telescope will be launched into Earth orbit, a mission which promises to revolutionize astronomy in general. And almost as a bonus, Voyager 2 will finally reach Neptune in August – its final planetary destination.

The Hubble Space Telescope highlights the fact that most planetary encounters are little more than snapshots of the worlds which have been observed. Hubble will allow greater long-term studies of the planets and moons in our Solar System, and may even reveal planets around other stars. In the 1990s, we can look forward not only to the first samples of the Martian surface but also cometary nuclei being successfully returned to Earth.

If the past 30 years of planetary exploration are anything to go by, there will be many more surprises in store. If nothing else, mankind's robotic envoys have revealed the planets as being far stranger and more exotic than even science fiction writers had imagined. A new generation of orbiters and landers will attempt to answer the mysteries that remain about the elusive worlds which populate our Solar System. It is a comforting thought that planetary exploration will pass through an even more golden age in the future.

THE GLOBAL VILLAGE: SPIN-OFFS AND APPLICATIONS

Telephone calls with relatives on the far side of the world: television reports from deserts a continent away: weather forecasts based on pictures from space. All are taken for granted today, and without satellites would be impossible. When the writer Marshall McLuhan described the space age world as 'the global village' he was not far wrong; in this regard, space exploration has truly come of age.

Yet ironically, the realization that space could bring about benefits is a recent development. The first mention of the commercial possibilities of space can be pinpointed to a 1945 issue of *Wireless World*. Arthur C. Clarke – then a relatively unknown science fiction writer – proposed that communications satellites be placed in orbit some 36,000 km (22,500 miles) above the equator where they could take advantage of the fact that they would complete one orbit in the same time that the Earth rotated. In other words, they would appear to hover over the same point on the Earth's surface. Today such 'geostationary' or 'geosynchronous' orbits are highly-prized and Clarke has wistfully commented that he wishes he had patented his idea.

The world's first telecommunications satellite, Telstar 1, was also the first to have a piece of music named in its honour. The satellite was covered by 3,600 solar cells which provided energy, while the helix-like antenna (below) was used to relay microwave signals. Its graphic demonstration of trans-Atlantic telecommunications ushered in the direct benefit from space.

INTERNATIONAL TELECOMMUNICATIONS

With the dawn of the Space Age, telecommunications from orbit, as predicted by Clarke, became a reality. The seventh American satellite, Score, in December 1958, tested the technology needed for transmission of radio signals from orbit. It was followed by Echo 1 in August 1960, a vast hydrogen-filled balloon used to bounce signals from continent to continent, but limited because it was a 'passive' reflector and, as a result, not very efficient. The first live TV broadcast between Britain and the United States took place in July 1962 by Telstar 1, which could also handle 600 telephone calls. But its elliptical orbit inclined at 45° to the equator limited its operational use.

The first communications satellite to operate from geostationary orbit was Syncom 3, which relayed TV pictures of the 1964 Olympic Games in Tokyo to the United States. It was a graphic demonstration of the power of telecommunications from space. Plans were already afoot for the setting up of an international organization which would operate a global network of communications satellites. Known by the name INTELSAT, it came into being in August 1964 with 15 founder member countries. From geostationary orbit, roughly one-third of the globe is visible, so theoretically satellites separated by 120° would be able to cover the whole of the globe. In reality, INTELSAT decided to station its satellites over the Atlantic, Pacific and Indian Oceans to handle intercontinental communications 'traffic'. Without the use of elaborate tracking equipment, geostationary satellites could be used for relaying TV, radio or telephone conversations more flexibly than underwater cables.

On 6 April 1965, INTELSAT's first satellite was launched and entered the history books as the world's first commercial communications satellite. Known as Early Bird, it could carry 240 telephone calls or one TV channel and was stationed over the Atlantic to service North America and Western Europe. Since then, INTELSAT has improved its service by developing ever more advanced technology, so that by the mid-1980s it was using 16 operational satellites to carry two-thirds of the world's inter-

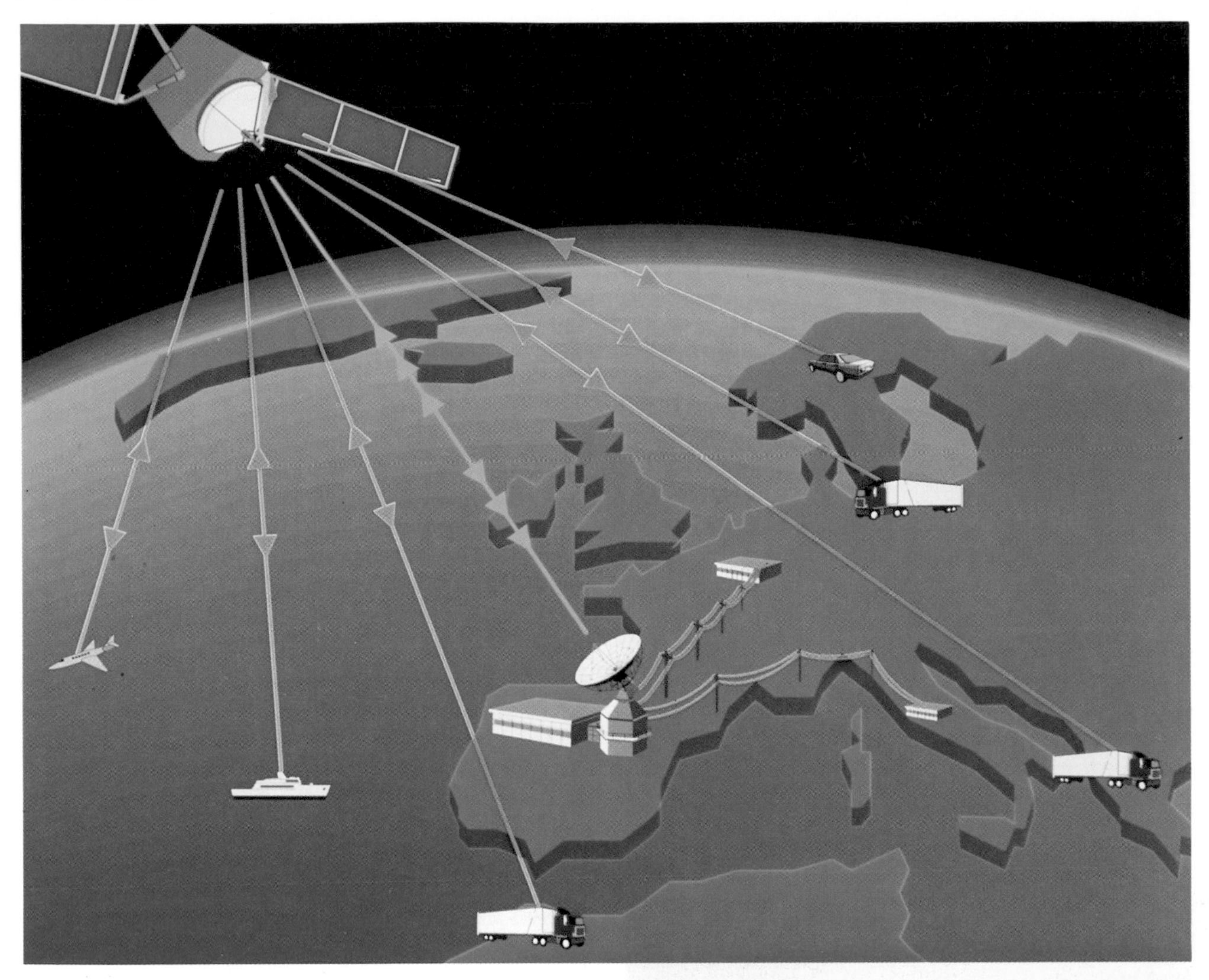

Left: Telecommunications capabilities are now more versatile than ever before. This diagram shows how the European ECS satellites can be used for communications between boats, trains and planes – as well as lorries and cars. Already carphones are a regular feature of the business community.

Below: ESA's PRODAT system is currently being tested by European lorry drivers. Information on the whereabouts, loads and 'health' of individual lorries can be returned to head office via the MARECS system. All that is needed for the driver is a small terminal on the vehicle's dashboard.

national telephone calls. The organization now numbers over 100 member countries, and operates over 800 ground stations. A new generation of satellite for the 1990s, known as INTELSAT 6, will carry 40,000 phone circuits and four TV channels: the first is due for launch by the end of the decade.

INTELSAT has acted as a blueprint for 'regional' communications groups: ARABSAT was formed in 1976 to cover the Middle East, and EUTELSAT was formed a year later in Western Europe. EUTELSAT now has four satellites, known as ECS (European Communications Satellites) in orbit, with 22 TV 'transponders' or individual TV transmission channels in operation. Rupert Murdoch's much-vaunted 'Sky' Channel would be unthinkable without the privately-owned Astra satellite, launched at the end of 1988 by an Ariane launcher. Australia now has AUSSAT to take care of its telecommunications needs, and Indonesia has operated Palapa communications satellites for a number of years.

For ship-to-shore radio communications and fax transmissions between ships, an organization called INMARSAT uses three satellites based on the European ECS – and known as Marecs. It now numbers over 50 countries as its members. In the United States, a whole range of companies operate 'domsats' or domestic communications satellites. The first of these was the Westar network, built and operated by Western Union, who sold its three satellites to the Hughes Aircraft Corporation in February 1988. There are so many television demands that 'transponder brokers' are now called in to lease time on the various commercial satellites to TV networks.

The Soviet Union, too, makes use of its own 'domsats' and is not a full member of INTELSAT, though it does send official observers to its meetings. An organization known as INTERSPUTNIK was set up in 1971 for Soviet Bloc countries, and makes use of a geostationary network of satellites known as 'Statsionar'. It now numbers 40 members, no longer restricted to countries behind the Iron Curtain. Telecommunications within the Soviet Union are hampered because the most remote areas which would benefit from 'domsats' are at higher northern latitudes, and signals relayed from geostationary orbit would be relatively weak and impractical. So since 1965, the Russians have used a series of Molniya satellites which are launched at an inclination of 65° to the equator and in highly elliptical orbits with their apogee (i.e. high point) above Siberia. This allows 16 hours a day of coverage over the Soviet Union, relaying TV broadcasts from Moscow to remote areas throughout the U.S.S.R.

AN EYE ON THE WEATHER

Another revolution in our everyday life has been in monitoring the weather and accurately forecasting it for many days ahead. Weather satellites in geostationary orbit have enabled there to be a global 'eye' on the weather, collecting and disseminating weather observations from aircraft, balloons and ships. Ever more powerful computers have allowed that data to be modelled and analyzed more accurately and quickly to produce highly reliable forecasts. Weather forecasts are not just about informing us if we should carry umbrellas in case of rain. They allow farmers to be warned of frosts well in advance, so that better harvests can be achieved. Ships can be re-routed to avoid bad weather and save fuel. Even lives can be saved – after the Chernobyl nuclear accident, predicting the motions of the radioactive fallout was of vital international importance. The UK Met Office has calculated that accurate weather forecasts already save the nation £50 million per year.

Today, we are more aware than ever about the environment in which we live. That awareness stemmed from the first pictures of the Earth in space, like this one returned by the crew of Apollo 17. Africa is seen in outline along with the familiar swirls of the clouds.

The first satellite devoted to weather observations was launched on 1 April 1960 by NASA, and known as TIROS – an acronym for 'Television and Infrared Observation Satellite'. It returned the first pictures of clouds from space, and although relatively primitive by today's standards, one meteorologist declared that weather forecasting had 'gone from rags to riches overnight'. More advanced satellites followed, such as the Nimbus series which returned higher resolution views as well as infra-red temperature measurements. In the United States, the National Oceanic and Atmospheric Administration operated the first American weather satellites, known as the Tiros Operational Satellites, in polar orbits. As they headed over the poles, the satellites 'covered' the whole of the Earth in a few days as it spun underneath them, so they could monitor changes in weather from place to place. Ever more advanced polar-orbiting U.S. NOAA and Soviet Meteor satellites regularly return data for the world's meteorologists.

For more accurate forecasts, however, meteorologists wanted the full Earth view provided by geostationary satellites. The World Meteorological Organization – the scientific body that coordinates weather research across the globe – proposed a remarkable plan in 1968 called the World Weather Watch which would use a network of such satellites to provide data across the globe. NASA launched a prototype series called the Synchronous Meteorological Satellites, which were later handed over to NOAA. The SMS series showed that visible and infra-red images of the Earth could return detailed information on cloud cover, wind motions and water vapour within the atmosphere.

By the 1980s, NOAA had provided two satellites called GOES (Geostationary Operational Environmental Satellites) as part of the World Weather Watch. Similarly ESA had launched its first two METEOSATS, and Japan its GMS satellites as part of the programme. Curiously, though the Soviet Union is part of the programme, its own GOMS satellite has yet to be launched. Nevertheless, the effectiveness of the data from even a slightly incomplete global network is shown by the fact that the UK Met Office, for example, has a 90 per cent chance of getting tomorrow's weather correct. The European Centre for Medium Range Weather Forecasts (ECMRWF) in Reading claims a 75 per cent accuracy in predicting the weather over the next six days. In November 1980, for example, its predictions that a cold snap would occur a week after a severe earthquake had taken place in Italy ensured that the homeless were provided with emergency blankets and shelters to avoid further casualties.

REMOTE SENSING

Another important benefit from space has now burgeoned into a multi-million dollar industry which comes under the general heading of remote sensing. As a science, remote sensing has its origins in the early 1930s, when aerial photography began in earnest. It was realized in the early 1960s that space technology could provide far greater coverage of the Earth, a promise hinted at by the results from the first weather and military reconnaissance satellites. The first remote sensing satellite was launched by NASA in July 1972 and was originally known as the Earth Resources Technology Satellite (ERTS). Based on the butterfly-shaped Nimbus weather satellites, ERTS-1 was later renamed LANDSAT 1, as its cameras returned valuable information about the land below.

Rather than using film, LANDSAT 1 used an instrument known as a scanning radiometer which produced 'digital' pictures of the Earth. Instead of a cumbersome film system, which either had to be retrieved or converted into numbers for radio transmission, a radiometer works by directing an image onto detectors sensitive to a wide range of wavelengths in visible and infra-red light simultaneously. They allow a multicoloured picture of rocks, soil, water, vegetation, etc. to be produced, as they each have their own characteristic 'signature' (i.e. appearance) at those wavelengths.

During the 1970s, remote sensing literally took off with the LANDSATs, which resulted in a greater awareness of rock formation, drainage patterns, marine pollution and ecological damage right across the globe. The destruction of tropical rain forests and the availability of crops in desert regions have been monitored with remote sensing satellites. Crop forecasts are a vital by-product: healthy plants appear brighter at infra-red wavelengths. So a multispectral image of a given area will allow a detailed assessment of how likely a good crop yield will be. Experienced geologists have been able to identify regions which are likely to contain mineral- and oil-bearing rocks. The way in which the rocks affect the vegetation can be inferred from multispectral images, and oil companies have saved valuable time and money in searching for new prospecting sites.

The Space Age has given meteorologists the 'whole Earth view' they so desperately needed. Continuous monitoring of the globe by networks of weather satellites enables ever more reliable forecasts to be made. This view of eddies over the Pacific highlights the reality of monitoring unusual weather patterns.

Data from remote sensing satellites is not just available as photographs. Digital tapes of 'raw' data which can be enhanced by computers have been made available so that enhanced pictures of certain regions can be produced there. A secondary explosion in these 'end-user' enterprises has occurred, so much so that by the mid-1980s it was hoped that further remote sensing satellites would pay for themselves. As a result, operations and marketing data from the fifth LANDSAT, launched in 1984, were handed over to a commercial company called Earth Observation Satellite (EOSAT) in 1986. This has meant that the U.S. government has withdrawn its subsidies to the LANDSAT programme, causing problems for EOSAT, and delaying the launch of the next satellite in the series, LANDSAT 6.

By the time it is launched, LANDSAT 6 will not be alone in the remote sensing stakes. Already, a French satellite called SPOT is returning pictures with far better resolution than LANDSAT 5. The European Space Agency's ERS-1 and

Canada's Radarsat will be launched early in the 1990s, followed in turn by remote sensing packages on the U.S.-led International Space Station, *Freedom*. The science of remote sensing has yet to reach maturity, with ever more varied uses of satellite data becoming apparent.

INTANGIBLE BENEFITS

The benefits from communications, weather and remote sensing satellites are amongst the rather more obvious products of the space age. Yet there are other, less immediate benefits, many of which resulted from the landing of a dozen men on the Moon as part of the Apollo project. All told, Apollo cost about $25 billion over slightly more than a decade, and it is instructive to realize that during that same time period, American women spent much more than that amount of money *per year* on cosmetics. Of more lasting importance have been the 'spin-offs' from Apollo. The need to develop and miniaturize computerized electronics resulted in the micro-electronics revolution of the 1970s. Digital watches and home computers that are commonplace today owe their origin to Apollo. The same knock-on effect occurred in engineering: the need to develop components which would not fail also led to better standards in high technology.

After the horrific fire that killed the crew of Apollo 1, the need for new standards of fireproofing led to many breakthroughs in fire safety. For example, a protective breathing system was developed out of the technology that had been used in the Apollo programme. It was lighter and more flexible, allowing fire crews to spend longer times fighting at the very heart of large, industrial fires. There have been more down-to-earth developments as a result of technological spin-offs from Apollo. These include Velcro, which was developed to keep objects fastened in zero gravity; and teflon, originally used in the heat shield of the Apollo Command Module, now used in non-stick frying pans.

Yet there are many who feel that one does not have to justify Apollo's $25 billion price tag in terms of non-stick frying pans. The science fiction writer Ray Bradbury has termed such justifications 'the tupperware syndrome'. He believes that human beings explore space for far more subtle reasons than the promise of new technologies. Humans are naturally inquisitive: space exploration fulfils the human need to explore new horizons. And though this reasoning suggests we explore space 'because it's there', it has undoubtedly produced some subtle benefits of its own. It is these 'intangibles' that add up to a far greater promise than the technological benefits that will accrue.

Many people have noted that the rise of the worldwide ecology movement in the late 1960s more or less coincided with the first views of the Earth as a globe, returned from astronauts on the Moon. In the phrase of biologist Norman Cousins, 'On the way to the Moon mankind discovered the Earth.' Seeing the Earth from afar represented a fundamental realization that our home planet was a fragile environment, dependent on the Sun for its power and limited to its own resources. An Apollo astronaut described it succinctly as 'a round and bounded globe that is utterly alone in the emptiness'.

SUMMARY

So three decades of space exploration have resulted in many benefits to date. The future looks hopeful as we have already entered the fourth decade of the space age, in which space *exploitation* as opposed to exploration will become ever more apparent. Within the next decade, permanently manned stations in Earth orbit – international in nature – will be producing new alloys and drugs in zero gravity. Vast 'farms' of solar panels will provide endless supplies of energy to the industrialized nations. And though these claims may seem like science fiction, there were many in the 1950s who classed journeys to the Moon in the same way. Even though it is often difficult to predict the future, one thing is certain. In the same way that explorers of the 16th century changed their world by their discoveries, so will future space travellers in the next.

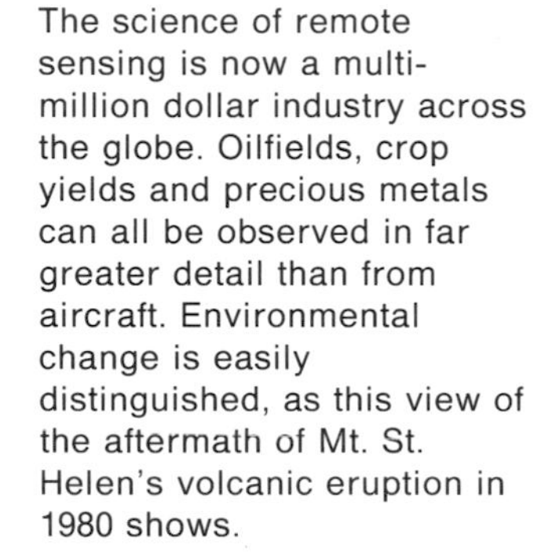

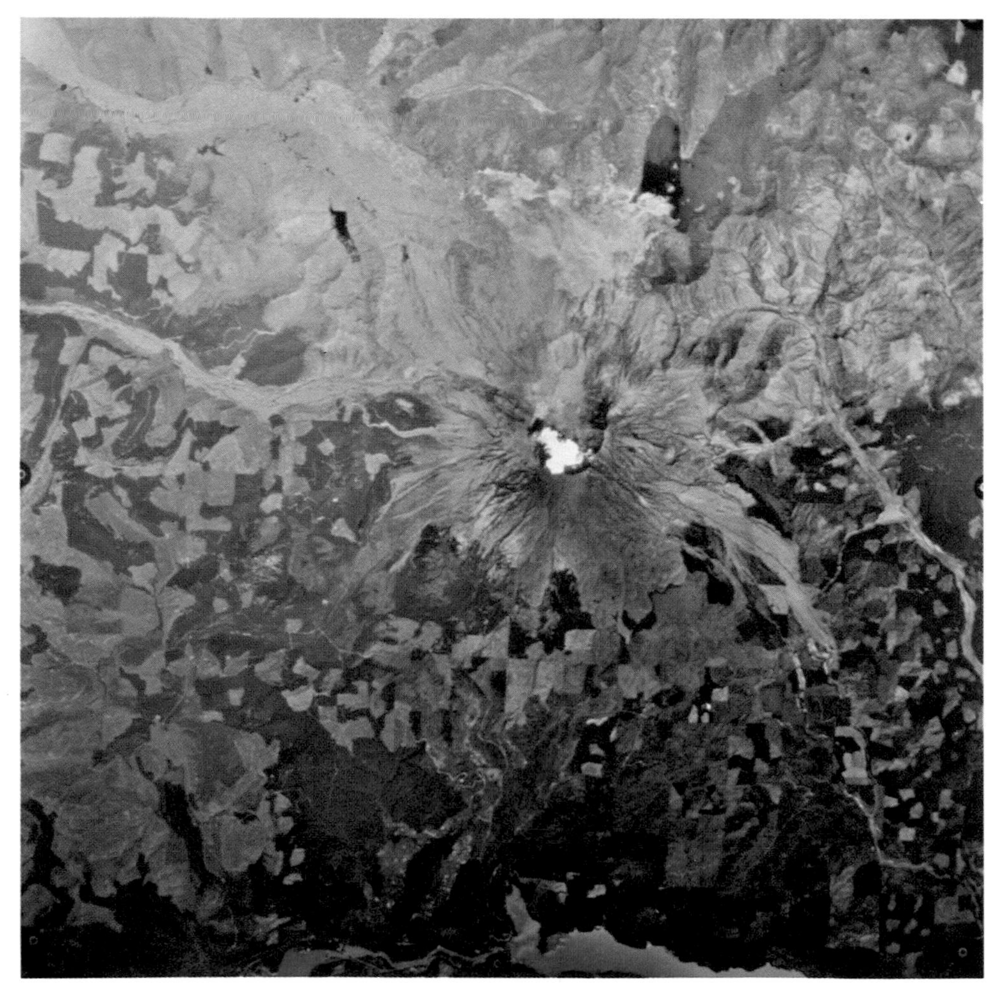

The science of remote sensing is now a multi-million dollar industry across the globe. Oilfields, crop yields and precious metals can all be observed in far greater detail than from aircraft. Environmental change is easily distinguished, as this view of the aftermath of Mt. St. Helen's volcanic eruption in 1980 shows.

BIGGER, BETTER, SOONER, CHEAPER: THE LAUNCHER BUSINESS

French Guiana in South America is an unlikely setting for a commercial space centre. Until recently, its only claim to fame was that it was the site of a notorious French prison and that the even more notorious Devil's Island is nearby. For French Guiana is hot and humid, made up for the most part of dense jungle, located near the equator. It is this last fact which makes it ideal for rocket launching. The speed of the Earth's rotation is at its greatest at the equator, a fact which means that rockets are given an extra 'shove' into space if they are launched there. It was for this reason in the 1960s that the French government developed its national rocket centre at Kourou on the coast. Even then, the French had their sights firmly on developing launchers independently of the superpowers – and marketing them. The French government realized early on the importance of space, becoming the first nation to launch its own satellite by its own launch vehicle in 1965. Its national space agency, CNES (Centre National d'Etudes Spatiales) expanded quickly, and developed a whole series of launch vehicles, including the Coralie rocket which later became part of Europe's first attempt to launch its own satellites.

EUROPA AND HER DAUGHTER

In 1964, seven countries joined together to form the European Launcher Development Organization (ELDO), primarily at the behest of the British government who wanted to make use of its Blue Streak missile to good advantage. An ambitious launcher, called Europa, was planned which would be made up of rockets developed by the individual nations. Blue Streak would be the first stage, Coralie the second and a German Astris booster the third. But problems beset Europa from the very start: when engineers came to mount the rocket stage together, it was realized that the British had used imperial measurements not compatible with the metric French! A test launch of a dummy satellite in 1970 from Kourou failed when Europa veered off course. Undaunted, the French decided to build an even more powerful version, adding a fourth stage for good measure. Despite the fact that the British government cancelled Blue Streak in July 1971, plans were well advanced for the test launch of the more powerful Europa 2. However, the test flight in November 1971 also failed. The rocket veered off course and had to be destroyed by explosive charges operated by a range safety officer. With a palpable feeling of 'enough is enough', ELDO soon disbanded; another Europa en route to Guiana was returned to France. The tattered remains of ELDO were later subsumed by the European Space Agency in 1973.

Europe's pre-eminence in the cut-throat world of commercial space launching has been achieved by the Ariane series. Seen here is the first flight of the Ariane 4 variant in June 1988 which successfully deployed three satellites. During the 1990s, Ariane will be the workhorse of the programme.

Despite the debacle, CNES decided to take the bull by the horns and develop a heavy launcher from scratch. While NASA was struggling with revolutionary designs for the Shuttle, the French decided that an old-fashioned launcher using proven 1960s technology could be produced in sufficient quantities to make space access economical. New generations of weather and communications satellites meant that there would be enough customers. After intense persuasion, the newly-formed European Space Agency adopted the French launcher as its first major project; and with no sense of irony was it named Ariane – in mythology, the daughter of Europa.

ARIANE

Described as '*le lanceur lourde*' or heavy lifter, Ariane was intended to compete with the Shuttle in the increasingly lucrative commercial launching sector. It was designed with requirements of launching satellites into geostationary orbit in mind. The basic design was a three-stage vehicle, with the first two stages using ultra dimethyl hydrazine (UDMH) as fuel and nitrogen tetroxide (N_2O_4) as an oxidizer. Such propellants are described as hypergolic – that is, they ignite spontaneously on contact. The third stage consisted of cryogenic tanks with liquid oxygen and hydrogen. This first version of the Ariane launcher, known as Ariane 1, was capable of lifting 1,850 kg (3,785 lb) into geostationary orbit. Ariane 2 used two solid strap-on boosters to lift 2,175 kg (4,795 lb) into geostationary orbit, and an improved version, the Ariane 3, could lift 2,580 kg (5,687 lb). By use of a special nose cone fairing, it was possible to launch two satellites together.

Until Ariane came into operations, ESA had to rely on NASA for all its launchings. Seen here is a Delta launch from the late 1970s. The *Challenger* accident and reliance on expendable launch vehicles has meant that the Delta is now being offered commercially to anyone who wants its services.

Ariane took eight years and £190 million to develop – 64 per cent of its funding came from France and 20 per cent from West Germany; Britain's contribution was a mere 2½ per cent. It was ready for its first test flight on Christmas Eve 1979 when it successfully launched a technological capsule into orbit. However, the second flight the following May broke up just under two minutes after launch. The cause was later determined to be a faulty fuel injection system with the first stage. Two more test flights in June and December 1981 were enough for ESA to declare Ariane 'operational' in early 1982. However, on the fifth flight, Ariane's third stage failed and two commercial satellites were lost. Cynics within the European Space Agency were quick to point out that if Ariane was successful it was a great day for France: if it failed, then it was a black day for Europe.

Despite the failures, Ariane reported brisk business. In 1980, a commercial company called Arianespace was set up to market the launcher. Its chairman was the Director General of CNES, which itself managed the production of the vehicles by prime contractor Aerospatiale. The 11-member ESA provided the finance, though from time to time there were grumblings, especially when Ariane 4 was approved in 1982. By using a combination of between two and four solid or liquid propellant strap-on boosters, Ariane 4 would be capable of lifting 4,200 kg (9,259 lb) into geostationary orbit. Many in the European space community were appalled by the continued development of high-cost, outdated technology in upgrading the new versions of Ariane. Many felt that a new cost-effective launch system developed from scratch was needed to minimize launch costs well into the future. On hearing of Ariane 4's go-ahead, the noted British engineer Alan Bond recalls thinking: 'There must be something better than that! You could buy better engines off-the-shelf even then.'

Problems with the third stage led to further failures in September 1985 and May 1986. In the 1985 failure, President Mitterand watched the range safety officer blow the rocket out of the sky to prevent it falling on Brazilian territory. More serious was the failure of the Ariane 2 variant on 30 May, when its third stage misfired and had to be destroyed 200 km (125 miles) above ground. Aerospatiale spent more than a year modifying its cryogenic engine on the third stage, testing the ignition system on eight different engines for a total number of 68 firings. When Ariane 3 was successfully re-launched in September 1987 with the Aussat K3 and ECS-4 communications satellites, the Ariane programme looked on course, though Arianespace was the first to admit that its planned eight launches a year were a long way off. 1988 saw the introduction of the more powerful Ariane 4 which successfully launched the Meteosat P2 satellite into orbit in June and the Astra TV satellite in December. At the end of 1988, Arianespace's order books contained 44 more satellites to be launched well into the mid-1990s.

Of far greater long-term significance was the decision to develop a fifth Ariane variant in 1985. Not only would it allow over 6,800 kg (15,000 lb) of payload to be launched into geostationary orbit, but it would also give the European Space Agency the ability to launch astronauts to the international space station in the French-built Hermes spaceplane. The decision to embark upon a European manned programme was not without its dissenters. In late 1987, when the Ministers from the ESA member countries met to discuss the agency's future in the Hague, the British government decided not to back Ariane 5 development or that of Hermes. The minister responsible, Kenneth Clarke, managed to alienate most of the attendees by his comment that 'the British taxpayer would not subsidize sending a Frenchman into space'.

THE UNITED STATES

1986 was the bleakest year in American space history. As well as the tragic loss of *Challenger*, a number of expendable launch vehicles were also lost. In April, a Titan booster launching a military satellite from the Vandenburg Air Force Base failed, as did a Delta with a weather satellite in May. In October 1986 NASA announced that it would no longer act as a space airline: only scientific and military payloads would be launched from the Shuttle. By early 1988, NASA announced that over the following seven years, its backlog of 135 satellites would be launched on a 'mixed manifest', meaning they would be launched by the Shuttle and expendable launch vehicles. New safety guidelines for the Shuttle would allow at most 100 launches over that seven year period, leaving a shortfall of at least 35 satellites to be launched by 'old faithful' expendables such as the Delta, the Atlas-Centaur and the Titan boosters. That there are also military payloads and commercial satellites needing to be launched gave the American aerospace giants who build expendables the impetus for keeping construction lines open and for active marketing campaigns to begin.

In 1988, McDonnell Douglas, for example, reported that its Delta launcher had at least 5 private customers, whilst Martin Marietta's Titan 3 had four. Yet the aftermath of *Challenger* saw the rise of smaller, specialized rocket companies keen to get a piece of the action. Space Services Inc. of Houston – with ex-astronaut Deke Slayton as its chairman – offered its Conestoga launcher on the market. Using rocket engines that had been developed for other boosters – a Minuteman missile engine, for example – the company promised to slash launch costs. The key to the undercutting offered by these 'small operators' is the fact that the rocket engines were originally developed at U.S. taxpayers expense. Rocket components available 'off the shelf' ensured that prices did not reflect the hefty initial investments needed for development. By the end of October 1987, SSI had eight firm reservations and 70 proposals. Perhaps the most bizarre was an offer by the Celestis Corporation of Florida to launch cremated remains of the recently departed in satellites that would have highly reflective exteriors. As the satellite gracefully soared past, their loved ones on the ground would see their memory preserved in the heavens!

Of far greater importance to any would-be satellite launcher was the question of where their boosters could be launched from. NASA was keen to offer its existing launch pads at competitive prices, but with the Department of Commerce overseeing the commercial licences, it was made clear that exclusive dependence on NASA was no longer *de rigeur*. So the possibility for commercial spaceports grew ever more likely – and these were not necessarily restricted to the mainland of North America.

Hawaii was particularly keen to get in on the act. The islands are no stranger to space-related activities as they are the location for a NASA tracking station and many of the world's largest astronomical telescopes. Technological consultants from the Arthur D. Little company were called in by the Governor's office to investigate the economics of creating a Hawaiian spaceport. Being close to the equator, advantage from the extra 'push' offered by its location obviously commended itself. In March 1988, the consultants reported that a spaceport comprising four launch pads for an average of ten launches per annum was economically viable. Seven possible sites were considered, with the most promising candidate being identified as Palima Point, on the southeast tip of the main island. But environmental concerns may delay the construction of a spaceport there because the noise and pollution from rocket launches would badly affect the breeding grounds for certain endangered species of turtles.

A Delta with a Thor upper stage is seen here ready to launch one of INTELSAT's first satellites in the 1960s. All the expendables being offered for launches by their parent companies derive from technology developed in ICBM research in the 1950s and 1960s. The U.S. taxpayer has effectively subsidized the competitors of the 1980s!

Another launching site has also been mooted – Cape York, in Northern Australia. A number of U.S. aerospace firms have expressed an interest in this site for commercial space launches. The most interesting plans have been advanced from a most unexpected quarter – the Soviet Union.

GLASNOST AND GLAVKOSMOS

The rise to power of Mikhail Gorbachev in 1985 has led to greater openness in Soviet society, or *glasnost* – an ideal which has extended to the country's space activities. Gorbachev has taken a great interest in space, and ensured that Russian space activities have been handed over from the bureaucratic Ministry of Medium-Size Machine Building to Glavkosmos, a new 'civilian' space agency. Within months of its formation, 1986 saw Glavkosmos hawking its wares in the West, offering the Proton launcher for commercial purposes. Proton has the distinction of being the first Soviet launcher not derived from a military missile, and was originally designed with the manned lunar programme in mind. It first came into operation in 1969, but it was not until the end of 1984 that pictures were released in the West when it was used to launch the VeGa probes to Halley's Comet.

Very soon, Glavkosmos was showing its customers how Proton worked – and how it could be launched at cut-price costs. However, the spectre of 'technology transfer' – the U.S. ban on allowing advanced technology to fall into the hands of the Eastern Bloc – cast its gloomy spell over the early optimism. Though India and Iran were keen on using Proton for launching their satellites, many Western companies had been more or less barred from launching their satellites by legal constraints in the United States. Despite earnest Soviet claims that they would not inspect the satellites to see inside, only India had used Proton for a 'promotional' launch, when a communications satellite was launched in a blizzard in March 1988. Nevertheless, it was reported at a meeting in Hawaii in November 1988, that Glavkosmos was actively seeking Australia as a partner. The fact that the Proton boosters would be shipped to Cape York and assembled by Australian technicians was an obvious way of avoiding the technology transfer issue. In mid-1989, Australia announced it had given the go-ahead for launching in the 1990s.

PACIFIC RIM

Indeed, commercial space activities may eventually centre on the Far East, and observers expect to see the rise of the 'Pacific Rim' – countries located within the Pacific Ocean – in the development of commercial space activities.

Japan and China became the fourth and fifth nations to launch satellites independently of the superpowers in February and April of 1970 respectively. Both have developed sophisticated space programmes, and are keen to offer their launchers commercially. In April 1984, a so-called Long March booster successfully placed China's first geostationary communications satellite into orbit, which gave the Chinese the confidence to market the three stage booster in the West. Three versions are outlined in very detailed 'User Manuals' which the Great Wall Industry Corporation based in Beijing offers to potential customers. The third, and most powerful Long March variant can boost a payload of 1,400 kg (3,086 lb) into geostationary transfer orbit.

There has been much interest in the West: China's 20th satellite launched in August 1987 carried a French-built materials science experiment, the first 'foreign' payload. In December 1987, it was also reported that McDonnell Douglas was discussing the possibility of allowing one of its boosters to act as a third stage for Long March 2.

China was hoping to launch 12 Long Marches per year, and had received 26 provisional book-

Above: The Soviet space agency is actively marketing its Proton launcher all around the world. Glossy illustrations accompanying slick PR handouts have surfaced as a result. The Proton has been in use since the late 1960s, and was the first launcher to be developed after Korolev's death in 1966. Proton vehicles have launched the Salyut and Mir space stations, as well as Venera, Mars and Phobos probes.

Right: To date, only India has taken up the Soviet's offer of post-*glasnost* commercial launchings. SRS-1A, India's first remote sensing satellite, was launched by an SL-3 booster on 17 March 1988. The satellite is shown here being mated to the booster. ISRO is the Indian Space Research Organization.

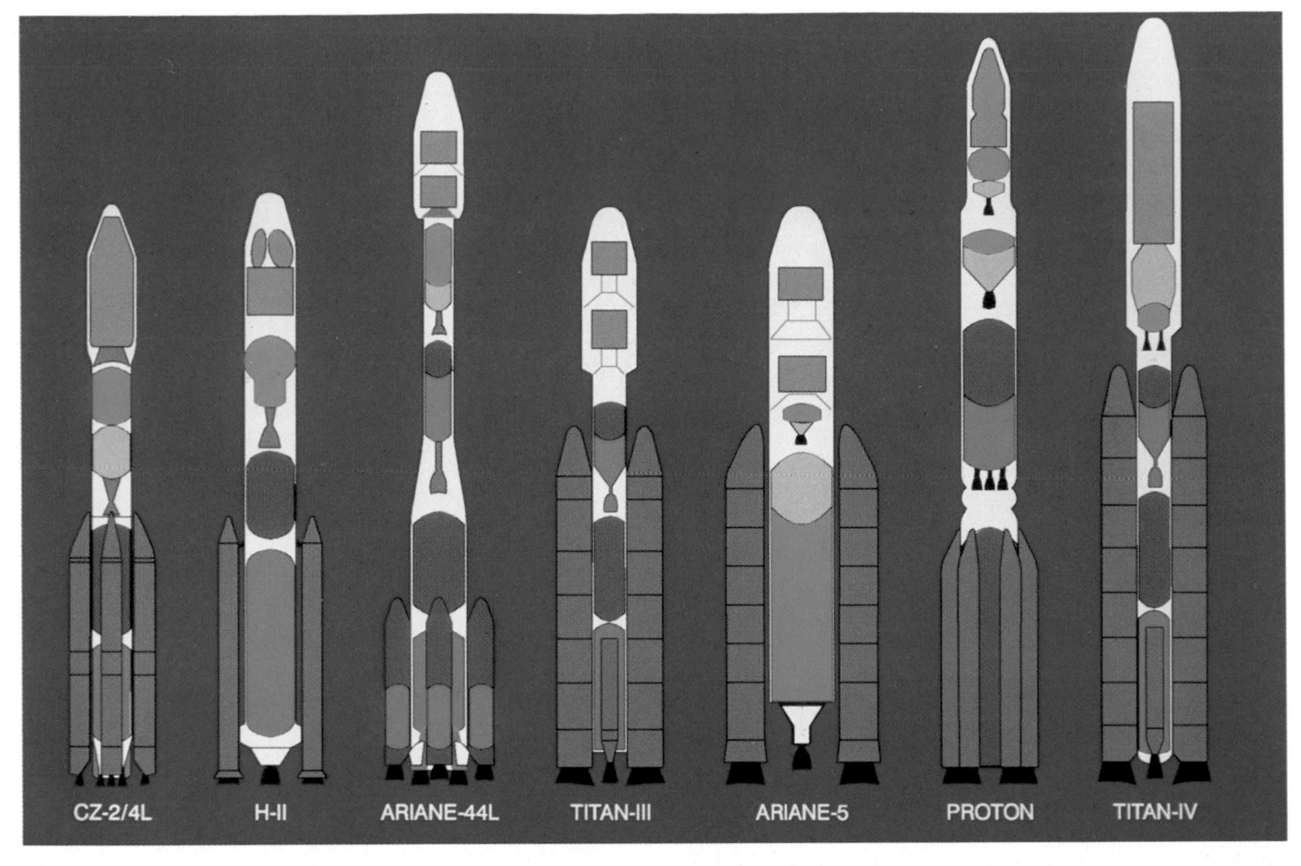

How the competition compares. From left to right: China's Long March; Japan's H-2; ESA's Ariane 4; Martin Marietta's Titan-III; ESA's Ariane 5, due to start service at the end of the 1990s; the Soviet Proton booster marketed through Glavkosmos; and Martin Marietta's Titan IV. To date, Ariane 4 has the biggest number of booked customers.

ings by U.S. and Swedish users in 1988. However, the events in Tiananmen Square obviously cast a blight over the use of Long March.

Japan's interest in cornering the lucrative launcher market can be gleaned from the fact that one-third of its annual space budget is devoted to the development of its H-2 launcher. With a liquid oxygen/hydrogen propellant first stage, to which solid rocket boosters can be added, H-2 will allow 2 tonnes of payload to be deployed in geostationary orbit. Due to come into operation in 1992, it will also be used to launch the Hope spaceplane towards the international space station, *Freedom*. Though still in the development stage, analysts point to the fact that with Japan's manufacturing costs being about half those in the West, a 'cut-price' launcher service could soon be set in motion.

SUMMARY – A NOTE ON COSTS

Trying to predict who will corner the greater share of the market in the lucrative business of launching satellites is virtually impossible. But it is obvious that more and more competitors will be vying for their share of the market.

An equally vexing question is that of launch costs – trying to work out what it costs to launch a given satellite is a difficult matter. Factors such as whether one or more satellites are launched together, development costs, launchpad costs and so on make it extremely hard to define which launcher is cheaper in the long run. Comparing Ariane with, say, the Long March launcher is not comparing like with like. And like high-tech car salesmen, each company makes extravagant claims to prove that their launch system is the best.

Before the *Challenger* accident, NASA was offering to launch a communications satellite at $75 million. Critics pointed out that the real cost which would include the development footed by the U.S. taxpayer was $235 million. An Ariane launch will cost between $55 and $75 million depending on the satellite and whether a dual launch facility is available. Included in that price for U.S. customers, Arianespace will fly representatives to Paris first class on Concorde! The Proton and Long March vehicles are being marketed in the West for around $30 million. A Delta launch costs in the order of $60–70 million while the more powerful Atlas Centaur will cost $100 million.

Nevertheless, these costs are still prohibitively high for most countries and it is clear that a revolution in space will require a dramatic decrease in launch costs. With the need to make space more accessible, a number of countries are developing totally re-usable launch vehicles. The British spaceplane HOTOL – with its revolutionary air-breathing engine – was the first to be announced in 1984. Even the most conservative estimates suggest that a satellite launch will be in the region of $15 million if its development costs are taken into consideration. But for the customer, a payment of only $5 million will allow a satellite to be launched. The potential that such vehicles offer is discussed in the final chapter of this section.

RED STAR IN ORBIT: FROM *SALYUT* TO *KOSMOGRAD*

Largely unnoticed in the West, the Soviet Union developed a whole series of ambitious space stations in the 1970s. Salyut 6, photographed here in 1978, was one of the most successful. The experience gained led to the *Mir* station launched in 1986 in which cosmonauts have lived for up to a year in space.

The idea of space stations is one of the most potent in the development of space, an idea first mooted by Tsiolkovskii in the early part of this century. Tsiolkovskii even went so far as to describe Kosmograd – the city in the sky – which he envisaged could be built in Earth orbit as a self-contained colony, the obvious product of developing space stations. Such ideas, he suggested, would not be put into practice until the 21st century.

In this, the great savant was wrong. The world's first permanent station is already a reality. Known as Mir, it has acted as home for a number of Soviet crews who have established long duration records, and the Soviets have talked about modifying a Mir station and sending it to Mars. That the Soviets are ahead in manned spaceflight is astonishing enough, the more so for the hazards and dangers that they have faced in building that lead.

Perhaps more astonishing is the fact that Mir forms the very heart of the Soviet Union's future space plans which will lead to the industrialization of space. Kosmograd is becoming an ever more likely possibility.

THE FIRST STATION

After Apollo 8 circumnavigated the Moon, it was clear that NASA was well and truly ahead in the Moon race. By erring on the side of caution the Soviets lost the chance of beating the Apollo 8 crew, and so abandoned their plans for manned Moon missions. The Kremlin decided that it would be more practical to concentrate on space stations in low Earth orbit. Thereafter Soviet officialdom, with well-feigned amnesia, was quick to point out that this had been their grand scheme all along. Space stations and long-term endurance records became Russia's ultimate goal in space in the 1970s.

So, in 1970, while modified Soyuz craft took to the skies, a station was prepared, its launch planned as a commemoration of the tenth anniversary of Yuri Gagarin's flight in April 1961.

Named Salyut 1 – Salyut meaning a salute to the pioneering cosmonaut – it was launched on 19 April 1971 and, though only about the size of a small house, it provided the first relatively luxurious living space for Soviet cosmonauts. A three-man crew aboard Soyuz 10 was despatched to become its first occupants, but though they docked with the station, they did not enter it. No explanation was forthcoming, such was the Soviets' embarrassment.

Nevertheless, the Soviets tried again with Soyuz 11 in June 1971. Its trio of cosmonauts docked successfully, and spent a record-breaking 24 days aloft. With Apollo winding down, and the Skylab station still two years away, Soviet space scientists could feel some measure of satisfaction: the future looked rosy. The crew were busy during their three weeks, training telescopes on remote galaxies, watering plants in zero gravity and experimenting with furnaces. Their return on 30 June elicited little comment, and rescue helicopters were soon hovering over the landing site. When the rescue teams opened the hatches, jubilation quickly turned to horror: Georgi Dobrovolski, Vladislav Volkov and Viktor Patsayev were dead. They were buried in the Kremlin Wall along with the other fallen cosmonauts, producing one of the largest crowds that Moscow had ever seen. A wave of emotion, similar to that which was felt in the United States after the *Challenger* accident, gripped the Soviet people.

The Soyuz 11 accident was perhaps the cruellest blow that befell the Soviet space programme during the early 1970s. It would be another two years before a modified Soyuz took to the skies, carrying a two-man crew equipped with spacesuits. The Soyuz 11 crew had perished when a pressure valve had blown on their capsule, leading to decompression and, almost instantaneously, to their deaths. Future Soyuz crews would be protected by spacesuits, but this meant that extra life support systems would be needed. The third crew couch would have to be removed. With no-one to visit it, Salyut 1 tumbled out of orbit in late 1971, an empty relic of the promise that had so cruelly been snatched from the Soviet space programme.

PROBLEMS AND PROGRESS

By 1973, the Soyuz craft had been extensively modified as a ferry craft to transport a pair of cosmonauts to new Salyut stations. Whereas the earlier Soyuz craft had been designed with the possibility of travelling to the Moon in mind, the new version would only be used in Earth orbit for two days at most. Solar panels were removed, chemical batteries were added to provide the craft with power, and the fuel tanks were decreased in size. Both developments would have serious repercussions on the Salyut ferrying missions.

Seen here in training for their stay aboard Salyut 1 are the Soyuz 11 crew who lost their lives. From left to right: Viktor Patsayev, Georgi Dobrovolski and Vladislav Volkov.

By May 1973, NASA was preparing to launch Skylab, its first ever space station. Perhaps in an effort to upstage the Americans – still a motivating force in the Soviet space effort – Salyut 2 was launched on 3 April. All seemed to be well until 14 April when its onboard engines were fired to boost it into its final orbit. The station became unstable and broke up into several fragments. In an even greater last ditch attempt, another Salyut was launched three days before Skylab on 11 May; it, too, broke up in orbit and was named Cosmos 557 by embarrassed Soviet officials.

It wasn't until September 1973 that Vasili Lazarev and Oleg Makarov took Soyuz 12 aloft for a two-day test which went some way to restoring Soviet faith in their technology. It also appeased NASA, preparing for detente's crowning glory, the joint U.S./U.S.S.R. Apollo–Soyuz Test Project scheduled for 1975, when an Apollo would dock with a Soyuz craft. In the meantime, June 1974 saw the next Salyut launch. Salyut 3 entered a slightly lower orbit than before and remained there until January 1975. Two further Salyuts were launched in December 1974 and June 1976, so that the three were in orbit more or less concurrently. What was not realized until much later was the fact that the first and last (Salyuts 3 and 5) were military stations, while Salyut 4 was civilian. By the time Salyut 5 re-entered in August 1977, analyses of the missions showed this beyond dispute.

Salyuts 3 and 5 were launched into orbits which were centred on 260 km (162 miles) above the Earth. Salyut 4 operated at around 350 km (220 miles). That extra 90 km allowed better photo-reconnaissance for military purposes. Unlike the U.S., Russia still relied on relatively simple film camera techniques to spy on its military opponents. After the last crews left Salyut 3 and 5, recoverable capsules were ejected which presumably carried the fruit of their photographic labours. Routine monitoring of conversations between crews aboard Salyuts 3 and 5 by Western intelligence sources showed that they were using different frequencies, and often speaking in code. That most of the crews were military officers and that very few pictures of the stations and their activities were ever released more or less corroborates the military theory. By comparison, Salyut 4 was trumpeted by the Soviets, though all three stations suffered a staggering variety of problems and were not completely successful.

MIXED RESULTS

Vostok veteran Pavel Popovich and Yuri Artyukhin spent 16 days aboard Salyut 3 in July 1974, where they made observations of specially-built targets around the Baikonur Cosmodrome. This we know from the U.S. military, whose satellites also spotted them at the same time! Sadly, the mission of Salyut 3 was blighted when the crew of Soyuz 15 failed to dock in late August 1974. The Soviets had made much of the improved computer and radar tracking system on the Soyuz ferry craft, but over the next decade, the system would fail on four further missions. The tracking and radar ranging data were not executed correctly by the computers, and Soyuz 15 was burning too much fuel in trying to chase Salyut 3. The design changes after the Soyuz 11 accident meant that the craft had less fuel and power, so had to return home. So Soyuz 15 landed in an emergency at night, after which the Soviet official spokesman claimed that 'practising emergency landings' had been the aim of the mission all along. Salyut 5 was abandoned and left to re-enter in December 1975.

Salyut 4 was launched in late December 1974, and in January 1975 was visited by the crew of Soyuz 17 who spent nearly a month on board. Lazarev and Makarov, who had tested the Soyuz 12 craft, were the next crew in line, planning to spend at least two months aboard Salyut 4. They were nearly killed in the process when they became victims of what the Russians still refer to as the '5 April anomaly'. After a seemingly successful lift-off on 5 April 1975, the upper stage of the SL-4 booster failed to separate from the first stage containing the clustered engines. The Soyuz 18 crew were sitting atop a rocket that was tumbling haphazardly, and they fired the emergency escape system at 120 km (75 miles) above the Earth. They sustained forces as great as 18G and blacked out in the process. Thankfully, the parachute opened and they landed near the Chinese border. However, the Soyuz capsule landed on the flanks of a mountain and began to roll down its slopes heading towards the edge of a high cliff. Luckily, the parachute snarled in some trees, and they emerged battered and bewildered. Their main fear was that they had come down in China, and would be made show hostages at a time of worsening relations between the two countries. However, they were inside the Soviet borders and were rescued some time later. They were severely bruised and battered: Lazarev never flew again. With the ASTP only two months away, a Soviet

Seen here being shepherded around a space exhibit are the participants in the Guest Cosmonaut programme. From left to right: Arnaldo Mendez (Cuba), Sigmund Jähn (East Germany), Vladimir Remek (Czechoslovakia), Miroslav Hermasziewski (Poland), Georgi Ivanov (Bulgaria), Bertalan Farkas (Hungary) and Pham Tuan (Vietnam).

official tried to placate NASA by saying that the Soyuz 18 booster had been 'checked less thoroughly than the ASTP booster would be'.

In May, Pyotr Klimuk and Vitali Sevastyanov were launched on just such a more thoroughly checked Soyuz booster and spent 61 days aboard Salyut 4. They were in orbit at the same time as Leonov and Kubasov in ASTP, and in the course of communications with them they mentioned a problem which had begun to plague them a month into their stay. Salyut 4's humidity control system had broken down, with the result that the station began to resemble a greenhouse. The windows fogged up and a green mould began to spread across the walls – very soon it covered most of the station's interior. Klimuk and Sevastyanov complained about the mould and the unpleasantness of sleeping in damp sleeping bags, and a week after Leonov and Kubasov returned, they got their wish to return home too.

Salyut 4 was home to no further crews, but did receive an unmanned Soyuz 20 in November 1975 which docked automatically. This was a prototype for a cargo vessel which came to be known as the Progress series and would play an important part in long-duration flights. Soyuz 20 undocked after three months, showing that a craft could survive that long without harm. This was important for the return home of long duration crews.

Salyut 5, launched in June 1976, was plagued with problems, and was home to only two Soyuz crews (21 and 24). Boris Volynov and Vitali Zholobov spent 48 days in the station from July onwards, but in late August they left it in a hurry. An acrid smell in the station was the reason, but this was presumably fixed from the ground in time for Soyuz 23 which was launched on 14 October. Once again, the computer failed and Soyuz 23 had to return home in an emergency. This time, they came down in a blizzard late in the evening of the 16th and became Russia's first unscheduled landing in water. The Soyuz 23 capsule came down in a freezing lake and it took rescue helicopters and frogmen nearly five hours to winch them to safety. All involved were given bravery awards and, with no sense of irony, it was revealed that before becoming a cosmonaut, Soyuz 23 engineer Rozhdestvenski had been a diving instructor. Before the flight he had said that his training would come in handy – he had no idea how true his words would be! It was left to the crew of Soyuz 24 to power down Salyut 5 in February 1977 before the station re-entered the Earth's atmosphere in August of that year.

BREAKING THE RECORDS

When Salyut 6 was launched at the end of September 1977, observers expected it would see new developments in the Soviet manned space programme. Ironically, the 60th anniversary of the Bolshevik revolution (and the 20th anniversary of Sputnik 1) was marked by failure when Soyuz 25 failed to dock with the station. Vladimir Kovalyonok and Valeri Ryumin attempted four dockings, but to no avail: they had to return home with no spare fuel in their Soyuz craft. This inauspicious start to the occupancy of Salyut 6 actually heralded a new era in the Soviet space programme: over the next five years, cosmonauts would notch up ever greater stays aboard the station. The Soviets were quick to emphasize the fact that Salyut 6 was a new generation vehicle: it had a second docking port at the rear which would allow Progress transporters to replenish the station with supplies and fuel. Unlike the American Skylab, the Salyuts had a propulsion system which could be fired to maintain them in orbit for many years.

In December 1977, Yuri Romanenko and Georgi Grechko were launched to Salyut 6 in their Soyuz 26 capsule and docked successfully. Because of the Soyuz 25 failure, it was feared that something was wrong with the forward docking port, so they docked at the rear. On 20 December, they donned Extra-Vehicular Activity (EVA) suits to investigate the front docking port for themselves. Grechko used handrails on the outside of the station to move to the front of the station and inspect the docking unit at close range. He remained there for over an hour and found very little wrong, so he returned to the rear airlock from whence he had come. During this time, Romanenko had been standing in the rear airlock porthole, and as Grechko returned, he moved out of the porthole to greet his comrade. But he had

Looking vaguely disconcerted, Vladimir Remek, the first non-superpower spaceflier, experiments with a 'Chibis' pressure suit on Salyut 6. Such suits were successful in minimizing the effects of weightlessness. Georgi Grechko (left) and Alexei Gubarev (right) watch with detached amusement.

The dramatic backdrop of the Earth provided Salyut crews with glorious sights as a counterpoint to their work. During rest periods, Soviet crews would spend hours gazing at the planet below.

not attached his tether and floated away from the station; luckily Grechko caught him in time. However, their adventures were far from over when they found that the valve which would repressurize the airlock seemed to be jammed open. They were stuck in space with no way of getting back into the station! Whereas most mortals would have panicked, Grechko and Romanenko merely hoped that the valve pressure instrument was wrong and attempted to repressurize the station. Thankfully it worked.

They stayed aboard Salyut 6 for 96 days and returned home in March 1978 after surpassing the final Skylab record of 84 days in space. During that time they settled down to a routine which became a blueprint for later Salyut crews, involving the replenishment of supplies and short visits by other cosmonauts. Soyuz 20 had been the test of the Progress capsule, and on 20 January 1978, the first in the Progress series proper was launched. Essentially a modified Soyuz, it contained supplies and fuel, the latter transferred by the cosmonauts after an EVA. Of more pressing concern to the crew, however, were letters from loved ones and food parcels contained inside the Progress vehicles. After unloading, the vehicles were filled with rubbish and they automatically undocked before burning up in the Earth's atmosphere.

GUEST COSMONAUTS

In early March, Grechko and Romanenko received their second set of visitors (from Soyuz 28) whose number included Vladimir Remek, a Czech. As the first 'guest' cosmonaut, his eight-day flight set the pattern for later visits from Soviet bloc cosmonauts. Conscious that NASA would be inviting foreign payload specialists to fly on the Space Shuttle, Intercosmos, the agency which oversees Soviet bloc space activities, made available flight opportunities for its own native cosmonauts. They would be trained with Soviet crews for over a year and spend eight days on Salyut before returning back to Earth. By returning on a Soyuz that had docked previously, they would effectively 'recycle' the ferry craft to ensure the long duration crew had a reliable vehicle in which to return home.

Vladimir Remek had been chosen because he just happened to be the son of the Deputy Defence Minister in the Czech government. In August 1978, Miroslav Hermaciewski joined the second long duration crew aboard Salyut 6 (Kovalyonok and Ivanchenkov) who succeeded in pushing the long duration record to 140 days. The fact that he was the brother of a Polish General was glossed over by official announcements. In turn, East German, Hungarian and Rumanian cosmonauts took their place aboard Salyut 6.

The crew of Soyuz 32 in 1979 extended the duration record to 175 days, and prepared to welcome a Bulgarian cosmonaut in April of that year. Cosmonauts Vladimir Lyakhov and Valeri Ryumin watched in horror as Soyuz 33 with veteran Nikolai Rukavishnikov and Bulgarian Georgi Ivanov overshot Salyut 6. The onboard computer had failed again, and they were ordered home. The engine fired for too long, and they descended on a far steeper trajectory than had been planned, encountering 10G forces on re-entry. It wasn't until June 1988 that a Bulgarian flew in space. Because of fears that their Soyuz 32 might decay in orbit, an unmanned Soyuz 34 was launched to bring the long-term crew home in August 1979.

Among the more exotic guests who visited Salyut 6 were North Vietnamese, Cuban and Mongolian cosmonauts. In July 1980, Pham Tuan visited Salyut where he made a special study of the effects of defoliants on his country. The Soviets were quick to point out that he had been a fighter pilot during the Vietnamese war against the American aggressors. Given that the superpowers were hardly on speaking terms during the 1980 Olympics, it was all that Moscow could do to emphasize their guest cosmonaut's political pedigree. Arnaldo Mendez (Soyuz 38, September 1980) became the first black to fly in space. Twenty-five years earlier he had been a shoe-shine boy in Havana. The last Salyut 6 guest cosmonauts included a Mongolian by the name of Jugderdemidiyn Gurragcha, whose name provided such a hurdle to native Russian speakers that he was known as 'Gurr' by one and all. The final guest to visit Salyut 6 was a Rumanian named Dimitru Prunariu in May 1981, who flew with Leonid Popov in Soyuz 40, the last in the old style Soyuz craft. An improved version with better computers and all-important solar panels known as the Soyuz T (Transport) had already been introduced by then. Control electronics had been miniaturized, so there was room for a third cosmonaut.

By the end of the Salyut 6 occupancy, the long-duration stay time had been increased to 184 days. Crews had learned to live in the strange environment of weightlessness, and performed many hours of experiments in materials science and Earth resources. The effects of weightlessness on the bones and muscles had been mitigated by exercise and a variety of experimental suits which provided pressure on the body. Psychologists had experimented with lighting levels and sounds to remind the crews of home. In turn, Salyut crews had been quick to pick up FM radio transmissions over the United States, showing a marked preference for heavy metal.

The last visitor to Salyut 6 was a strange vehicle called Cosmos 1267, twice as large as a Soyuz and referred to as a 'Star Module'. It boosted the station to a higher orbit, and undocked before Salyut 6 re-entered in July 1982.

SALYUT 7

Nearly a year passed before the final Salyut was launched in April 1982. Salyut 7 was similar to its predecessor, except it had improved computer and navigational equipment. A month later, Anatoli Berezovoi and Valentin Lebedev journeyed to it in Soyuz T-5, and increased the long-duration record to 211 days. It was an auspicious start to the final chapter of the Salyut programme, which was beset by triumphs and setbacks in almost equal measure.

During their stay, the T-5 crew were visited by two notable cosmonauts in the Soyuz T-6 and T-7 crews. In June 1982, French test pilot Jean-Loup Chretien became the first Western European to fly in space, and he amused his hosts – with typical test pilot bonhomie – with a Quasimodo mask among other items. His wish to bring good French wine and garlic to add some Gallic spice to the food aboard Salyut was banned because it was thought the cosmonauts could not cope with the former, and the air filtering system with the latter. (Chretien flew again in December 1988 when he performed the first spacewalk by a non-superpower spaceman.) In August 1982, Svetlana Savitskaya became the second Soviet female cosmonaut. NASA had already decided that Dr. Sally Ride would fly aboard *Challenger* in 1983, prompting the Soviets to steal some propaganda thunder. Savitskaya had two outstanding qualifications for her role as a cosmonaut: in her youth, she had been a sky-diving daredevil, and perhaps more importantly, she had a father who happened to be a Marshal in the Soviet Air Force. In December, Berezovoi and Lebedev landed back on earth at night in freezing conditions. Despite a hampered rescue, they were glad to be home with their families for the start of 1983.

The next visitor to Salyut 7 was another mysterious Star Module, Cosmos 1443 – heavier than

In 1982, Svetlana Savitskaya became the second women in space when she spent time aboard Salyut 7. Seen here with her are Igor Volk and Valentin Lebedev. Savitskaya flew again the following year: Volk is slated to pilot the first orbital test of the Shuttle *Buran*.

The last crew to visit the ailing Salyut 7 station were Leonid Kizim and Vladimir Solovyov aboard Soyuz T-15 in 1986. Seen here on an earlier flight (Soyuz T-10), Kizim performs one of many spacewalks to repair faulty equipment. Salyut 7 remains in orbit awaiting further visits by crews aboard *Buran*.

Salyut itself. The Russians remained notoriously coy about its purpose, leading Western military observers to suggest it was a military vehicle, a prototype battlestation no less. Described fleetingly as a 'space tug' or 'supply craft', the Soviets compounded the mystery by their reticence. They were, however, quick to announce in April 1983, that a three-man crew had been launched to take advantage of the Module's extra capacity. Alas, the radar tracking system on the Soyuz T-8 craft failed to deploy. Flying 'blind', it was impossible to estimate the distances in order to dock, so the crew returned home.

There was better luck for Vladimir Lyakhov and Alexander Alexandrov who were launched on Soyuz T-9 on 27 June 1983. Their luck held out until September, when a fuel line burst and then a solar panel failed to work. Temperatures within the station dropped to near freezing, and rather than abandon the station, it was decided to launch Vladimir Titov and Gennadi Strekalov to help them repair the station. On 27 September 1983, Titov and Strekalov were strapped into their Soyuz T-10 capsule when a fuel valve jammed in the SL-4 booster and there was an explosion on the pad. The emergency escape system came into its own, and the cosmonauts were pulled away from the pad to safety, suffering only slight bruises.

For the hapless Lyakhov and Alexandrov, the miraculous survival of their colleagues came as a relief, despite their own problems. Once the pad explosion became known, the Western media ran headlines which suggested the T-9 crew were stranded in space, as their Soyuz vehicle was well past its expected design lifetime. But the crew stayed on, valiantly fixing the problems with a spacewalk, and returned home in November after 150 days in space.

1984 brought better luck to the station, and a three man crew was launched on a successful Soyuz T-10 in February. They included Dr. Oleg Atkov, whose work into biomedicine ensured that the effects of weightlessness were minutely recorded. They extended the long duration record to 237 days, and were visited by an Indian guest cosmonaut as part of the Soyuz T-11 flight in April and Svetlana Savitskaya as part of the Soyuz T-12 crew in July. She performed a $3\frac{1}{2}$ hour spacewalk testing an electron welding device to do some running repairs on Salyut 7. That NASA's Kathy Sullivan was scheduled to perform the first U.S. female EVA in October of 1984 was probably enough to prompt Svetlana's second flight.

By early 1985, the rot had set in for Salyut 7 – or so it seemed. The water control system aboard the station was leaking, and the solar panels were out of alignment with the Sun. With no power, conditions inside the station were icy, to say the least. In June, veteran cosmonauts Vladimir Dzhanibekov and Viktor Savinykh were launched aboard Soyuz T-13 to perform the final rescue of the station. Conditions aboard had deteriorated so much that they spent most of their time in their Soyuz craft. They moved into the station after finally bringing the solar panels back on line. In September a replacement crew was brought up in Soyuz T-14, which involved the first direct swap of crews. Georgi Grechko returned home with Dzhanibekov, leaving Savinykh with Vladimir Vasyutin and Alexander Volkov to attempt to push the long duration record beyond 237 days.

In October, another Star Module (Cosmos 1686) was launched to Salyut 7, allowing even further experiments to be carried out. But the best laid plans of mice and spacemen were thwarted when Savinykh and Volkov noticed Vasyutin had become fatigued and listless. For the first time, a space flier had begun to suffer psychological illness, so there was little alternative to bringing the crew home. During 1986, Salyut 7 was visited by the two-man crew of Soyuz T-15, who effectively 'mothballed' it. However, the Star Module's propulsion engine fired Salyut 7 into a higher orbit, and in late 1988, the Soviets announced they would bring back the station piece by piece in the early 1990s.

MIR

On 20 February 1986, the Soviet Union launched its latest space station, christened Mir – meaning peace. Though roughly the same size as Salyut, it was designed as a modular station out of which a permanent complex would be built. Instead of having experiments inside the station, these were relegated to modules which were added at a later date. This meant more room for the crew, who also had creature comforts like their own personal sleeping cubicles, as well as improved toilet, washing and eating facilities. The Soyuz T-15 crew visited the station in earnest in 1986 before taxiing over to Salyut 7.

In February 1987, Yuri Romanenko and Alexander Laveikin became Mir's first long-term occupants, flying a Soyuz TM-2 (Modernized Transport) craft for the first time. Romanenko spent 326 days aboard the station with minimal physiological effects, so that the next long-term crew spent a year aboard. Musa Munarov and Vladimir Titov returned to Earth in December 1988, also in good physical shape.

Soviet doctors announced that there remained very few hurdles left to a flight to Mars, biologically speaking. The introduction of the Energia booster in May 1987 put that goal within the realm of possibility, given that it is the most powerful rocket ever built. On its second flight in November 1988, Energia carried the Soviet Space Shuttle Buran into orbit, the first unmanned test of the vehicle which promises to revolutionize Soviet space activities. Very similar in design to NASA's Shuttle, Buran will be used as a freighter to space stations like Mir in the 1990s. It will also be used to bring back the Salyut 7 station.

Ironically, 1989 has seen some remarkable developments in the Soviet space programme. In April two cosmonauts who were 'housekeeping' aboard Mir were brought home early, the reason being that they were spending most of their time reading faulty equipment. Also, two new modules were not ready on the ground.

As these words are written, it is becoming increasingly clear that *glasnost* has wrought the most staggering changes. Greater openness has led to greater accountability within the Soviet space programme. It is possible a 75 percent cut has been approved in funding manned spaceflight.

Yet, Tsiolkovskii's ideals remain like a bright light guiding the long-term plans for his descendants. Despite current difficulties, the experience with Salyut and Mir – coupled with the potential of Energia and the Buran vehicle – promises to fulfil Tsiolkovskii's dream rather sooner than he would have dared imagine.

Buran, looking remarkably similar to the U.S. Shuttle, is seen atop the Energia rocket. After a test launch in November 1988, *Buran* is set to launch a new phase of the Soviet space programme in the 1990s. Budgetary problems, however, will delay the first flight into 1991.

CAT AND MOUSE: WAR IN SPACE

by **Frank Miles**

Sometimes referred to as 'the dark underbelly' of space, military research in Earth orbit is an ever-present, often frightening, aspect of modern space programmes. Seen here is a NAVSTAR global positioning satellite, which will form part of the Pentagon's communications network in the 1990s. The U.S. armed services spend nearly three times as much on space than NASA does.

If a major war ever broke out the first blow would be struck in space and probably *from* space. Modern armies are so totally dependent upon satellites for reconnaissance, surveillance, communication, navigation and early warning of enemy attacks that space offers prime targets. In order to nullify an enemy's effectiveness to wage war, the first act would be to try to knock out its satellites, either by destroying them completely or by 'jamming' their communications systems and sensors. Adding to this the planned use of space-based lasers or particle-beam weapons to destroy the enemy's missiles before they reach their targets, space becomes a new theatre of war – the aptly named 'Star Wars' scenario.

The Pentagon hopes that by the end of 1990, its forces will have carried out tests in space of five different missile interceptor systems. Since 1983, when President Reagan first publicly revealed the plan to research space-based anti-ballistic weapons (the Strategic Defense Initiative or SDI) the Soviet Union has condemned it. However, the Soviets have themselves been running such space-based experiments for many years.

A NEW THEATRE OF WAR

The military on both sides have always been the biggest users of space although their activities have, of course, been largely shrouded in secrecy. Of the Soviet Union's 100 or so satellites launched

every year, about 80 are for military purposes. While the Americans launch fewer than the Russians do (with the Shuttle back in operation it is expected to average around 20 a year), in one sense they are equal since the American satellites have a much longer life. (The Soviet Union's often last for only a few weeks, or even days, the Americans' many months.) The loss of the American Space Shuttle *Challenger* in January 1986 and the subsequent virtual grounding of their space programme for two years was of more concern to the Pentagon than to NASA. And the much-vaunted Space Station which should have been in the course of assembly by now will almost certainly be used by the Department of Defense for espionage and/or 'Star Wars' research. After all, for years now the Russians have apparently made use of their own Salyut and now Mir space stations in this way, although they deny it.

Both major powers acknowledged the vital role of space in warfare right from the birth of the Space Age. Having said they had a rocket capable of carrying a nuclear bomb, the U.S.S.R. proved it with the launch of the first satellite, Sputnik 1, in October 1957. The rocket used was an ICBM known to the Russians as the A-1, and classed by Western military sources as the SS-6. It soon became obvious that the Moon would be an ideal base from which to launch such rockets. To destroy them would be virtually impossible: it would take an enemy's anti-ballistic missile three days to cover the quarter of a million miles to the lunar launch sites. This was one of the hidden reasons behind the race to the Moon in the 1960s.

What was horrifyingly clear even before the Moon was conquered was that both major powers had the technology to put nuclear bombs in orbit around the Earth – ready for almost instant delivery. This fact scared both sides so much it resulted in a resolution unanimously adopted by the United Nations in December 1963 that no nation would put nuclear weapons in orbit – although it took another four frightening years for that resolution to become a formal UN Treaty to ban such weapons.

So most of the 'aggressive' use of space has rested with spy satellites, anti-satellite satellites (ASATs) and research into space weapons – all of which will be covered here.

SKY-HIGH EYE SPY

Extraordinary and sometimes extravagant claims are made about the amount of detail that can be seen in photographs taken from space. It might be possible to read a newspaper headline from space, but it is more likely that the best resolution at the moment is down to about 30 cm (12 in). This means one could, for instance, see the number on the side of an aircraft or count the number of weapons under its wings. In many cases, military strategists need to see the overall layout of military installations, so too much detail in resolution is useless in this regard.

Warships and submarines form a vital part of a country's defence. Because 70 per cent of the World's surface is covered by ocean, their theatre for warfare is almost limitless. Seen here is a data link from a U.S. Navy Destroyer to the Pentagon via satellite. Without such links their effective deployment in times of emergency would be unthinkable.

Spy satellites fall into two categories, surveillance and reconnaissance. Surveillance is the regular monitoring of a potential enemy's activities – such as studying the building of new runways, missile sites, submarine pens, factories, and so forth. Satellites used for this role are often sun-synchronous, which means that they pass over the same territory at the same time each day and so can register precisely any changes in the scene below. Reconnaissance involves sending up often short-life satellites to study specific targets such as military manoeuvres, the movements of a fleet of ships or even a single warship, the deployment of tanks and troops, the exact size and location of a weapons factory to verify reports from agents on the ground. There are also satellites that can intercept telephone conversations or 'sniff' exhaust gases of rockets to determine what fuel has been used.

All conflicts over the past 30 years – the Arab-

One of the more pressing customers for the U.S. Space Shuttle is the Department of Defense. Had it not been for the *Challenger* accident in 1986, DOD Shuttle missions would have been launched out of the Vandenburg Air Force Base in California. Seen here in a 1985 training exercise is a guard, one of many overseeing a roll-out ceremony.

Israeli wars, the Iraq-Iran war, the battle for the Falklands, Afghanistan, etc – have been closely monitored by spy satellites launched by both the U.S.A. and the U.S.S.R.

THROUGH THE KEYHOLE

One of the best known spy satellites was the Big Bird series in the 1970s, America's main surveillance satellite which was also used for reconnaissance. It was built by Lockheed, launched by a Titan 3C and earned its name by its size – roughly 15 m (49 ft) long and 3 m (10 ft) in diameter. It also had a 6 m (20 ft) dish as one of its many aerials. Each satellite had two cameras, one for general surveying, the other of extremely high definition for close-up work. Satellites in the series were sometimes put into extremely low orbits to pass a target as close as 160 km (100 miles) above them. There were two ways of sending back pictorial information: by ejecting pods of film to be recovered by aircraft or by processing the film on board, scanning it with lasers to turn it into electronic signals and then transmitting the pictures back to USAF bases around the world. At the end of their useful life each Big Bird satellite was destroyed by triggering an explosive device on board.

Big Bird has been replaced by the Keyhole series of satellites, KH-11 and KH-12, built by TRW and which have a lifetime of about two years. The KH-11 is over 19 m (62 ft) long and though it orbits at higher altitudes than Big Bird did, it is believed to have superb digital imaging systems as well as infra-red and radar sensors (to 'see' through clouds). The pictures can be relayed back to Earth immediately and, being digitized, they provide extreme clarity. Infra-red sensors can detect such things as well-camouflaged tanks or aircraft on the ground, by 'reading' the heat from their engines. The multi-spectral sensors can detect normal camouflage from natural surroundings (although the Russian military forces have developed new camouflage techniques which might not be penetrated by Keyhole satellites). The KH-12 is bigger and designed for launch by America's Space Shuttle, but the loss of *Challenger* in January 1986 and the subsequent delay in Shuttle flights seem to have postponed the KH-12 going into operational use.

In addition to the imaging satellites there are others known as ELINT (Electronic Intelligence) satellites, such as the American Rhyolite and now its successors, Magnum and Aquacade, which are highly secret but presumed to be more powerful and sensitive and even more resistant to jamming. The ELINT satellites, often referred to as 'ferrets', monitor radio and radar transmissions (these last to measure pulse widths and frequencies to help determine what the radar is being used for) and eavesdrop on telephone conversations. The U.S. military also make use of White Cloud clusters of ELINT satellites – 3 satellites in each of 3 clusters – which carry infra-red and millimetre wave sensors that can pinpoint surface vessels at long distances. Other highly-secret ELINT satellites are able to reveal the positions of nuclear submarines by detecting the warm water trailing behind them as they move under the oceans.

THE ELITE OF ELINT

The Soviet Union has been using ELINTs since 1967, launching them, as they do all their civilian or military satellites, simply under a 'Cosmos' number. Cosmos 1603, launched in 1984, was the biggest military satellite they have ever launched. They have on several occasions set up constellations of six ELINT satellites spaced at 60 degree intervals to detect military radio stations and by using co-ordinates from three or more satellites, pinpoint the location accurately.

Since 1962, the Russians seem always to have a spy satellite ready for launch. As soon as a conflict breaks out anywhere in the world they have a close-observation, short-life satellite launched to investigate it. They are almost all photographic reconnaissance satellites, either returning to Earth with the film or jettisoning pods of film to be retrieved by aircraft. The earliest spy satellites had a lifetime of one to two weeks; later satellites, weighing almost 2 tonnes and believed to be based on the Soyuz manned capsule design, also featuring solar panels, were highly manoeuvrable and carried out missions which lasted 60 days.

The spy satellites that go into near-polar orbits, the Earth spinning beneath them as they orbit it, can photograph the whole world within a 10–11 day period. Others are sent on highly eccentric orbits allowing low passes, down to about 140 km (88 miles), over a particular target on the ground. They have used such satellites to monitor the Iraq-Iran war, Lebanon, The Falklands, and the rebel emplacements in Afghanistan.

Electronic surveillance from orbit is no longer the preserve of the superpowers. It was revealed in 1987 that the British Ministry of Defence had developed an ELINT satellite system called ZIRCON, to monitor Russian electronic messages. Hitherto they have relied on co-operation from the U.S.A. Costs for ZIRCON, it was alleged, were hidden under the development costs of the UK's Skynet system of military communications satellites. It is also known that the French will be launching an ELINT and photo-reconnaissance satellite system known as Helios in 1993. There will eventually be four of these orbiting at 850 km (530 miles).

NUCLEAR-POWERED DANGER

Quite a few spy satellites, on both sides, are nuclear-powered as radioactive decay provides higher power outputs than solar panels. The Russian RORSATs (Radar-equipped ocean-reconnaissance satellites) have, since 1967, been launched in pairs, to identify and track warships. The delay between the pass of the first and the second satellite enables the Russians accurately to determine the speed and direction of any ship. The U.S. Space Command believes these

Britain's Ministry of Defence (MOD) has developed a compact communicator system for deployment on the European battlefields in the event of hostilities. Re-routing commands via the Skynet system, battalions could receive up-to-the-minute data about their strategic manouevring. In the U.S., handheld systems have been pioneered.

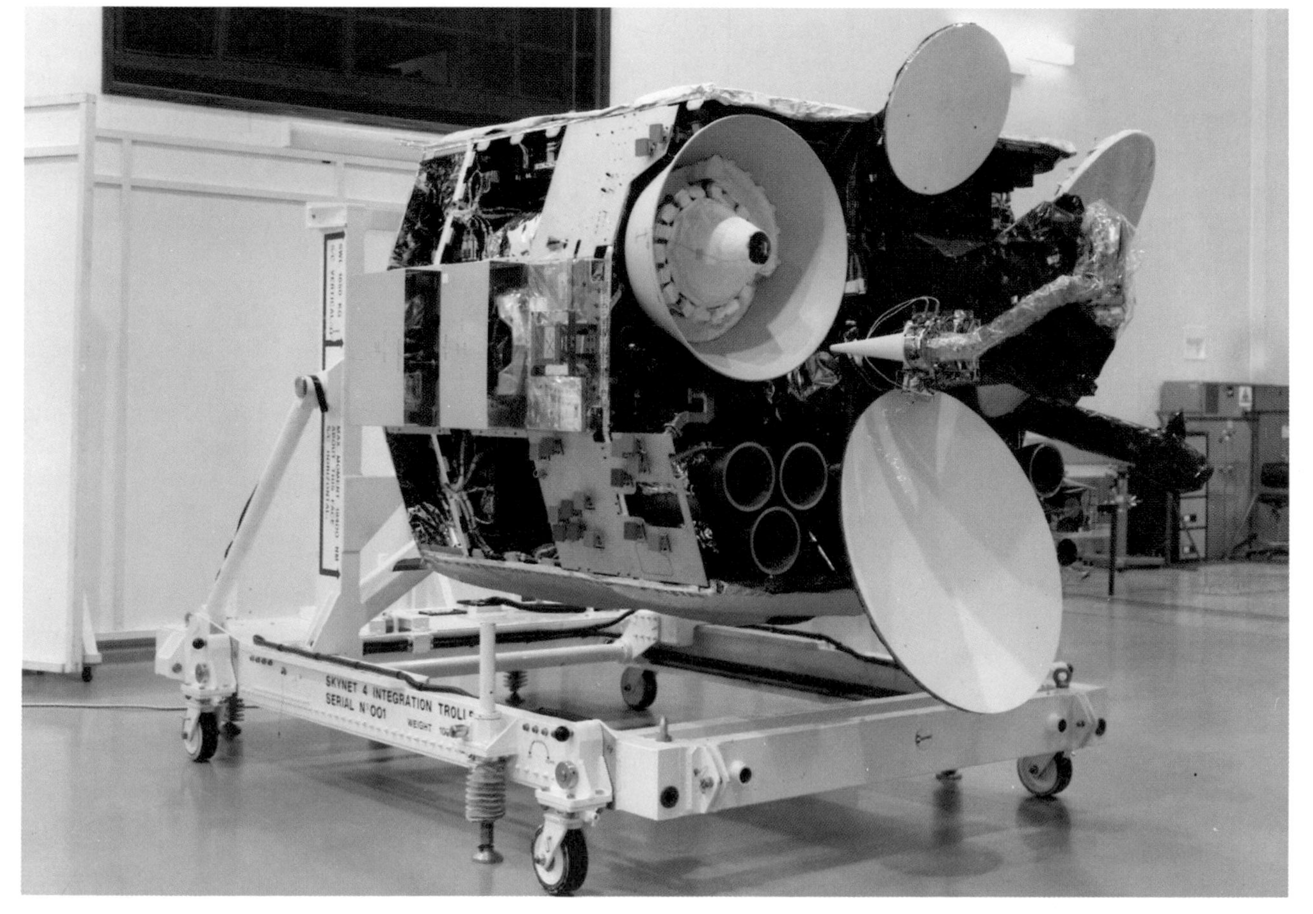

As with espionage on the ground, military activities in space are not without their controversies. Seen here is a Skynet 4B satellite being built by Marconi, the same company which was allegedly involved with the ZIRCON affair. The satellite shown here was launched in tandem with the Astra satellite in December 1988 by an Ariane 4.

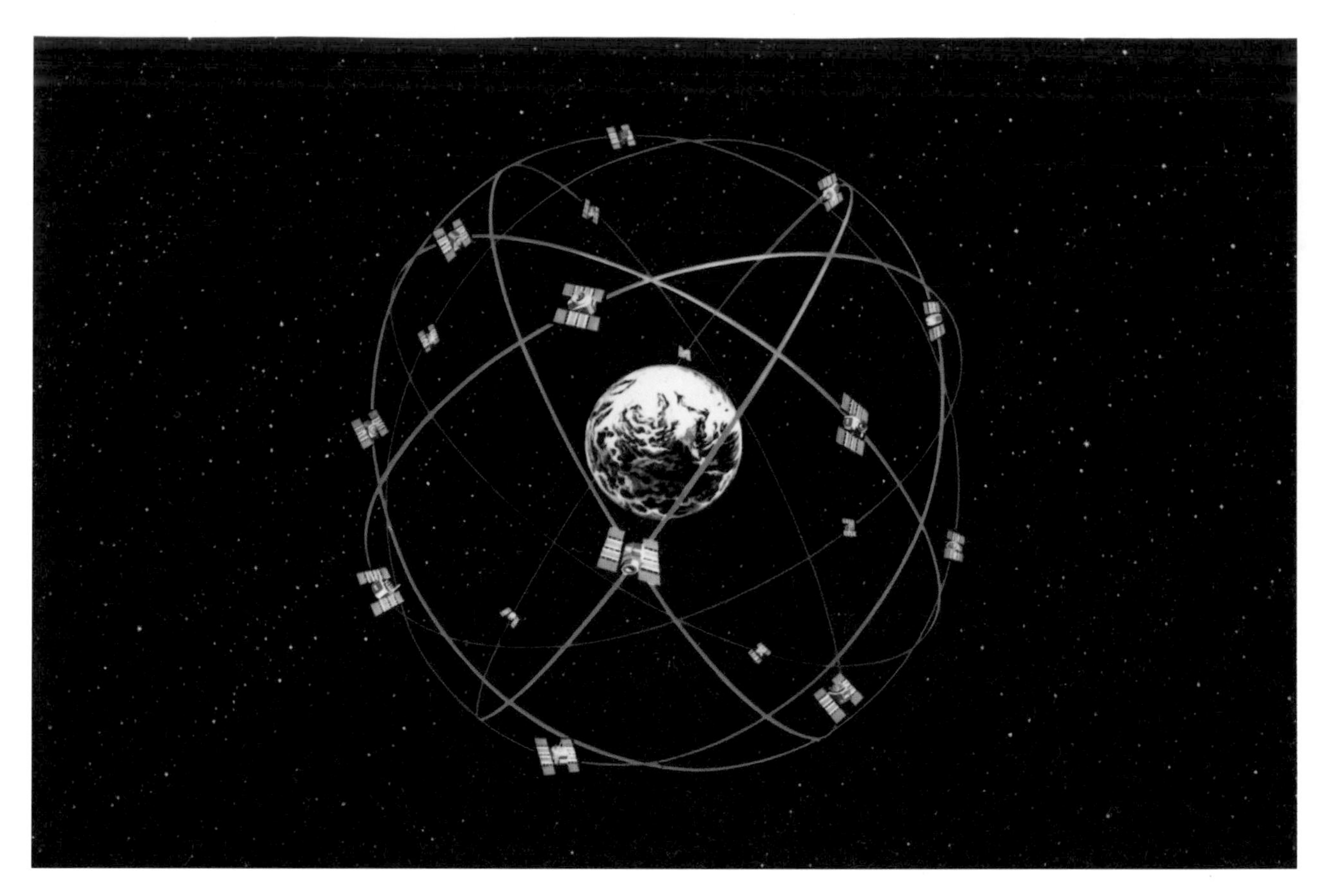

The NAVSTAR system is shown here in outline, a constellation of six satellites in circular orbits that will enable users across the globe to work out their exact position, velocity and local time at any time under any weather conditions.

RORSATs to be so effective they could neutralize the U.S. Navy in wartime.

The realization that these satellites had nuclear-power plants on board came when, at the end of their two-month missions, each of the pair would in turn be raised to a 900 km (562 mile) circular orbit. The reason for this higher orbit is to allow some 500 years to pass before they fall back to Earth – leaving the problem of lethal radioactive debris for some future generation to worry about.

Three of these RORSATs, Cosmos 954 in 1978, Cosmos 1402 (launched in 1982 to study Britain's Task Force operating in the Falklands war), and then Cosmos 1900 in 1988, failed to go into the higher orbit and fell back to Earth. The worst result was with Cosmos 954 when radioactive debris was scattered over part of Canada. The debris of the others fell into the Indian and Pacific Oceans despite the addition of safety features after the Cosmos 954 incident.

In all there are 50 nuclear-powered satellites at present orbiting the Earth at various heights.

KEEPING IN CONTACT

The jamming of military communications satellites was, for a long time, considered an easy matter, yet even so their use in peacetime became so valuable that the armed forces of most developed nations now depend upon them. This resulted in moves to make them 'jam-proof' which is becoming possible by 'hardening' the spacecraft systems (i.e. making them less electronically vulnerable) and by using Extremely High Frequency (EHF) band transmissions.

America's DSCS (Defense Satellite Communications System) comprises 16 satellites, each of which has 1,300 voice channels. It is possible to 'steer' the beams from them at will and it is also possible to shift their positions quite easily to meet defence requirements in any part of the world. The hold-up in Shuttle flights due to the *Challenger* mishap seriously upset the launching of replacements for two DSCS satellites that showed signs of wearing out.

The U.S. Navy's FLTSATCOM Ultra High Frequency (UHF) series of communications satellites are used in the control of the U.S. nuclear forces around the world, acting as links between warships, submarines, U.S. Navy aircraft and a thousand USAF aircraft with Strategic High Command. The President also uses these satellites to communicate with his commanders. Some of the FLTSATCOMs will eventually be replaced by the LEASAT and Syncom satellites which again use UHF bands, although there were problems in deploying some of them in orbit from the Space Shuttle.

SDS, or Satellite Data System, is used by America's nuclear forces stationed in polar regions (and SDS is also used to relay data from Keyhole spy satellites). They are top-secret satellites which, it is believed, have actually been damaged by ground based lasers in the Soviet Union. TDRSS, the Tracking and Data Relay Satellite System, used principally for Shuttle operations, is also used by the military.

The Soviet Union uses military versions of its

Molniya communications satellites, eight of them, for its forces. There are also more than 40 short-life (two years) Cosmos satellites apparently being used at any one time for medium-range VHF and UHF military communications. The frequency of replacing such large numbers means the 'dead' satellites and their rocket stages are causing one of the chief concerns about the amount of dangerous orbiting space debris.

NAVIGATION

The United States is establishing a whole constellation of NAVSTAR/GPS (Navigation, Timing and Ranging/Global Positioning System) satellites. There will eventually be 21 of them, which will provide 'three-dimensional information' to enable not only every ship and plane but even individual soldiers to determine their precise position on the surface of the globe. The individual soldier would use a lightweight backpack comprising a receiver and computer which, without him needing to transmit anything, could determine his position to within 15 m (50 ft). These satellites can also enable an anti-missile unit to measure the precise velocity of an ICBM to less than a tenth of a millimetre a second. NAVSTAR satellites, which are Shuttle-launched, are being 'hardened' against enemy attack. The system was to have been fully operational by now but Shuttle grounding has caused delays and target date has been put back to at least 1991. NAVSTARs are also intended for civilian use.

The Soviet Union has been using GLONASS (Global Navigation Satellites System) navigational aids, similar to the American NAVSTARs, although fewer in number – probably 12. Also like NAVSTAR, although GLONASS is primarily a military system, these satellites could be used by civilians. The Russians began launching them in 1982.

EARLY WARNINGS

America's Space Based Surveillance System (SBSS) involves the use, primarily, of two types of early-warning satellites capable of giving the U.S. early warning of the launching of missiles and of the deployment of military satellites. IMEWS (Integrated Missile Early Warning Satellites) replaced an unsatisfactory, earlier system called MIDAS, and embraces both the detection of missiles within seconds of their launch (even from submarines) and also detection of nuclear weapon tests carried out in the Soviet Union or elsewhere. One IMEWS satellite stationed over Panama watches over the Atlantic and Eastern Pacific oceans, while a second over the Indian Ocean watches the other half of the world.

DSP (Defense Support Program) uses satellites in geostationary orbit to work in conjunction with ground-based BMEWS (Ballistic Missile Early Warning System) stations to monitor Soviet missile launches. The satellites use camera-like telescopes which use infra-red sensors to detect the heat of missile engines soon after launch. There are critics of the efficacy of these satellites and it would seem that IMEWS is the more favoured of the systems.

The Soviet space-based early warning satellites, which first became operational in 1976, seem flawed: the Russians launch many more satellites for this role than do the Americans, indicating the Russian version is proving less efficient. They are put into extremely elliptical orbits of between 600 and 40,000 km (375–25,000 miles) – the apogee enabling them to study American missile sites for several hours of each 12-hour revolution.

HUNTER-KILLERS

It is thought that the Soviet Union is further advanced in a system of anti-satellite (ASAT) weaponry than is the United States. The Russians have tested their low Earth orbit systems since the mid-1960s and have shown an impressive ability to launch hunter-killer satellites, rendezvous close to a target satellite and then destroy both by exploding the hunter satellite. The Americans, on

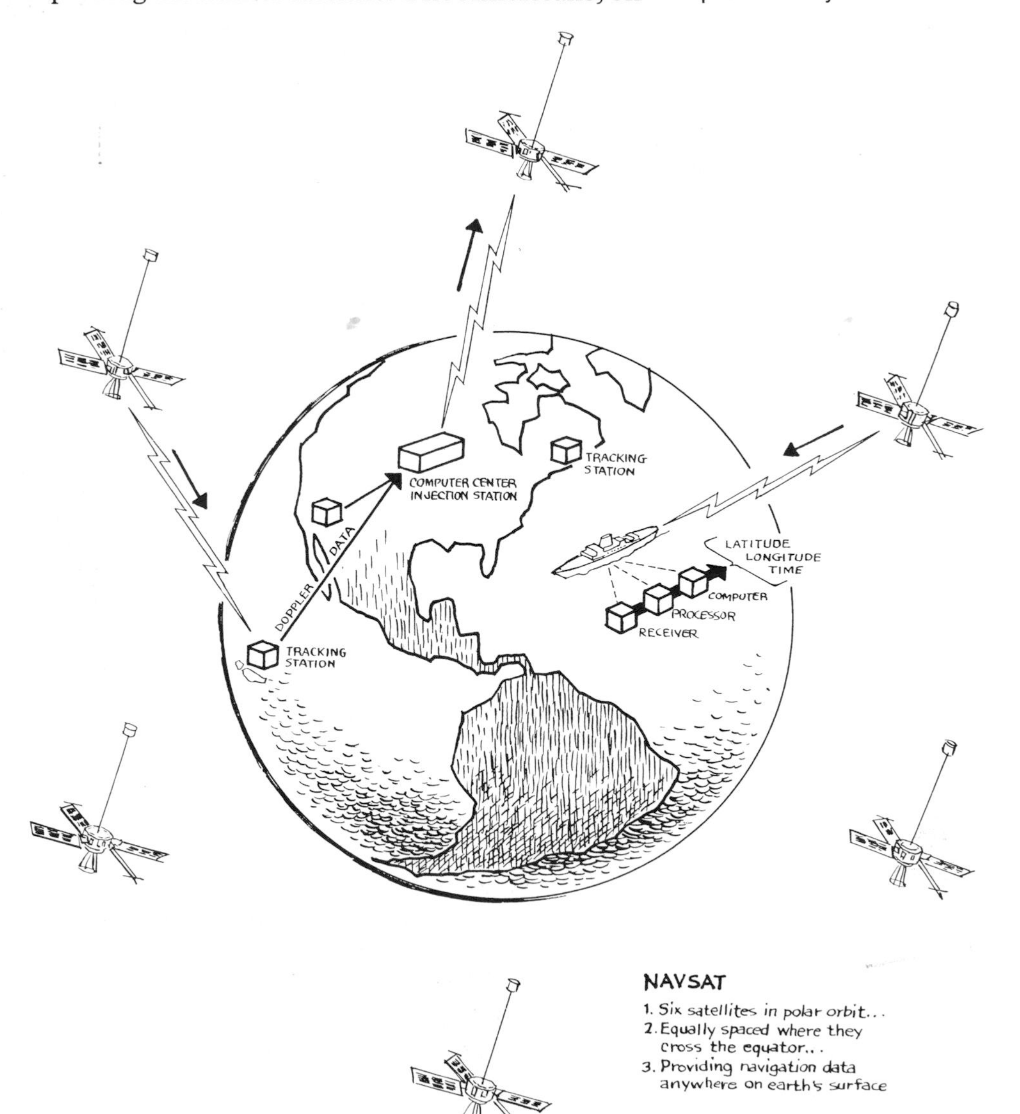

An earlier system used by the Pentagon called NAVSAT, points an interesting aspect of military space systems. In use during the 1970s, its technology is now obsolete. The systems used in the field have now increased in capacity, requiring new and improved relays.

the other hand, who also started working on ASAT projects virtually from the start of the Space Age, have changed methods several times without reaching a conclusion of which to adopt, but now hope to have an effective system by the early 1990s.

America's first successful test was in 1959 with an air-launched missile called Bold Orion, fired from a B-47 bomber against an early Explorer satellite that had finished its useful life. This seemed successful but was abandoned in favour of the development of another system. That, too, was cancelled, in 1962, while still under development. Within two years the U.S. had yet another ASAT system known as Programme 437 with missiles launched by Thor rockets. Ten years later this was scrapped and yet another system, which they hope to have operational within a few years, was undertaken.

The Russians were developing an ASAT system in 1962 and held hunter-killer satellite tests between 1968 and 1983, using infra-red homing devices and conventional explosives which resulted in a blast of metal fragments. Having apparently perfected a system against satellites, mainly those in Low Earth Orbit but some orbiting as high as 5,000 km (3,250 miles), they then called for a ban on the development of all space weapons.

It is not known whether either side has a proven ability to tackle the all-important geosynchronous satellites (those for communications and navigation). However, there has been much speculation that the U.S.S.R.'s new massive Energia rocket could, with strap-on boosters, develop the power needed to launch an anti-satellite weapon up to the same height – 36,000 km (22,500 miles). In fact, such an ASAT would be in a slightly higher orbit which is termed 'retrograde' – it would be orbiting in the opposite direction to the rotation of the Earth and therefore to the geosynchronous satellites. The value here is that such an ASAT would pass close to every satellite at that height in just 24 hours and so could selectively 'zap' all an enemy's communication and navigation satellites in that time.

The next arena for warfare will be space. Seen here is an artist's impression of an ICBM interceptor tested by the U.S. Army over the Pacific in June 1984. Five years later, the SDI programme is over budget, decreased in scope and unlikely of ever being completed as originally envisaged. Perhaps technical limitations will ensure that mankind's instinct for warfare will not continue indefinitely.

WAR AND PEACE

As with most technological endeavours, the military and strategic implications of space have acted as a spur to its development. There are many who fear that the 'hidden' space programmes of the world will lead to a further escalation in the arms race – this time into space. Yet others feel that the strategic importance of space is negligible for the simple reason that military satellites are little more than 'sitting targets'. And despite the vast sums of money poured into particle-beam weapons as part of SDI or 'Star Wars' research, many scientists believe that such systems will never operate efficiently because of the vast amounts of energy needed to power them. Pentagon spokesmen no longer refer to SDI as an effective 'peace shield' which will defend the United States from any missile attack. At best, the project will minimize the number of warheads which will impact on the American mainland.

At the very heart of discussions of military uses of space is an interesting paradox: as space surveillance becomes more sophisticated, it is less likely that the 'other side' can effectively hide its forces which it could be amassing for a surprise attack. In the late 1950s, the U.S.S.R. possessed only a handful of operational tactical missiles. If surveillance from orbit had been possible at that time, the United States would not have embarked upon lessening the so-called 'missile gap' which Soviet leaders claimed was stacked heavily in their favour.

Military surveillance satellites effectively mean that it is possible to estimate the armed capability of potential enemies. The fact that the watchers are being watched themselves is an important factor in arms limitation talks and agreements. It is perhaps a further paradox that such systems make the world a safer place in which to live.

EYES ON THE UNIVERSE: ASTRONOMY FROM SPACE

The Space Age has revolutionized the science of astronomy. Our knowledge of the cosmos has literally extended beyond vision, for telescopes in space have revealed objects at wavelengths which cannot be seen from the surface of the Earth. As a result, our picture of the Universe has undergone as dramatic a revision as that brought about by Galileo's introduction of the first telescope in 1610. Many new objects and processes have been found, leading astronomers to wistfully talk of ever stranger additions to the 'cosmic zoo'. Yet the story is far from over: further telescopes are planned for the next decade, and another quantum leap will doubtless take place in the 1990s. At that time our perception of the Universe will change once again, almost certainly in ways which we cannot predict.

DARKNESS VISIBLE

On a clear night, around 2,500 stars are visible to the naked eye, and many more become apparent with telescopes and binoculars. But even with the largest telescopes, that is not the end of the story; there is a multitude of astronomical objects which we cannot see directly. Our eyes are only sensitive to a spread of colours generally known as visible light. White light itself is made up of individual colours, a fact observed by Isaac Newton in the late 17th century when he passed sunlight through a prism. The rainbow-like spread of colours which resulted – the spectrum – was the birth of the science of spectroscopy.

In the early 1880s, William Herschel repeated the experiment and showed that the temperature of a thermometer with a blackened bulb increased when placed beyond the red part of the spectrum. This led to the realization that infra-red radiation – literally, 'radiation beyond the red' – existed. At around the same time, it was noted that paper soaked in salt was darkened when placed beyond the ultraviolet part of the spectrum obtained from sunlight. Similarly, the notion of ultraviolet radiation came into being.

The development of physics in the 19th century led to the notion that infra-red, visible and ultraviolet radiation were all similar. The discovery of radio waves and X-rays in the laboratory led physicists to realize that the radiation was caused by waves of energy associated with electric and magnetic fields, which also had the property of being able to travel in a vacuum (unlike sound waves). These different electromagnetic radiations had different frequencies, yet all travelled at the speed of light. Radio waves had the lowest frequencies, while the highest frequencies belonged to gamma rays: physicists classed the different types of radiation as best they could, but ultimately some of the distinctions were arbitrary.

Astronomical research from space will be revolutionized with the launch in 1990 of the Hubble Space Telescope. Shuttle *Atlantis* and its crew will deploy the telescope to begin its work in pushing back current frontiers of knowledge.

In the same way that light could be split into colours, each with their own frequency, spectroscopy could be extended to the other electromagnetic radiation by use of the same technique. As electromagnetic radiation is essentially the result of an electric charge being accelerated, spectroscopy was a powerful technique in understanding the physics of different objects. The motion of electrons surrounding atomic

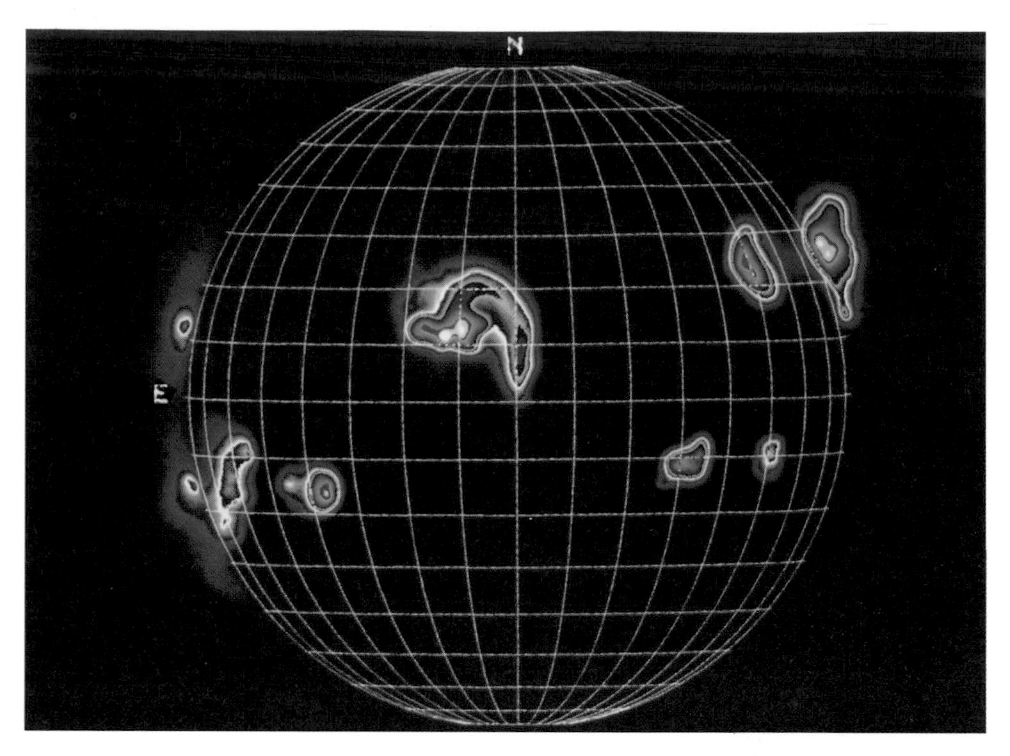

The full range of the electromagnetic spectrum is shown in this schematic diagram. The 'rainbow spread' of the optical window, to which our eyes are sensitive, forms a small part of the spectrum. The highest energy regions are seen to the right of the diagram.

nuclei between different energy levels gave rise to specific electromagnetic radiation. As the number of electrons varies for different elements, it was soon realized that each presented its own characteristic frequency through the spectroscope. Each element has its own unique spectral signature. In this way, for example, helium was identified in the Sun's spectrum before it had been chemically recognized on Earth!

Though the full range of electromagnetic radiation is emitted by astronomical objects, most of that radiation is absorbed by our atmosphere. Only certain radiations (such as visible light and some radio frequencies) are not blocked by the vast blanket of air which surrounds our world. Our view of the heavens was almost like that of a crab at the bottom of the sea, itself hampered by the vast ocean above it. It is hardly surprising that by the 1930s, many astronomers blithely claimed that they had more or less come to the end of the road for making new discoveries. It was just a case of mopping up a few new galaxies and trying to explain some of the remaining puzzles.

Yet, a new era was dawning that had its unlikely origins in a field in New Jersey. An engineer by the name of Karl Jansky, who worked for the Bell Telephone Company, was investigating troublesome static which interfered with telephone communications in 1931. Using a home-made aerial which was mounted on parts from an old Model T Ford, Jansky noted that radio noise was coming from the same portion of the sky as the Milky Way. This was the inauspicious birth of radio astronomy, which blossomed with wartime research into radar. After the war, the development of radio astronomy continued apace, revealing many new objects in the Universe.

It soon became obvious that other electromagnetic radiation could be detected if one could get above the vast bulk of our atmosphere. The sounding rockets that were developed from wartime missiles like V2s enabled astronomers to experiment in earnest. History records that the first attempt to observe ultraviolet radiation from space failed when a V-2 sounding rocket crashed in June 1946 and the ultraviolet film was destroyed. But in October of that year, ultraviolet spectra of the Sun were returned from the upper atmosphere. In 1948, X-rays were also detected from the Sun, and astronomers borrowed the experimental equipment devised by particle physicists to look at these very energetic emissions.

Because of its close proximity, the Sun was an obvious target for the initial sounding flights. Indeed, the development of solar astronomy from space has dramatically redrawn our portrait of our nearest star. Far from being the benign, life-giving Phoebe of the ancients, new techniques revealed that the Sun pulsates with scarcely-believable violence. It is a vast ball of hydrogen and helium which shines by the same nuclear transformation which powers hydrogen bombs: at its core, hydrogen is being converted to helium at the rate of 600 million tonnes per second. It would be impossible to catalogue all the results, but flotillas of interplanetary spacecraft such as the early Pioneers and Explorers revealed the existence of supersonic streams of charged particles – known as the solar wind – generated in the uppermost reaches of its atmosphere. X-ray astronomy by astronauts aboard the Skylab space station in the 1970s, and missions like the Solar Maximum Satellite in the 1980s, have told us more about unpredictable outbursts of energy – solar flares – equivalent to many thousands of thermonuclear explosions. Its significance lies in the fact that this radiation reaches the Earth's magnetic field and causes fantastic auroral displays which ultimately affect our weather and climate.

X-RAY EYES

The first X-ray sources beyond the Solar System were discovered by a test flight of an Aerobee rocket in June 1962. Ironically, Ricardo Giacconi's team were looking to see if X-rays were being emitted from the Moon by its interaction with the solar wind. The first X-ray object, Scorpius X-1, was duly discovered, believed to be a black hole into which matter is accelerating, giving off streams of X-rays as a result. Further flights had determined about 50 sources by the end of the 1960s, as well as a 'background' emission of X-rays which was hard to explain at that time.

The first dedicated X-ray astronomy satellite was not launched until the end of 1970. Known officially as Explorer 42 it was the first of NASA's Small Astronomy Satellites (SAS) and was launched by a Scout rocket from a converted oil rig off the coast of Kenya in December 1970. Operated for NASA by Italy, the San Marco platform saw the start of many space astronomy missions. The launch date also happened to be the 7th anniversary of Kenyan independence, so the pioneering satellite was quickly rechristened *Uhuru*, the Swahili word for freedom. Over the next few years, *Uhuru* added a further 300 objects to the catalogue of X-ray objects, including supernova remnants and globular clusters, some outside of our own Galaxy. And *Uhuru* introduced astronomers to exotic objects called X-ray binaries, in which a fairly normal star orbits a denser, older neutron star, resulting in matter spiralling into the neutron star and being heated to many thousands of degrees and generating X-rays.

In the early 1970s, further satellites gave astronomers new insights into X-ray processes which were both startling and significant. NASA's third Orbiting Astronomical Observatory (also known as *Copernicus*) observed the Crab Nebula and slowly rotating neutron stars. A British satellite called *Ariel 5* also provided interesting results. The *Ariel* program had started as a joint Anglo-American project to investigate upper atmosphere physics and, as a result, Britain holds the distinction of being the first nation to have a satellite launched for it, *Ariel 1* in 1962. By 1974, British X-ray astronomers had built their own satellite, the fifth in the series, which was duly despatched from San Marco in October of that year. Its instruments detected spectral lines which showed that the space between galaxies was filled with a hot, ionized gas at temperatures of around 100 million degrees. Two further small X-ray satellites, the Dutch ANS and NASA's SAS-

Some of the most striking objects seen at higher energies are associated with stellar explosions. The Crab Nebula resulted from a star that exploded in 1054, and at its centre is a rapidly-ticking pulsar. These objects can be seen at radio wavelengths, but remain enigmatic.

The European Space Agency's EXOSAT returned much-needed data about unusual X-ray objects. This object is known as Cassiopeia A, the remnant of a supernova explosion which is now a powerful radio source. The data have been colour-coded to show maximum X-ray emission brighter.

3, investigated X-ray bursters, objects which release more radiation in a few seconds than our Sun does over a few weeks.

By the mid-1970s it was clear that there was a multitude of X-ray sources in the sky, but the satellites to date had not been able to pinpoint exactly where they were located in the sky. So NASA planned the High Energy Astronomical Observatories (HEAOs) which would attempt to resolve these objects directly. The first, HEAO-1, was launched as a test of the instruments and spacecraft sub-systems in 1977. In August 1978, HEAO-2 was launched and unofficially renamed *Einstein*, in honour of the 100th anniversary of the great scientist's birth. Over the next two and a half years, it increased the number of X-ray sources to several thousand, using the first telescope capable of focusing the highly energetic radiation. Its most important discovery was that most stars emit X-rays, the cause of the unidentified background emission.

The most recent X-ray telescope was the European Space Agency's EXOSAT, launched in May 1983 atop an Ariane. Its two telescopes, though less sensitive and not capable of the resolution of *Einstein*, investigated many X-ray binaries, including an object called AR Lacertae, believed to be two Sun-like stars in which X-ray production seems related to starspot formation. EXOSAT's highly elliptical orbit around the Earth's poles allowed astronomers to make continuous observations for up to 90 hours, a distinct advantage over *Einstein*.

Further X-ray telescopes are planned for the 1990s. NASA's Advanced X-Ray Astronomical Facility (AXAF) will allow far greater resolution – by a factor of ten. ESA's XMM mission with a lower resolution will allow a catalogue of X-ray spectra to be built up at the turn of the century. And in the meantime, a Russian craft called Spectrum-X will be launched in the early 1990s on which European X-ray instruments will fly.

ON THE NUCLEAR TRAIL

The very highest energy radiation is associated with very violent processes, and is known as gamma radiation. Gamma rays are the most penetrating of radiation and in astronomy are associated with cosmic rays, the mysterious radiation which reaches the Earth from the furthest reaches of the galaxy. Cosmic rays interact with the Earth's atmosphere, causing showers of charged particles to be generated. Most people are aware of gamma rays in their most insidious guise – the fallout from nuclear explosions – and ironically, it was from this angle that gamma ray astronomy began.

The Partial Test Ban Treaty of the early 1960s banned unauthorized testings of atomic weaponry, and U.S. Intelligence wanted to check on any illicit tests by other nations. If a weapon was tested, it would release a shower of tell-tale gamma rays, and so from the late 1960s onwards, the Department of Defense launched the Vela series of satellites to monitor gamma rays. At any one time, two satellites separated by 180° ensured that the globe was covered, though the gamma rays it detected were of the cosmic variety. In 1969, the first 'gamma ray bursters' were discovered, though they were not announced until 1973. Since then several hundred bursts have been observed, all different from each other.

The first gamma ray surveys of the whole of the sky were made by NASA's SAS-2, which was launched in November 1972. Though it only operated for six months due to an electrical fault, its work was later supplanted by COS-B in 1975, one of ESA's first scientific satellites. COS-B showed that gamma ray emission was concentrated in the Milky Way, generally towards the centre of our galaxy. Though COS-B operated until 1982 and noted 25 discrete gamma ray sources, its angular resolution was not very good. A number of those sources were possibly pulsars, but because their position could not be well specified, they could not be correlated with known objects. So for example, the second brightest gamma ray object in the sky, known as Geminga, similarly remains a mystery.

NASA's HEAO-3, launched in 1979, monitored gamma ray emission, leading astronomers to conclude that most galactic gamma ray emission comes from cosmic rays. It also monitored a peculiar binary system called SS433 from which vast jets of material, like a celestial lawn sprinkler, are flung outwards at a quarter the speed of light. The HEAO's instruments detected a gamma ray burst from the Large Magellanic Cloud, a satellite galaxy to our own, which was also observed by instruments on eight other spacecraft. The next step in gamma ray astronomy will be NASA's Gamma Ray Observatory in the mid-1990s which will map the whole sky and improve the resolution of individual objects.

TENACITY AND TRIUMPH

In the energy ranges nearer to those of visible light, space astronomy assumes a familiar guise. In the ultraviolet reaches, not many new objects have been discovered, though the technique of spectroscopy has come into its own. For many of the physical processes in the universe, spectral lines are stronger in the UV than in the visible regions of the spectrum. For example, in the material between the stars – the interstellar medium – most atoms have their strongest spectral lines. Ultraviolet spectroscopy has allowed astronomers to observe gas flows around hot stars, the gases in the outermost reaches of planetary atmospheres and the hottest parts of cooler stars.

The first satellites to carry ultraviolet instruments were NASA's Orbiting Astronomical Observatories (OAOs) in the mid-1960s and early 1970s. Though only two of the four satellites worked, their discoveries included the detection of a hydrogen cloud around a comet and ultraviolet radiation from a supernova remnant. The third in the series was renamed *Copernicus* after the great Renaissance astronomer.

By far the most successful ultraviolet observatory has been the International Ultraviolet Explorer (IUE), a joint project between Great Britain, NASA and ESA. Launched in January 1978, it became the first major international

The Infra-Red Astronomical Satellite (IRAS) was another unqualified triumph for international space research. It is seen here over Europe – mission control was the Rutherford Appleton Laboratory in Oxfordshire. The large cutaway at the front was a sunshield designed to protect the satellite from excess heating by sunlight.

Measurements from IRAS enabled this 'whole sky' picture to be assembled from its data. The bright central region corresponds to our galaxy, rendered white here to show its relative warmth. The circular dark band (mid-picture) was caused by no scanning data being returned by IRAS of these particular regions.

astronomical satellite and is still working a decade after launch, a triumphant vindication for British astronomers who envisaged the project in the 1960s. The vagaries of international cooperation and budgetary cutbacks saw NASA unwilling to go it alone with the project, and the European space science community unconvinced of its scientific merit in the 1960s. But with British backing, the project went ahead and became the first telescope in geostationary orbit. With control centres on both sides of the Atlantic – another first – a greater degree of flexibility in operations has been possible, allowing it to be available to astronomers throughout the world. Over 3,000 scientific papers have resulted from IUE and its spectroscopes were instrumental in determining which star had exploded as Supernova 1987A, the first supernova naked eye object for three centuries, in February 1987.

The region known as extreme ultraviolet has not been well-observed to date, with only an instrument carried as part of the Apollo–Soyuz Test Project in 1975 devoted to this part of the spectrum. However, NASA is planning to launch its Extreme Ultraviolet Explorer (EUVE) in 1991 which should bring about new insights into ultraviolet astronomy.

HEAT AND DUST

Another quantum leap in astronomical research has been in the infra-red. The world of infra-red astronomy is a curious one, where the darkest objects appear brightest, and the sky is bright whether it is day or night. Infra-red radiation essentially corresponds to 'heat' radiation, and so it enables astronomers to examine colder bodies than those seen at X- and gamma-ray detectors. During the 1970s, the first simple infra-red detectors were flown on the Salyut space stations, but the first true infra-red satellite was the Infra-Red Astronomical Satellite (IRAS). A joint project between NASA, Holland and Great Britain, IRAS was launched in January 1983 and spent 300 days producing the first all-sky survey at infra-red wavelengths. The results shed light on many mysterious processes in the universe.

Until fairly recently, the process of star birth had remained enigmatic. It was generally known that stars were born in nebulae, vast regions of dust and gas, but the exact details remained hazy. The dust in question was mainly made of carbon, a kind of celestial soot, which absorbed light and re-radiated it as infra-red radiation, easily observed by IRAS. In visible light the dust had tended to shield the regions where stars were being born, and so IRAS has enabled astronomers to get a look into the very heart of stellar kindergartens. When enough dust and gas has amassed, the forerunner of a star undergoes thermonuclear transformation. When it does, vast streams of radiation are given out which can disrupt the formation of stars in the vicinity. After this violent period of birth, the remaining dust and gas may clump together to eventually form planets like those in our Solar System. Dust characteristic of star formation was seen around a number of stars by IRAS, leading researchers to conclude that planet formation is commonplace.

IRAS essentially consisted of a liquid-helium cooled telescope which minimized thermal interference from the spacecraft itself. As soon as the helium ran out, IRAS could no longer make observations, but by the time this happened in November 1983, the telescope had amassed over 200,000 objects in data banks back on Earth. Its results merely whetted the appetites of astronomers, so a new generation of infra-red telescopes is planned. ESA's Infra-Red Space Observatory (ISO) (due for launch in 1992) will be joined at various times by NASA's Shuttle Infra-Red Telescope Facility (SIRTF).

THE NEXT STEPS

Noticeable by its absence has been discussion of optical astronomy from orbit. To date, advances in optical astronomy have come to bear on advances by ground-based telescopes. And while a new generation of advanced Earth-based telescopes is planned, two space astronomy projects which promise to revolutionize astronomy are about to come to fruition.

The first is perhaps the less spectacular of the two, perhaps because it is known by an unwieldy acronym. HIPPARCOS – the High Precision PARallax COllecting Satellite – will be the first satellite devoted to the science of astrometry, the accurate measuring of the positions of astronomical objects. ESA seriously began considering the mission in the mid-1970s, at a time when other space agencies said the very accurate measurements required were beyond the range of even the most advanced technology. HIPPARCOS will increase the accuracy of specifying positions by a factor of ten, measuring angles as small as that made by a man on the Moon as seen from the Earth! Over a period of at least two years, HIPPARCOS will build up a picture of the sky, enabling astronomers to accurately specify the positions of over 120,000 stars. A better understanding of a star's position will enable its distance to be measured, which in turn allows its luminosity, mass and temperature to be calculated. So HIPPARCOS will allow astronomers to produce a stellar population census which promises to open a new era in astrophysics.

The other instrument is the Hubble Space Telescope (HST), the first optical telescope to be sent into space. Its 2.4 m (7 ft 10 inches) diameter mirror is not revolutionary by today's standards, but its location certainly is. Far above the turbulence of our atmosphere, HST will enable astronomers to get far, far better views of the universe. It promises to answer important astronomical questions and seemingly trivial ones with equal ease. Hubble will allow astronomers to look even farther into the universe, towards its earliest stages of development. It will also allow clearer observations of objects closer to the Earth, and may observe planets around other stars directly. In fact, astronomers are beside themselves with excitement as to what the five instruments which will use the main mirror may find. Coffee mugs specially produced for project scientists boast a legend which says it all: 'In conscious expectation of the unexpected'.

No finer summary of the philosophy behind space astronomy to date – and the new era which Hubble will introduce – can be found.

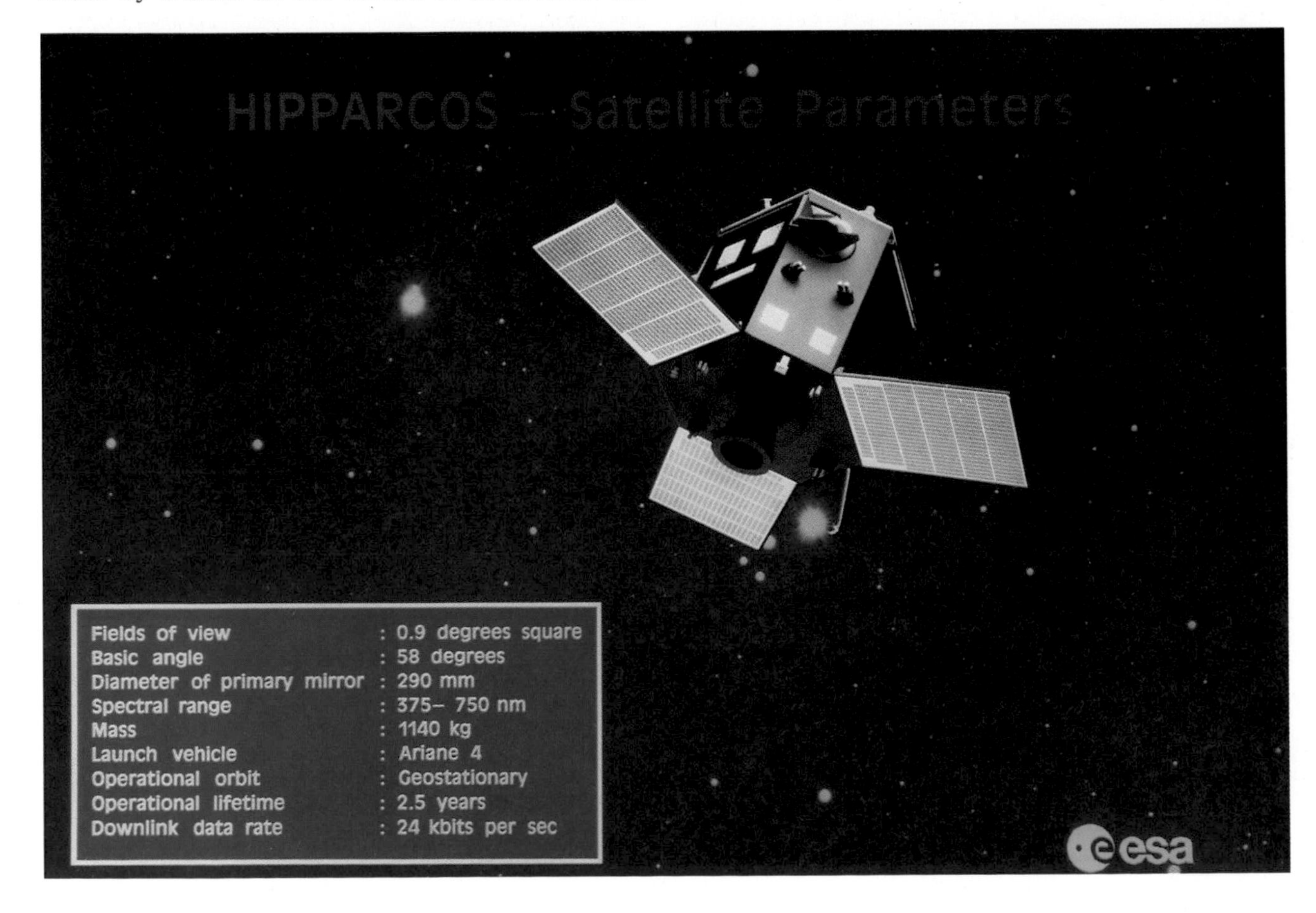

The European Space Agency's HIPPARCOS has been praised as a fine example of a 'small but beautiful' mission. Launched in 1989, its intention was to help astronomers to build up the most detailed picture of the stars yet attempted. Technical details of how the satellite builds up its sky views are shown in the box.

FREEDOM: THE INTERNATIONAL SPACE STATION

by **Peter Beer**

By the start of the 21st century, the international space station *Freedom* will have been built up in low Earth orbit. Over 40 separate Space Shuttle flights will be required to build up the station, to allow international crews to take up residence in the central, pressurized modules.

The idea for a permanently occupied space station is not new. Russian pioneers of the early 1920s described this as an essential step in the future conquest of space, a notion that was eloquently echoed by a German engineer named Guido von Pirquet. In his 1934 book *Rockets Through Space – The Dawn of Interplanetary Travel* he observed that: 'A spaceship would be able to escape from the Earth with a plentiful supply of fuel. The ship would arrive at the outward station with its fuel supply all but gone. But after refuelling, it would be in a position to proceed spacewards with the expenditure of comparatively little fuel.' With uncanny prescience, he had realized that stations in Earth orbit could act as staging posts for manned exploration of the Solar System.

FORERUNNERS

There have already been some remarkable forerunners to a permanently-occupied space station. As early as 1970, NASA engineers made a brilliant improvisation by adapting the empty fuel tank of a Saturn V booster to provide working and living space for an astronaut crew. During three missions, they achieved a total of 171 days of manned presence in space. Presence, moreover, in an environment that was spacious if a bit spartan – the Skylab space station. A more recent venture by America and Europe was Spacelab, which provided more sophistication. However, each Spacelab flight is limited in duration to ten days or so by the fuel and life-support capability of the Shuttle, in which it is carried.

The Soviet Union, meanwhile, quietly got on with its own programme for prolonged manned spaceflight, using the Salyut/Soyuz combination (see p. 64). The results achieved have been remarkable by any standard, with cosmonauts surviving with little harm for over 200 days in the extremely cramped confines of the Salyut – about

the size of a modest holiday caravan trailer. Things got a little more comfortable – but scarcely more spacious – with the launch of Mir in 1986. Small though it is, its design, which is on a modular basis, means that it can, and will be, added to in the near future.

Despite the occupancy of Mir since early 1987, there are many who feel that these pioneering efforts aren't 'proper' space stations. A large, wheel-shaped assembly spinning elegantly in space (usually to the tune of the Blue Danube) remains one of the more enduring romantic images of space travel. And though such vehicles are unlikely to be built within the foreseeable future, the first purpose-built space station will become a reality in the quite near future. The project is within a few years of seeing the first metal being cut.

In fact, designs for an International Space Station to be built by the United States and other Western nations are far advanced. By the turn of the next century, a spidery framework festooned with aerial dishes and solar panels 400 km (250 miles) above the Earth promises to usher in a new age of space exploration. Long, cylindrical modules at its heart will house astronauts from a variety of countries.

Working with 'EASE'. In December 1985, astronauts Jerry Ross (left) and Woody Spring conducted the first experiments in assembling space station components. The so-called 'Experimental Assembly of Structures in EVA' bodes well for construction proper in the 1990s on the *Freedom* station.

STATIONS IN THE SKY

But why aim at a permanent laboratory in the sky – a true space station? Mainly because the scientists and engineers knew that there were discoveries to be made and work to be done that simply could not be done on Earth. Many of the things they would like to do would not be at all practicable on a short flight. Long-term astronomical and ionospheric observations; growing huge crystals in the microgravity environment; making new materials; long-term biomedical research – these were all high on the list. The materials science applications, starting with work on the laboratory scale, could eventually lead to whole factories in space, close to and controlled from the Space Station. And – while many of the operations of such factories could certainly be done automatically – breakdowns and repairs of equipment in space have made it abundantly clear that there is still no substitute for the flexibility of the human hand and brain – and may never be!

As with most large projects, there are detractors. Though construction experiments were successfully achieved at the end of 1985 by Shuttle astronauts, there are many who feel that basing the station's construction on the Space Transportation System will lead to inevitable delays. NASA itself admits that it will not meet President Reagan's deadline of 'within a decade' by early 1994 – the hiatus after the *Challenger* accident being to blame. Other critics, including eminent scientists like James van Allen, believe that the station will become another 'white elephant' like the Shuttle because of cost overruns. Others seriously question the necessity of spending $20 billion on a laboratory for research into the effects of weightlessness on the human frame and on materials.

Many of these arguments fall within the 'manned versus unmanned' label which has beset space research from the very beginning. To the outsider, such arguments seem at best like squabbling among colleagues about whose work is the more important. There are some things which unmanned probes can do that those carrying humans could not do, and vice versa. But it is fair to say that if mankind's ultimate aim is to permanently occupy space, then a purpose-built space station will provide an invaluable foothold. The interest shown by the Japanese, Canadians and ESA highlights the importance which they attach to the space station. A space station will provide the focus for their own spacefaring aspirations.

WHO DOES WHAT – THE INTERNATIONAL PARTNERS

The limitations of Spacelab and the spur of Soviet progress have lent impetus to NASA's plans. During the early 1980s, a whole series of design studies was started by NASA and the American aerospace industry to define more closely both the ideal requirements and the most achievable means of meeting them. When President Reagan an-

By the end of the 1990s, the European Space Agency and Japan will be using their own spaceplanes to dock with elements of the station. Here ESA's *Hermes* is seen docked with one of the 'free-flyers' which are separate from the main station.

nounced the project in 1984, it was almost a foregone conclusion.

Some of the earlier designs have been dropped because of expense and the difficulty their construction would pose. The 'twin keel' concept had emerged as the favourite, but it became clear that it would be too costly to start with. The *Challenger* accident led to a total design review in terms of safety aspects. A major budget review in mid-1986 finally led to the present single-keel layout. NASA believes that completion by the mid- to late-1990s is a realistic target, with a total cost in the region of $20 billion. Though this is a huge sum it is considerably cheaper than the Apollo programme, even allowing for inflation.

A project on such a grandiose scale would be too vast for any one laboratory, or even a single country, to handle adequately. NASA has divided its own development effort into four 'work packages' assigned to its major establishments such as the Marshall Space Flight Center in Huntsville, Alabama, and the Johnson Space Center in Houston. Study contracts have been given to all the major industrial companies with aerospace experience. Much of the main system design has been done by Boeing, working closely with its NASA neighbour in Huntsville.

The story concerning the manufacture does not stop there. From the outset, the Americans have actively sought international participation, partly to spread the immense costs more widely, but also for political reasons. A closer involvement of the international space community would obviously be very desirable. After years of negotiation, an agreement was signed by which Europe (through ESA), Japan and Canada would all design and build substantial parts of the station. By a happy coincidence the signings took place on the day when *Discovery* successfully returned to flight operations after the hiatus caused by *Challenger*'s destruction. A few months earlier, it had been decided to name the whole of the station *Freedom*.

Japan will produce a sophisticated module to be attached to the central part of the main *Freedom* structure. The Canadians will develop a 'Mobile Repair Facility' to provide servicing ability anywhere on the main framework. ESA's contribution is Columbus, one of the four main modules that form the heart of the Station. Two separate 'free fliers' which will be serviced by astronauts are also part of the Columbus project. Laboratories all over Europe are helping to design and build the Columbus components, which will finally be assembled and tested at ESA's European Space Technology Centre (ESTEC) in Noordwijk, on the coast of Holland.

DESIGN OUTLINE

What will the final *Freedom* station look like? As all the drawings show, it will be a huge lattice structure, extending a hundred metres (330 ft) or more in length and breadth – an improbably lightweight construction at first glance. But it is important to recall that in Low Earth Orbit the gravitational field is only about one-millionth of that on the Earth's surface. Structures that would buckle under their own weight on the ground can be built up with ease in space. This gives the planners almost unlimited flexibility.

Accordingly, the reasons for the big framework are simple: to give more room for manoeuvre by visiting spacecraft; to give enough space for the necessarily large solar power arrays; and to keep some experiments well-separated from the disturbing effects of the central modules. Space dust and debris, local electric and magnetic fields, and even the gravitational influence due to the station itself can all disturb sensitive experiments. It would even be possible – though this is not planned at present – to have a small nuclear power source attached at a safe distance from the inhabited modules. If even greater separations were ever needed, space tethers – long cables holding a unit up to a few kilometres away – could be deployed.

Attached to this lattice framework are the main solar panels, radio aerials for communications with Earth and spacecraft, and a movable repair facility – something like a travelling crane in a factory roof. At the centre of it all are the main modules (four of them, initially) in which the astronauts will live and work. The modules are interconnected by multi-part connecting units, onto which supply craft can be docked, as well as providing storage units for supplies and waste materials. Flexibility of the design has played an important part in the station's development. Only time and cost considerations will limit the attachment of more modules as they become desirable, and this will almost certainly happen in time. It would, in fact, not be too far from the mark to call the station 'a giant Lego set in space'.

LIVING IN SPACE

A six to eight person crew will probably have a three month tour of duty in space. Their life must be made as comfortable as is reasonably possible to achieve. Though not huge, the modules will nonetheless provide far more space than anything since the pioneering days of Skylab. Each module is something like the size of a railway car and although they are cylindrical in outer form, they are fitted out with equipment bays cunningly designed to leave a roughly rectangular working space at their centre.

One module is designated as the living quarters. Each crew member has his or her own miniature 'cabin', not very much larger than a superior telephone booth, which holds the sleeping-hammock – a kind of sleeping-bag hung on the wall – storage for clothes and personal effects. Here in their off-duty moments crew members can relax when they seek that most important human need, a little privacy. At other times, they can sit in the communal dining or recreation areas, watching television or looking down towards the Earth (perhaps a little wistfully) through the windows thoughtfully provided to lessen their sense of isolation. Or they might take their daily turn on one of the exercise machines which are essential to keep muscles in good tone in the weightless conditions.

Unobtrusive but vital (and costly) elements of the station are the life-support systems. It will hardly be practicable to ferry up tankfuls of water and cylinders of oxygen; that would be far too costly in resources. Water supplies alone would require a Shuttle to be filled with water and launched every three months: merely to maintain other supplies, that would mean adding an extra Shuttle to the support fleet. The answer has to be in a complex set of recycling systems whose functions will include the filtering and distillation of waste water, and chemically re-converting carbon dioxide from the station atmosphere into pure oxygen. Only the irreducible minimum of solid wastes will be collected into containers for return to earth on a home-going supply craft. No more pollution of space by simply dumping wastes into the vacuum will be permitted.

There will also be quite advanced workshop facilities on board to enable even quite ambitious running repairs to take place. As for interior equipment – the experiments, for example – each equipment bay is hinged to pivot out from the wall, and be removable entirely in minutes. The hatchways in the station are large enough to allow a whole bay to be removed, for storage elsewhere

The central pressurized modules will be individually provided by the United States, Europe, Japan and Canada. They will allow unhindered passage for shirt-sleeved crews. Note the docking attachment with the Shuttle's airlock at the right of the picture.

Stations in low Earth orbit will allow relatively easy – and efficient – access to Mars and the Moon. In this artist's impression, one of the first elements for the station has been transported to the lunar surface. Technology developed for *Freedom* could easily be adapted.

and later return to Earth and replacement with a spare or improved unit.

In all but the most extreme emergencies, crew members will have to await the next Shuttle flight for return to earth. Although these are planned to take place at regular intervals (probably every six to 12 weeks) an earlier flight could sometimes be arranged, but not always. Shuttle flights will never be possible at less than three to four week intervals. Both the Japanese and ESA hope to introduce their own spaceplanes to service the station in time (Hope and Hermes respectively). Even so, they will not come on line until the end of the century and cannot be launched at will.

So something will have to be done in case a real disaster strikes – mechanical, or even medical. A ruptured appendix cannot wait six weeks for treatment. Studies are being made for a simple return capsule to be kept permanently attached to the station – a kind of 'space lifeboat' – though no final designs have been made.

ORBITAL CONSTRUCTION

A structure of complex as a space station which will weigh almost 200 tonnes when it is completed, clearly has to be assembled in space. Once again its modular nature and construction-set design make the whole task possible. Every major unit is designed to fit inside the cargo bay of the Space Shuttle, which will be NASA's workhorse for the station's assembly and then for its supply when in full service. Spacesuited astronauts will carry out the assembly in a long series of spacewalks, a process which they have already been rehearsing on earlier Shuttle missions and practising in the huge diving-tanks at the Marshall Space Flight Center. Present plans call for the first assembly flights to begin in 1995 or 1996. From then on, they will continue every few weeks until, after a total of around 20 flights, all the essential units are in place.

But the station will not have to wait two to three years until it receives its first crew. Current planning allows for a 'presence in space', probably by a skeleton crew of two or three men, to be established from perhaps the sixth or seventh assembly flights. At that time, there will be at least one living module and the first of the vital solar-power panels. So if all goes to plan, soon after the middle of the 1990s, America will finally have its first 'permanent' space station in embryonic form at least. Thereafter the rest of the assembly will become considerably quicker.

GATEWAY TO THE PLANETS

Over the next few years, it seems likely that *Freedom* will bring a new sense of pace and purpose to NASA's plans and a focus for other Western nations' space programmes. In NASA's case, it is a sense of direction that has been sadly lacking over the last decade. And though the *Challenger* tragedy casts its gloomy shadow over planning too optimistically for the future, there is now a greater awareness of the safety factors involved in such plans. A disaster of *Challenger*-like proportions will not be allowed to blight the space station.

As the station takes shape, it will stimulate new developments in every area of space technology. The need for new systems of all kinds, orbital manoeuvring vehicles (the 'Space Tug'), and above all better and cheaper replacements for the ageing Shuttle, will drive technology forward. The work that will be done on the station itself will open new windows in materials science and other manufacture-related studies. A space-led 'industrial revolution' is a very real possibility in the not too distant future.

Yet the real promise of the station must be as a staging post in our further exploration of the Solar System. A number of far-reaching plans for setting up bases on the planets have been proposed by NASA. In 1988, the agency was given $50 million by the retiring Reagan administration for development of technologies to send human beings to the Moon or Mars – or both. Though no concrete plans have emerged, all scenarios for manned flight to the Moon and beyond use a space station as a forward base at which the long-range interplanetary spacecraft can be assembled and fuelled, crews can be changed, and the most opportune moment for launch chosen. A Mars mission would require a craft considerably bigger than anything launched to date.

The experience gained in building a station like *Freedom* would be invaluable for sending human beings to the planets. In this regard, a permanent station will assume a significance which was predicted by Konstantin Tsiolkovskii: 'When the Solar System is conquered, man will have three dimensions.'

‘AUSTRALIA IN 45 MINUTES’: THE NEW GENERATION OF SPACEPLANES

by **Frank Miles**

As we head towards the next century, the most important challenge facing future space programmes, worldwide, is to find the best way to cut the cost of getting into space and thus make it more widely accessible. It is a goal which has been aimed at before – but in most cases, the promised revolutions have failed to materialize. Yet the establishment of permanent space stations in Earth orbit, requiring the ferrying of crews and cargo to and from space, is giving impetus to the development of cost-effective and economical launch systems. Such systems will require dedicated investment over many decades, a timespan that will appear particularly attractive to politicians or financial institutions. How likely is routine access to space going to be?

LAUNCH COSTS

To answer that question, it is necessary to make a quick survey of space exploration to date. Current high launch costs stem from a basic error built into civilian space programmes right from the start. In the 1950s a lot of national prestige hung on winning the race to become the first country to launch a satellite, so both the U.S.S.R. and U.S.A. chose simply to adapt their existing military weapon-launcher technology to get hardware onto civilian launchpads. With weapons programmes it has never been thought necessary to recover the launcher: the overriding consideration has always been the accurate delivery of the nuclear warhead, rather than cost.

But for peaceful space exploration that built-in high cost of using expendable launchers has meant many would-be users cannot afford space. Unfortunately, other nations (Europe and Japan in particular) who set out to compete with the U.S.A. and U.S.S.R. to launch satellites, etc., likewise chose to follow the only known technology and so also went for costly expendables.

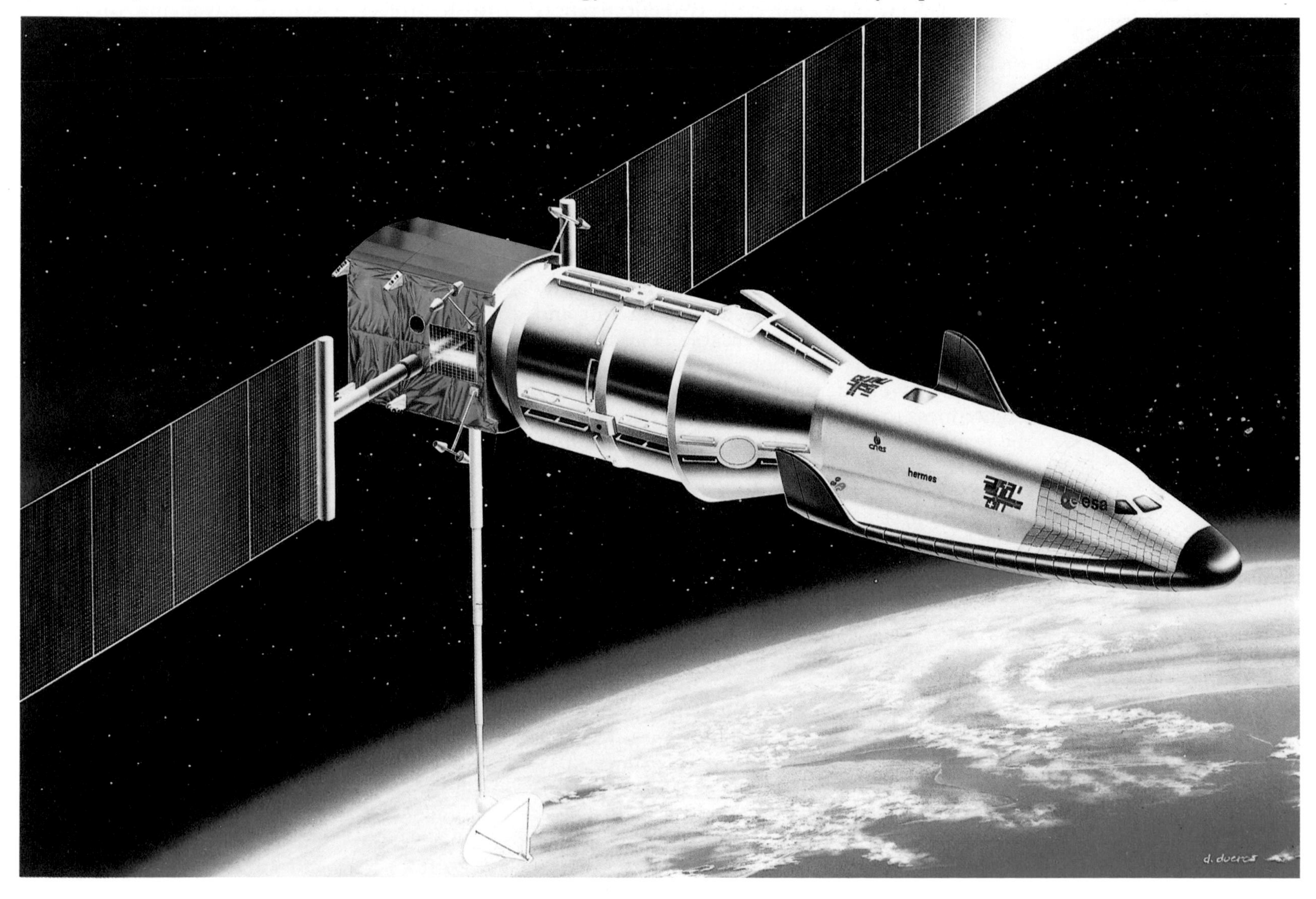

The European spaceplane *Hermes* is due for launch in the late 1990s. It will enable ESA astronauts to have autonomous access to space. This artist’s impression shows the mini-Shuttle docked with the Man-Tended Free Flyer element of the *Columbus* programme.

Britain's HOTOL was the first single-stage-to-orbit vehicle to be revealed. Though it has undergone many design changes since, the basic outline is shown here. Note the large atmospheric oxygen intake underneath the vehicle.

But the 1980s saw the first attempts to economize with the development of the partially-reusable American (1981), and now Russian (1988), Space Shuttles. As we saw earlier, the cost effectiveness of the Space Transportation System failed to materialize. Only time will tell if the Soviet Buran vehicle and the Energia booster will become cost effective into the next century.

Clearly, the next step is for all spacefaring nations to develop much more economic launchers. Already, several are planning completely reusable, or largely reusable, craft. The most economic, from an operational point of view, will undoubtedly be single-stage-to-orbit (SSTO) launchers able to take off and land horizontally with the flexibility of conventional aircraft and which will be completely reusable. The set-back at the moment is enormous development costs. Very few governments have the resources or far-sightedness to back projects whose benefits won't be reaped until decades later. No better example of this attitude is shown than by Britain's HOTOL (Horizontal Take-Off and Landing) craft, the first totally reusable spaceplane on the drawing boards.

With the need to cut launch costs in mind, the noted British aerospace engineer Alan Bond began to consider reusable spaceplanes in the early 1980s. Very quickly he developed an idea for a revolutionary engine that could operate with atmospheric oxygen or else oxygen in the form of a liquid propellant. The possibility of flying to Australia in about 45 minutes from Britain was an attractive by-product of the drive to reduce costs. Original plans were for a first flight by the mid-1990s. But after initial backing the British government decided not to continue giving any financial assistance to the developers, British Aerospace and Rolls Royce.

It was a decision that puzzled the space world, especially when other nations – alerted by the British project – started advancing their own plans to build reusable spaceplanes. It now looks as if the British will be overtaken in the race towards a fully reusable spaceplane. Nations now known to be competing include the United States, Japan, France, West Germany, India and (it is presumed) the U.S.S.R. That most of the work is being done in secret shows how important it is.

HEARTBREAK HOTOL

The concept of HOTOL developed from a highly secret design of engine (the RB-545) to use oxygen from the air, instead of liquid oxygen (LOX), to combine with liquid hydrogen (LH_2) in initial stages of flight. This would mean a massive weight saving since LOX is extremely heavy (the U.S. Space Shuttle has to lift 600-tons of it during launch). HOTOL would need a comparatively small amount of LOX on board – enough for only the final stages of launch, when the craft reaches high altitudes where there is insufficient oxygen in the atmosphere.

In fact, the HOTOL vehicle requires only the one engine to operate in both flight modes. It is this ability to use atmospheric oxygen and then perform a smooth switch-over to LOX that makes this invention unique. HOTOL's design is dominated by the large LH_2 fuel tank which is an integral part of the load-carrying structure. The LOX tank constitutes only one-eighth of the volume. Both tanks are covered by an aeroshell of metal panels backed by high-temperature insulation to avoid LH_2 boil-off and also to withstand re-entry heating. The payload bay is 7.5 m by 4.6 m (25 ft × 15 ft) and could accommodate very large payloads, such as larger communications satellites.

ORIENT EXPRESS

In America, the development of a National Aerospace Plane (NASP), also referred to as the X-30, is being undertaken jointly by the Department of Defense and NASA. It is the biggest experimental aircraft project ever undertaken in the United States. The federal government has allocated $3.3 billion to the project with the emphasis on slashing launch costs. George Keyworth, President Reagan's science adviser, declared that NASP's launch costs would be 'not five times cheaper, but a hundred times cheaper than they are today' before a hearing of the Committee on Science and Technology of the U.S. House of Representatives in July 1985.

Rockwell International, General Dynamics and McDonnell Douglas are in the process of producing competitive designs. In addition, studies are being undertaken by Rocketdyne (Rockwell's engine division) and by United Technologies (Pratt and Whitney) of air-breathing propulsion systems to allow a craft to take off and land horizontally and accelerate to Mach-25 (i.e. 25 times the speed of sound) and reach low earth orbit. The pace of technology is being maintained to a point that by 1990 it should be possible to give the go-ahead to build a demonstration spaceplane for test flights within five years.

The NASP or X-30 project is conceived as a military launcher for the so-called 'Star Wars' programme but it is sometimes referred to as 'The Orient Express' since it is envisaged there could eventually be a civilian version to fly passengers from New York to Tokyo in two hours. America's Advanced Manned Launch System (AMLS) studies also include: a fully reusable orbiter with a payload pod on top of the fuselage; a partially reusable spacecraft with a flyback booster; an expendable large core vehicle with a recoverable orbiter on top (similar to the French Hermes-Ariane 5 configuration and the Japanese Hope designs); and an air-breathing expendable rocket to take-off horizontally, using NASP technologies.

THE RISING SUN

Japan seems to have the most ambitious plans for cutting launch costs. There are separate project studies for a spaceplane being made by three major Japanese companies – Kawasaki, Fuji and Mitsubishi – in conjunction with Japan's National Aerospace Laboratory (NAL) and the National Space Development Agency (NASDA). The designs are for SSTO hypersonic manned spaceplanes with 'aircraft flexibility' in operation weighing around 385 tonnes. The government's target is to have such a spaceplane flying by the year 2006 and they are prepared to spend $1.8 billion over seven years to get one. (Since Japanese production costs are only half what they are in the U.S.A. this means, in effect, they are

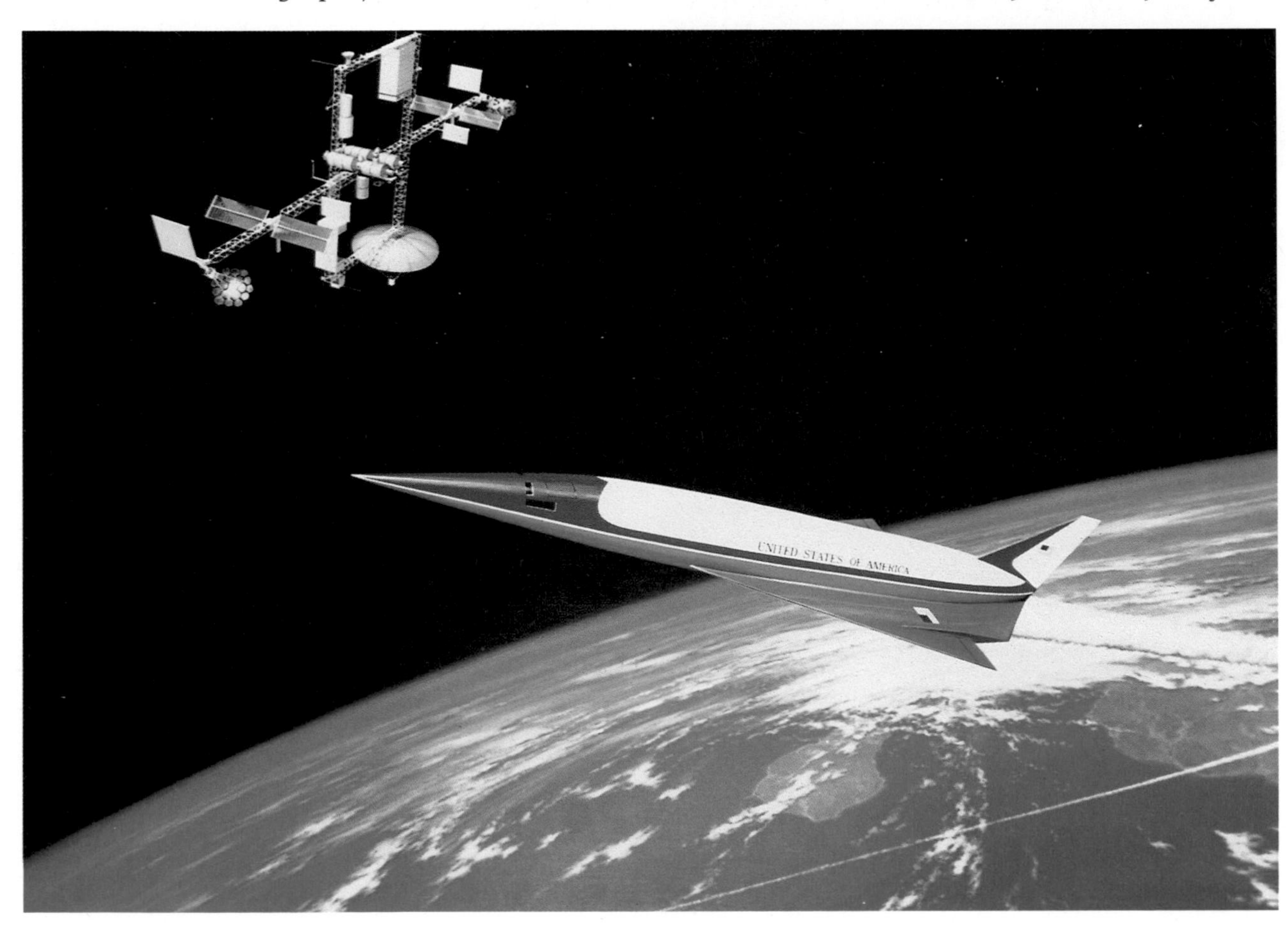

America's planned rival to HOTOL is variously known as X-30, NASP (National Aerospace Plane) or 'Orient Express'. This artist's impression shows the vehicle approaching the international space station as it may appear by 2010.

prepared to spend more than America is on a spaceplane).

All the early Japanese designs – for a craft that could carry a crew of four to orbit at 500 km (312 miles) – are known to envisage air-breathing engines for atmospheric flight with a switch over to separate liquid hydrogen/liquid oxygen rockets for later stages of flight. Work on liquid dynamics for air-breathing engines is going on at Tokyo and Ngoya universities. Kawasaki have designed a craft 66 m (216 ft) long, with a wingspan of 30 m (98 ft) and featuring twin rudders: so does the Fuji design which is longer – 75 m (246 ft) – but has a narrower wingspan of 25 m (82 ft). Mitsubishi's design has a single rudder, and is the longest at 87 m (285 ft) but again with a fairly narrow wingspan of just over 25 m (82 ft).

As an interim measure to cut launch costs, Japan is also developing a two-stage-to-orbit unmanned, winged, shuttle-like orbiter that could go into space as early as 1996. It is named HOPE (H-2 Orbiting Plane) and would be launched on an H-2 booster with strap-ons. The H-2 should itself be operational by 1992. In addition to all this NAL are thinking of a 67 m (220 ft) long aircraft capable of Mach-7 using air-breathing engines to carry a crew of two to an altitude of 150 km (94 miles).

France's STS-2000 programme currently envisages three types of vehicle. From top to bottom: a single-stage vehicle with both air-breathing and conventional rocket engines; a two-stage air-breathing rocket derivative; a single-stage rocket vehicle, derived from *Hermes*.

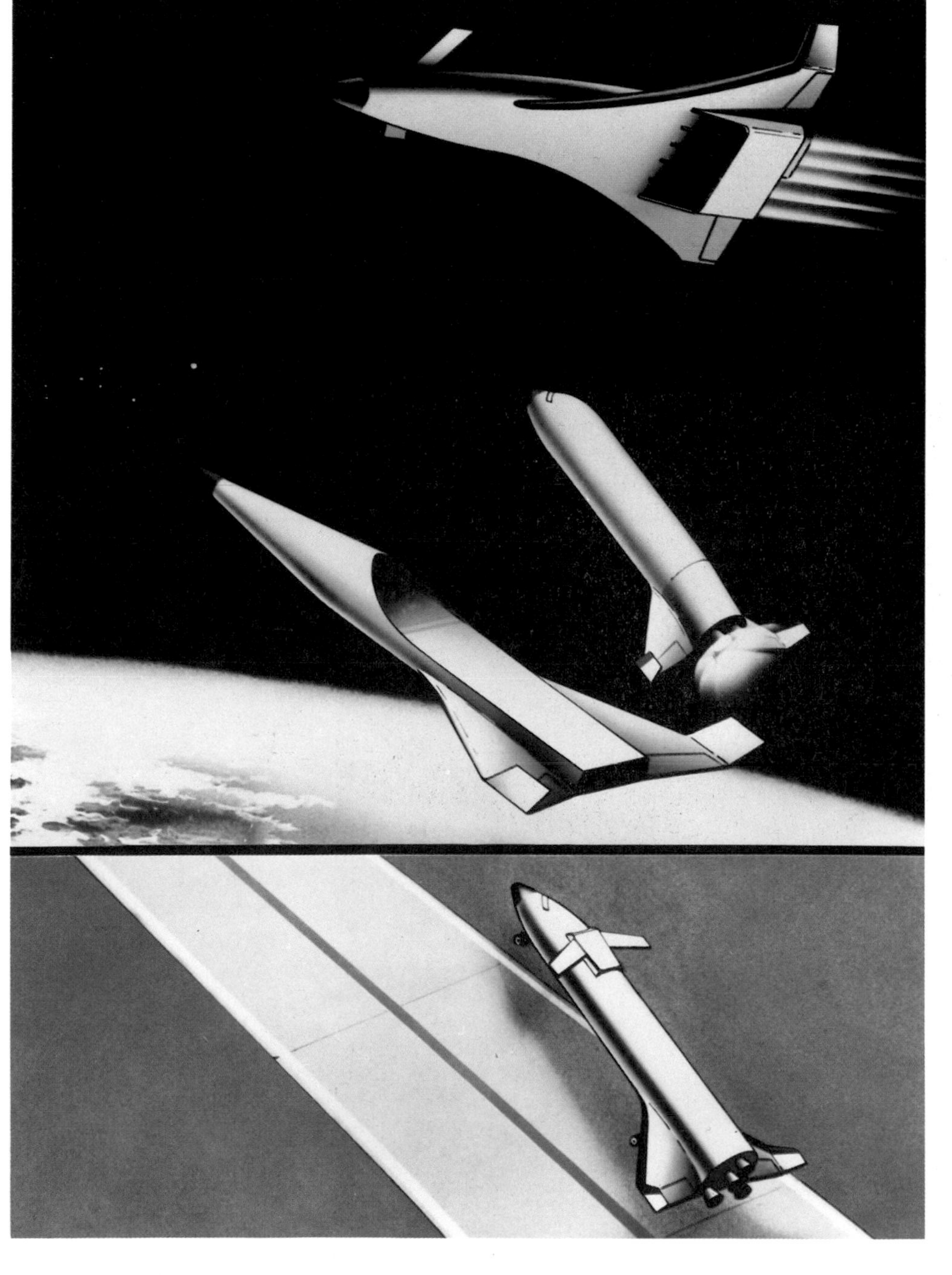

BEYOND HERMES

The French have already persuaded the European Space Agency (ESA) to build Hermes, a recoverable manned spacecraft which will be launched on top of an Ariane 5 rocket. The first test flight is due in 1996 and it should become operational the following year. Hermes would orbit at 400 km (250 miles) for flights lasting up to 90 days. It is 18 m (60 ft) long with a wingspan of 10 m (33 ft) and has a pressurized crew compartment designed to accommodate between two and six people. Hermes should carry out a wide range of missions, the main one to service a 'man-tended free flyer', the main element of Columbus, the European contribution to *Freedom*.

Hermes will have onboard engines for use at the end of the ascent phase (when it is separated from the Ariane rocket) to manoeuvre it into its planned orbit and later to perform the 'braking' de-orbit burn prior to re-entry and landing on a runway.

France (through Aerospatiale and Dassault) is now known to be planning at least two other types of spaceplane to replace Hermes by the year 2010, calling the combined project 'STS-2000' (Space Transport System for the next century). Development of a new spaceplane will take place over the next ten years (from 1989). The French are concentrating on two vehicles, one to lift payloads of up to 40 tonnes and the other a lighter 'space taxi' for rapid, flexible use to shuttle men and materials between Earth and a manned orbiting space station, with a payload capacity of between 7 and 10 tonnes. Where the French differ from the British, American, Japanese and Indian concepts is that they include recoverable vertical take-off rockets in their plans for entirely reusable launchers. There are also studies of single and two-stage versions of a project labelled AGV, an aero-cruiser for testing propulsion units, but little is known of this project.

VORSPRUNG DURCH TECHNIK

The Germans have a major project known as Sanger II – a two-stage winged space-transportation system for manned flight – originally conceived during the Second World War but cancelled in 1943. It was resurrected by the aerospace manufacturing company MBB-ERNO when work on Britain's HOTOL was revealed.

The Sanger concept's first stage, equipped with turbo-ramjet engines, is seen by the Germans as an advanced version of a jet airliner, a forerunner of a future hypersonic global passenger carrier to replace conventional airliners. Initially, however, it would be developed as a space launcher, to carry on its back a rocket-propelled spaceplane named Horus that would separate from the mother craft at a speed of Mach-6 and at an altitude of 35 km (22 miles). This second stage, using high-performance rocket engines (not air-breathing engines) would take a crew of up to 12 plus a small, 4 tonnes payload into low earth orbit – 300 km (187 miles). Total launch mass would be around 500 tonnes.

The fully-recoverable twin craft would be launched from, and then land separately on, conventional runways. In fact MBB-ERNO claim it would be the only system to allow launches from European airports. The West Germans also have a project known as LART (Luft Atmender Raketen Trager) or Air Breathing Rocket Carrier Vehicle. It is said to be a single-stage-to-orbit HOTOL lookalike but the engine, unlike HOTOL's, relies on pre-coolers and is closer to what the Japanese are working on.

PASSAGE FROM INDIA

India is planning a Hyperplane (which they regard as an acronym for Hypersonic Platform for Airbreathing Ascent to Near-Earth Orbit). Studies are being carried out jointly by the Scientific Defence Research and Development Laboratory and Bharat Dynamics Limited at Hyderabad. The Hyperplane would be a fully-reusable SSTO craft, 50 m (164 ft) long with a take-off weight of 100 tonnes and using a high specific-impulse air-breathing engine (present configuration has the air-intake on top of the airframe), to orbit at 300 km (187 miles) and launch both communications and earth-resources satellites at low cost. The payload weight envisaged is 16.7 tonnes.

What has been published of the plan so far states that: 'The plane would take off from a conventional aircraft runway, then cruise at constant acceleration while collecting air and liquefying oxygen required for the rocket phase of flight.' This suggests a sort of in-flight propellant factory making LOX on board and so is unlike HOTOL. Indeed, India claims a basic difference between their Hyperplane and Britain's HOTOL and America's NASP in that it would incorporate 'both air breathing engines and air-collection technology' and they modestly claim launch costs would be reduced by 'at least half what they are today'.

BACK IN THE U.S.S.R.

When the Soviet Union revealed their own version of a Space Shuttle in 1988 it proved not unlike America's in appearance. It is known that Buran will be an important part of the Soviet plans to industrialize space. To date, the U.S.S.R. has given no information about plans they are believed to have to develop a SSTO fully-reusable spaceplane. One clue was that they announced at an international space conference in Bangalore in 1988 that studies are being carried out of variable cycle engines for boosting launch vehicles – presumably spaceplanes.

No illustrations of the next generation of Soviet spaceplanes are available in the West. This schematic diagram of a two-stage hypersonic vehicle from the 1970s shows that the U.S.S.R. has been considering the idea for many years.

CONCLUSION

Whichever of the competing nations wins the race to build the first totally reusable spaceplane, one thing is clear. Space travel will soon become accessible to the average man in the street once the launch costs are down to a minimum. Such a grandiose claim may seem, at best, like science fiction. A sense of *déjà vu* pervades any talk of minimizing launch costs, as projects like NASA's Space Shuttle showed only too well in the 1970s. Predicted to become a 'DC-3' of the space age, the complexity of the Shuttle system has rendered its flight operations to a bare minimum of flights per year.

Yet history provides a telling precedent. In the early 1930s, flights across the Atlantic were the preserve of aviating daredevils. Only 40 years later more than 25 million people cross the Atlantic per year. We are at roughly the same stage of development in space as aeronautics was during the 1930s. Given the impetus that permanently-occupied stations is giving to the development of space, perhaps it isn't too far fetched to assume that fare-paying passengers will actually be commuting from continent to continent in 40 years' time.

MANNED MISSIONS

APOLLO

The Apollo programme, which lasted for just a decade, was America's most successful space enterprise. In essence, each Apollo spacecraft consisted of three elements: the Saturn launch vehicle, the Command and Service Module in which the astronauts were transported to the Moon, and the Lunar Module in which two of them descended to the Moon's surface. They are considered here separately.

THE SATURN LAUNCH VEHICLES

The family of Saturn launch vehicles used in the Apollo project was the brainchild of Wernher von Braun; it is a testimony to his skills and those of his team that none of them ever failed in flight. The first launch vehicle, designated 'Saturn 1', was used to test the rocket components of the series, many of which had been developed in the earlier Redstone and Jupiter boosters. It was a two-stage rocket and the power needed for launch was generated by clustered engines. A more powerful variant known as the Saturn 1B was used to launch the first Apollo crew (Apollo 7) and later the crews to Skylab and the Apollo-Soyuz Test Project.

The Saturn V was a true leviathan: with the Apollo spacecraft firmly mounted at the top, the whole structure stood 110 m (363 ft) tall – 18 m (60 ft) higher than the Statue of Liberty. It was the most powerful rocket ever built, capable of lifting 142 tonnes into orbit, and has only been surpassed in capacity by the new Soviet Energia vehicle. The first stage (known as the S-IC) was responsible for lifting the whole Apollo/Saturn combination off the launch pad to an altitude of about 64 km (40 miles) above the Earth. Its total thrust of 7.5 million lb was generated by five F-1 engines using liquid oxygen and kerosene which burned for a total of 150 seconds.

The second stage (S-II) used five J-2 engines to generate as much power as 30 diesel locomotives, and by far the most powerful rocket engine ever built which used liquid oxygen and liquid hydrogen as fuel. Both these fuels (sometimes referred to as LOX and LH_2) are cryogenic: that is, they have to be maintained at temperatures many hundreds of degrees below the freezing point of water. The liquid oxygen was stored at temperatures of −182°C (−297°F) and the hydrogen at −252°C (−423°F). Liquid hydrogen is particularly difficult to store, mainly because it has a very low density and can explode on contact with air. However, its advantages are that the cryogenic propellants ignite on contact and produce a very great thrust per kilogram of weight compared to other fuels. Liquid oxygen/hydrogen engines are very efficient, which is one reason why von Braun developed them, despite the new technology and insulation required to harness their energy.

After burning for 359 seconds, the J-2 engines boosted the Apollo spacecraft to speeds of 22,400 km/h (14,000 mph), reaching an altitude of 192 km (120 miles). After this, the third stage (SIV-B) took over with its single J-2 engine to launch the spacecraft into Earth orbit. The engine would be re-ignited to speed the spacecraft towards the Moon.

COMMAND MODULE

At the very top of the Apollo structure was the Command and Service Module in which the three astronauts journeyed to and from the Moon. In case of emergencies at launch, there was a 10 m (33 ft) long solid-propellant emergency rocket which would have lifted the Command Module away from the pad. The Command Module (CM) itself

Above: The key to the success of the Apollo missions was the mighty Saturn 5 launcher. Shown here is the cryogenic second stage, 24.8 m (81.5 ft) tall and weighing more than 453,600 kg (1 million lb) when fully fuelled.

Opposite: After being undisturbed for millennia, the surface of the Moon now carries the mark of a dozen men. Perhaps more than anything else, this Apollo 11 footprint is one of the most striking images of man's exploration of the Moon.

This unusual view of a Saturn V at Cape Canaveral shows a test of the water control system used for cooling and quenching during lift-off. The scale of the 110.6-m (363-ft) high rocket can be discerned from the station wagon at top right of the picture.

was cone-shaped and, though only 322 cm (10 ft 7 in) high and 330 cm (12 ft 10 in) in diameter, it provided adequate living quarters for the crew. They spent most of their time on couches, but part of the centre couch could be removed to allow greater flexibility of movement. Two of the crew could easily stand and, in zero gravity, were able to float around quite freely. Food, water and clothing were packed into bays which lined the walls of the craft. The cabin was pressurized to around 0.35 kg/cm² (5 psi) (about one-third that at sea level) and temperature maintained at 24°C (75°F). Though the Apollo environmental control system allowed the crew to wear shirtsleeves for most of the time, spacesuits were worn during launch, re-entry, docking and crew transfer. The Apollo 204 fire in January 1967 resulted in a total overhaul of the spacecraft including new wiring and flame-proofing in the CM.

On returning to Earth, the Command Module was separated from the Service Module and encountered re-entry heating twice as great as that for Mercury and Gemini re-entries: temperatures in the range of 2,760°C (5,000°F). Its heatshield consisted of a reinforced plastic called phenolic epoxy resin. As the CM entered the Earth's atmosphere, the resin turned white, charred and then melted away, absorbing the fierce heat of re-entry and dissipating it away. At its thickest, at the CM's base, the shield was about 5 cm (2 in) thick.

THE SERVICE MODULE

The Service Module (SM) contained the main spacecraft propulsion system and supplied the crew with hydrogen, oxygen and water. It was attached to the CM until just before re-entry into the Earth's atmosphere. The SM contained the spacecraft's main engine (known as the Service Propulsion System) which was used to brake the spacecraft into orbit around the Moon and later send it homewards to Earth, as well as for mid-course corrections. The Service Module also contained electrical power, environmental and reaction control systems and part of the main communications system. In shape, the SM was essentially a 3.96 m (13 ft) long cylinder made up of 2.54 cm (1 in) thick aluminium panels supported by six radial beams. The cryogenic tanks containing liquid hydrogen and oxygen were used in the production of electrical power by fuel cells which also provided oxygen and water for the use of the crew. The insulation in the tanks was so good that it was estimated that if they were filled with ice and placed in a room at 21°C (70°F), it would take 8½ years for the cubes to melt.

Despite such exemplary engineering standards, Apollo 13 was crippled by an explosion in one of the liquid oxygen tanks. A heater switch inside one of them had welded shut when higher currents were passed through it before launch. After the Apollo 13 explosion, the SM was redesigned so that materials which were potentially combustible in close proximity to liquid oxygen were minimized. In the last three missions (Apollos 15–17), the individual Apollo craft were much heavier to allow more scientific work to be carried out over longer periods. A Scientific Instrument Module Bay (SIMBAY) was carried which had eight instruments on board including mapping cameras and spectrometers to take detailed pictures of the lunar surface. A small satellite was also deployed, which allowed geophysical measurements to be made from lunar orbit. En route to Earth after leaving the Moon, the Command Module pilot performed a spacewalk to recover film from the SIMBAY cameras.

LUNAR MODULE

The Lunar Module (LM) was contained in an adapter section which was located above the S-IVB third stage. The LM was 'retrieved' from its storage by docking with the combined CSM combination after leaving Earth orbit, before the main SPS engines fired the spacecraft towards the Moon. Because the LM never had to fly inside the Earth's atmosphere, it did not have to be aerodynamically-shaped. With its landing legs extended, the craft looked like a bug and was sometimes described thus by the astronauts. The LM came in two parts: the lower Descent Stage which contained the landing gear and the main engines; and the upper Ascent Stage, which was pressurized, and in which the astronauts stood. As the cabin inside the Ascent Stage was a mere 233 cm (92 in) in diameter, there was just enough room for the two crew to stand up – there were no seats. Though there were two forward-facing triangular windows, as the LM descended from lunar orbit the astronauts had to rely on the navigational equipment and radar, because the craft was pitched on its side with the main engines braking its motion towards the surface. The crew only got a 'forward view' within the last two minutes of descent, finally orientated upright for landing. The crew could make as many as four separate forays onto the lunar surface after depressurizing and pressurizing the cabin. When

This view of the Apollo 15 Command and Service Modules (CSM) shows the detailed structure of the 'SIMBAY' in which scientific instruments were located. The propulsion nozzle of the SM's main engine is prominent. Also visible is the docking drogue of the Command Module.

they lifted off the Moon's surface, the Ascent Stage was separated from the Descent Stage by four explosive bolts. The Ascent Stage engines were fired for nearly seven minutes in order to enable the craft to enter a transfer orbit with the CSM in a lunar orbit.

The Descent Stage was an octagonal-shaped base in which the main landing engines were housed. Surrounding them were the propellant and water tanks as well as the scientific equipment. The four landing legs remained stowed away until the Apollo spacecraft arrived safely in lunar orbit. The landing gear was then explosively deployed so that the legs were locked in place by a mechanism. The Descent Stage contained a number of self-contained scientific experiments which were deployed on the lunar surface, and known as the Apollo Lunar Surface Experiment Package (ALSEP). Each contained a variety of instruments performing seven scientific experiments ranging from magnetic and thermal studies of the lunar surface to the Moon's interaction with the solar wind and interplanetary environment. A seismometer was included as part of the instrumentation, and in later flights, just after the crew were safely inside the CSM before returning back to Earth, the discarded Ascent Stage was dropped onto the lunar surface. The seismometers recorded 'Moonquakes' of such ferocity that one of the scientists referred to the fact that 'the Moon rang like a bell'.

LUNAR ROVING VEHICLE

In the final three Moon landings (Apollos 15–17), a Lunar Roving Vehicle – more commonly referred to as a 'Moon Buggy' – was carried to the Moon's surface, stowed away inside the Descent Stage of the LM. Its function was to allow the

The Lunar Roving vehicle is seen here against the dramatic backdrop of the Hadley Apennines region where Apollo 15 landed. The mesh aerial at the front of the vehicle was used for voice and TV communications with mission control.

crew to journey further from the LM to collect a greater variety of surface rocks. On landing it was unfurled by pulling a number of lanyards and was then 'rolled' onto the Moon. The vehicle was 3 m (10 ft) in length, and just over 1.8 m (6 ft) wide, allowing both astronauts to travel. Attached to the central chassis were four wire-mesh wheels which were powered by silver zinc batteries and had specially designed hydraulic shock absorbers. It could reach a top speed of 14 km/hr (9 mph) and was eagerly tested by Dave Scott and Jim Irwin on Apollo 15. Scott reported: 'I've never been on a ride like this before, oh boy!' They found the ride quite 'bumpy' and so seat belts were added on Apollos 16 and 17. On Apollo 16, John Young accidentally backed into a sensitive heat flow experiment, and on Apollo 17, a mudguard came loose, which resulted in the crew being showered with lunar soil. A temporary mudguard made of maps and lighting clips saved the day.

APOLLO 7

This was the first manned flight of the Apollo spacecraft Command and Service Modules (CSM). Though a Lunar Module was not carried, the crew practised the manoeuvres required to remove the Lunar Module from its adaptor above the SIV-B engine stage. The craft made 163 orbits of the Earth and was recovered by the U.S.S. *Essex* in the Atlantic Ocean. Commander Schirra was the only astronaut to fly the Mercury, Gemini and Apollo spacecraft.

APOLLO 8

The first manned flight of the Saturn V booster saw Apollo 8 become the first manned flight to the Moon. The successful firing and re-firing of the main SPS engines allowed the crew to make 10 orbits of the Moon before returning back to Earth and a recovery by the U.S.S. *Yorktown* in the Pacific Ocean.

APOLLO 9

Apollo 9 saw the first flight of a Lunar Module (named 'Spider' by the crew). Schweickart made a 37-minute spacewalk just before the CSM and LM undocked for the first time. McDivitt and Schweickart then separated over 180 km (112 miles) and fired the ascent stage engines for the first time before returning to dock with the CM 'Gumdrop'. Altogether, the crews made 151 orbits of the Earth before splashing down in the Atlantic near to the U.S.S. *Guadalcanal.*

APOLLO 10

The crew of Apollo 10 took their Lunar Module 'Snoopy' to within 15 km (9 miles) of the lunar surface, after swooping down to the surface on two occasions. 'Snoopy' spent a total of four orbits separated from CM 'Charlie Brown' out of a total of 31 orbits both made of the Moon. The crew were safely recovered by U.S.S. *Princeton* in the Pacific ocean.

APOLLO 11

Neil Armstrong and 'Buzz' Aldrin became the first human beings to land on the Moon, and spent a total of 21 hours 36 minutes and 21 seconds inside the LM 'Eagle' on the surface. The crew performed surface EVAs for a total of 2 hours 31 minutes and 40 seconds. Mike Collins made a total of 30 orbits around the Moon in the Command Module 'Columbia' and his colleagues returned with 22 kg (48.5 lb) of lunar soil. They

APOLLO LOG

The abbreviations are: Cdr, Commander; LMP, Lunar Module Pilot; CMP, Command Module Pilot. From Apollo 11 onwards (excepting Apollo 13 which did not land) the Cdr and LMP were the crew members who ventured onto the Moon's surface. Apart from Apollo 7 which was launched by a Saturn 1B, the other missions were launched by the Saturn V. The duration is measured in hours, minutes and seconds.

Mission	Crew		Date	Duration
Apollo 7	Cdr:	Walter Schirra	11–22 October 1968	260:09:03
	LMP:	Walter Cunningham		
	CMP:	Donn Eisele		
Apollo 8	Cdr:	Frank Borman	21–27 December 1968	147:00:42
	LMP:	William Anders		
	CMP:	James Lovell		
Apollo 9	Cdr:	James McDivitt	3–13 March 1969	241:00:54
	LMP:	Russell Schweickart		
	CMP:	David Scott		
Apollo 10	Cdr:	Thomas Stafford	18–26 May 1969	192:03:23
	LMP:	Eugene Cernan		
	CMP:	John Young		
Apollo 11	Cdr:	Neil Armstrong	16–24 July 1969	195:18:35
	LMP:	Edwin Aldrin		
	CMP:	Michael Collins		
Apollo 12	Cdr:	Charles Conrad	14–24 November 1969	244:36:25
	LMP:	Alan Bean		
	CMP:	Richard Gordon		
Apollo 13	Cdr:	James Lovell	11–17 April 1970	142:54:41
	LMP:	Fred Haise		
	CMP:	John Swigert		
Apollo 14	Cdr:	Alan Shepard	31 January–9 February 1971	216:01:57
	LMP:	Edgar Mitchell		
	CMP:	Stuart Roosa		
Apollo 15	Cdr:	David Scott	26 July–7 August 1971	295:11:53
	LMP	James Irwin		
	CMP:	Alfred Worden		
Apollo 16	Cdr:	John Young	16–27 April 1972	265:51:05
	LMP:	Charles Duke		
	CMP:	Ken Mattingly		
Apollo 17	Cdr:	Eugene Cernan	7–19 December 1972	301:51:59
	LMP:	Harrison Schmitt		
	CMP:	Ronald Evans		

Charlie Duke, Lunar Module Pilot on Apollo 16, adjusts his communications carrier or 'Snoopy helmet'. This process of suiting up took place in a special room known as the 'clean room' at Cape Canaveral.

made a successful Pacific splashdown near to the U.S.S. *Hornet*.

APOLLO 12

Despite their Saturn V booster being hit by a lightning strike at launch, the Apollo 12 mission pressed on regardless. Conrad and Bean spent nearly eight hours on the lunar surface and returned a total of 22 kg (48.5 lb) of surface rock as well as deploying the first Apollo Lunar Surface Experiment Package. They also returned parts of the Surveyor 3 spacecraft which had landed nearby. Their total stay time on the Moon was 31 hours 31 minutes. After making a total of 45 orbits of the Moon, the crew were recovered by the U.S.S. *Hornet* in the Pacific Ocean.

APOLLO 13

The mission had to be aborted when one of the SM 'Odyssey's' liquid oxygen tanks exploded, severely damaging another. The lunar module 'Aquarius' effectively became a lifeboat, as its oxygen and power supplies were used just until the time of re-entry. The crew survived their ordeal and were recovered by the U.S.S. *Iwo Jima* in the Pacific Ocean.

APOLLO 14

Under the command of Alan Shepard – the only Mercury astronaut to walk on the Moon – the Apollo 14 LM 'Antares' successfully landed at Fra Mauro, which had been the planned landing site for Apollo 13. It was the first landing in the rugged lunar highlands in which the crew spent 33 hours 31 minutes on the surface. The crew made use of a Mobile Equipment Trolley (MET) to carry equipment to investigate a crater called Cone Crater some 2 km (1.25 miles) away, but had to turn back such was their exhaustion. Roosa, in the CM 'Kitty Hawk', made a total of 34 orbits around the Moon at a lower orbit than ever before. Forty-four kg (96 lb) of samples were returned by the crew before a splashdown in the Pacific Ocean near to U.S.S. *New Orleans*. The Apollo 14 crew were the last to spend three weeks in quarantine after landing on the Moon.

APOLLO 15

Scott and Irwin successfully landed in the Hadley Appenine region of the Moon in the LM 'Falcon', the first landing site well away from the equatorial regions of the Moon. One of 'Falcon's' landing legs was lodged on the rim of a small crater which caused an inconvenient tilt of 10 degrees when unfurling the first Lunar Roving Vehicle (LRV). They spent 66 hours and 55 minutes on the lunar surface, and performed three EVAs for a total of 19 hours 8 minutes outside the Lunar Module. Worden made 74 orbits of the Moon in the CM 'Endeavour' and the whole crew and their 77 kg (170 lb) of samples successfully returned for a Pacific splashdown. The landing was not without drama, as one of the three parachutes failed to open, but the crew reported little more than a 'heavy' landing.

APOLLO 16

Young and Duke successfully landed LM 'Orion' in the Descartes region of the Moon, terrain which was believed to be produced by volcanic activity. They spent a total of 71 hours 2 minutes on the Moon and made three separate EVAs. After orbiting the Moon 64 times, Mattingly performed a 1 hour 24 minutes EVA to retrieve film from the SIMBAY of SM 'Casper'. A total of 77 kg (170 lb) of soil was returned after the crew landed in the Pacific and were recovered by U.S.S. *Ticonderoga*.

APOLLO 17

The last Apollo flight was the first night-time launch, a spectacular sight as the Saturn V lit up the Florida coast like an artificial Sun. A landing site in the Taurus-Littrow valley was chosen to allow Jack Schmitt – the only trained geologist in the programme – to locate samples indicating volcanic activity. Three EVAs lasting longer than seven hours each resulted in a record 110 kg (243 lb) of samples being returned. As with Apollo 16, the U.S.S. *Ticonderoga* recovered the crew. Gene Cernan was the last of 12 men to walk on the surface of the Moon – to date!

ASTP

The Apollo-Soyuz Test Project (ASTP) came about in the climate of detente between the superpowers in 1972, and stands as a curiosity in the development of space. It was heralded as the start of a great breakthrough in space collaboration, but the vagaries of superpower politics intervened in later years with increased hostilities following the Soviet invasion of Afghanistan. Nevertheless, ASTP saw the first direct collaboration between 'old rivals', and shed some light in the West onto the workings of the Soviet space programme.

After an agreement signed by Premier Kosygin and President Nixon in March 1972, it was agreed that a Soyuz capsule would dock with an Apollo in July 1975. Earlier discussions had suggested that they both dock with a Salyut, but this was deemed too complicated. Nevertheless, the docking of two spacecraft with different atmospheric composition could cause problems. Apollo's normal atmosphere was pure oxygen at 0.35 kg/cm² (5 psi), whereas Soyuz contained a mixture of oxygen and nitrogen at 0.98 kg/cm² (14 psi). If the two craft just docked, the American crew would suffer from diver's bends because of the nitrogen if they transferred. So a 'Docking Module' was built by NASA as a technological one-off, 3.15 m (10 ft 4 in) long, and 1.4 m (4 ft 6 in) at its widest. It was essentially an airlock which would allow crews to pass from one craft to another. For ease of transfer, the Soyuz cabin pressure was reduced to 0.70 kg/cm² (10 psi). The most important mission rule was that one crew member always remained inside his native craft.

THE CREWS

The choice of crews for ASTP was important, given the diplomacy and mutual training that would occur. The Soyuz Commander was no less a person than Alexei Leonov, whose larger than life bonhomie ensured he was well appreciated by his American hosts. His companion was Valeri Kubasov, veteran of Soyuz 6, who had trained with Leonov for an earlier Salyut mission but never flown as a result of the Soyuz 11 accident. The American crew was a mix of the old and the new. Commander was Tom Stafford, whose experience and mid-Western honesty impressed the Russians. The astronaut in charge of the docking manoeuvres was Vance Brand, a rookie in his 30s whose boyish good looks led to his being called the 'Robert Redford of the Astronaut Corps'. And finally getting a flight at the age of 51 was Mercury astronaut Deke Slayton. After spending years as Director of Crew Operations, Slayton assigned himself the last chance to fly – no-one seemed to begrudge him. After reaching orbit he was to exclaim: 'Man, that was worth waiting 16 years for'.

ASTP allowed both countries to view and compare their respective space progammes. Mutural training lasting for many months took place: back-up crews were announced by both sides, an unprecedented step for the Russians. Romanenko and Dzhanibekov, who later played key roles in Salyut programmes, were first introduced at this time. Nevertheless, the Russians tried to limit the amount of information they could give away by flying the Americans into Star City that surrounds Baikonur late at night, minimizing their view of the vast sprawling site. The NASA astronauts found that their rooms were bugged after remarking there weren't enough clothes hangers, only to find more the next day. Another unexpected hurdle was the plethora of

ASTP CSM/DOCKING MODULE

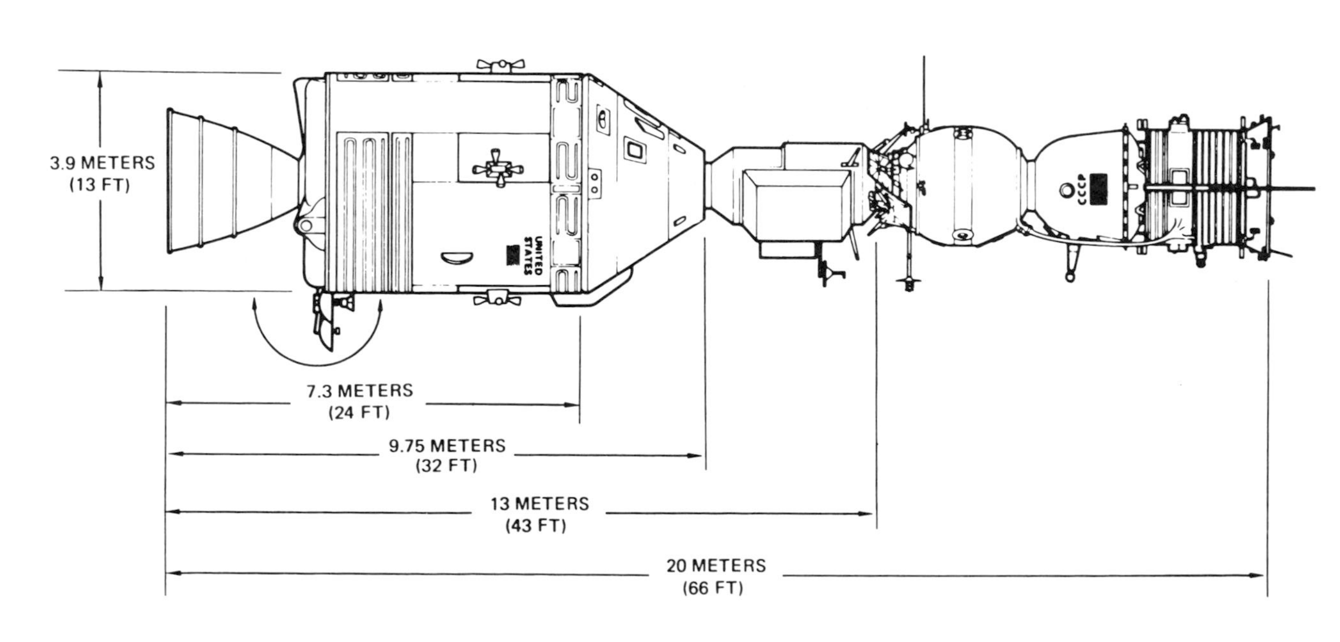

The first – and to date, only – joint manned mission between the superpowers involved a docking between Apollo and Soyuz spacecraft. Seen here in diagramatic form are the Apollo CSM (left) and the Soyuz (with solar panel). A special docking module was needed, carried by Apollo which 'chased' Soyuz after launch.

The much vaunted 'handshake in space' was a terribly confused affair. Seen here in the docking module are Tom Stafford and Alexei Leonov, the respective mission commanders. This view was taken by Valeri Kubasov in the orbital module of Soyuz.

official receptions which both crews were required to attend. Given the Russians' – including Leonov's – fondness for alcohol, it was decided unofficially amongst the Apollo crew that one of their number match his prodigious quaffing of alcohol lest they appeared rude in declining their hosts' vodka!

THE MISSION

Soyuz 16 was a dress rehearsal for the flight, and went some way to reassure NASA after the loss of the crew of Soyuz 11. Soyuz 19 was launched first on 15 July, followed by the Apollo craft $7\frac{1}{2}$ hours later from Cape Canaveral atop a Saturn 1B. In fact, Apollo was launched into its lowest orbit ever, only 320 km (200 miles) above the Earth, and effectively chased Soyuz over the following two days. By noon on 17 July, they were 50 km (31 miles) apart, and by early evening were within a few hundred metres. They docked west of Portugal at just after 4 o'clock Greenwich Mean Time, and prepared to open the hatches. Because of reports of an acrid smell in the docking module, the main hatch opening was delayed for over three hours. Given that instructions in Russian and English were being relayed to the crew, the much-vaunted handshake in space was confused, but eventually Stafford and Leonov shook hands with the former saying, 'Glad to see you'. Slayton joined Stafford and the Russians for dinner in Soyuz, where they received congratulatory phone calls from Premier Brezhnev and President Ford. It was another mission rule that in orbit the Americans should converse in Russian and vice versa: Stafford's accent was soon christened 'Oklahomski' by his hosts.

The crews spent a further day in each other's spacecraft, performing joint experiments and enjoying each other's company. After the final undocking, Apollo was photographed from Soyuz artificially eclipsing the Sun. They redocked and then separated. Soyuz landed on the 21st and the final stages of its descent were relayed around the world by TV cameras from their rescue helicopters. It was the perfect end to the Soviet side of the mission, leading Soviet Mission Control to cheer wildly. The crew emerged from their craft jubilant and were whisked away to Baikonur for a heroes' welcome.

Apollo spent a few more days in orbit, running a limited number of experiments, and allowing the crew to savour flying in space. Though they had no idea at the time, it would be another six years before the next American astronauts entered space on the Space Shuttle. On the 24th, Apollo re-entered the Earth's atmosphere to be picked up by the U.S.S. *New Orleans* in the Pacific. What was not realized at the time was how close the crew came to being killed – within minutes of recovery, they were in intensive care. During their descent, the parachutes had failed to deploy automatically, and Brand had fired them manually. Inadvertently, the capsule's reaction control thrusters started to fire to maintain the capsule's orientation and after trying to switch them off, fumes fron the thrusters entered the craft as a pressure equaliser valve opened. Nitrogen Tetroxide fumes filled the cabin, and the crew were nearly asphyxiated before they could don emergency oxygen masks. The long-term effects of the gas were enough to bleach their lungs, and they were flown to a military hospital in Honolulu.

Such a near-tragic ending could not detract from the achievement of the mission, and the crew remained good friends. Ten years later, they formally reunited in Washington DC at a conference where joint exploration of Mars was mooted.

GEMINI

Named after the heavenly twins in classical mythology, NASA's Gemini series had a two-man crew. These missions developed the techniques needed for flights to the Moon, and many of the astronauts in Gemini later flew on Apollo. Notable firsts of the Gemini programme included rendezvous and docking with target vehicles as well as the first American spacewalks. All in the series were launched by the Titan II booster, with the first two in the series being unmanned tests. The final five in the series involved Agena target vehicles which were 'hunted' in orbit for rendezvous and docking.

THE SPACECRAFT

Though it was also bell-shaped like the earlier Mercury capsule, the Gemini spacecraft was twice as heavy and far more advanced. It was the first spacecraft to utilize fuel cells which produced electrical power by electrolysis of oxygen and hydrogen. Drinking water was a useful by-product of this process. A miniature computer was carried which weighed just under 23 kg (55 lb) and was able to perform 7,000 operations a second: it enabled Gemini astronauts to calculate the thrust required to change from one orbit to another. Each Gemini capsule was made up of a pressurized crew compartment and an adapter module. Though the crew compartment had 50 per cent greater volume than the Mercury capsule, it was still a cramped environment, leaving about the same space inside as a VW Beetle. Gemini did not have an emergency escape rocket, but rather ejection seats which could be used up to 18,000 m (60,000 ft) altitude. Toilet facilities were rudimentary, consisting of little more than plastic bags with adhesive edges!

The adapter section contained the fuel cells and the cryogenic fuels for attitude control thrusters. Small thrusters were also dotted around the crew compartment and used for changes in pitch, yaw and roll for tricky rendezvous manoeuvres. A radar system was installed in the front end of the Gemini craft which automatically measured the range and position of the rendezvous target. The spacecraft's heat shield consisted of a complex mixture of glass fibres sandwiched together.

GEMINI LOG

'Cdr' refers to Commander; 'Pil' to pilot.

Mission		Crew	Date	Duration
Gemini 3	Cdr: Pil:	Virgil Grissom John Young	23 March 1965	04:53:00
Gemini 4	Cdr: Pil:	James McDivitt Edward White	3–7 June 1965	97:56:11
Gemini 5	Cdr: Pil:	Gordon Cooper Charles Conrad	21–29 August 1965	190:55:14
Gemini 7	Cdr: Pil:	Frank Borman James Lovell	4–18 December 1965	330:35:31
Gemini 6	Cdr: Pil:	Walter Schirra Thomas Stafford	15–16 December 1965	25:51:24
Gemini 8	Cdr: Pil:	Neil Armstrong David Scott	16 March 1966	10:41:26
Gemini 9	Cdr: Pil:	Thomas Stafford Eugene Cernan	3–6 June 1966	72:21:00
Gemini 10	Cdr: Pil:	John Young Michael Collins	18–21 July 1966	70:46:39
Gemini 11	Cdr: Pil:	Charles Conrad Richard Gordon	12–15 September 1966	71:17:08
Gemini 12	Cdr: Pil:	James Lovell Edwin Aldrin	11–15 November 1966	94:34:31

GEMINI 3

The mission was a three-orbit demonstration of the Gemini spacecraft. While on its first pass of Texas after launch, Grissom used the hand-held computer to change orbit for the very first time in a manned spacecraft. The spacecraft landed about 80 km (50 miles) uprange from the U.S.S. *Intrepid* in the Atlantic ocean.

GEMINI 4

This four-day flight saw Ed White become the first American to walk in space – officially known as 'Extra-Vehicular Activity' or EVA. He spent about 20 minutes outside the spacecraft, enjoying the sensations of weightlessness and manoeuvring with a hand-held oxygen gun. Mission control had to force him back into the capsule, and the hatch was open for a total of 36 minutes. After 62 orbits of the Earth, the crew landed 80 km (50 miles) uprange from U.S.S. *Wasp* in the Atlantic.

GEMINI 5

This 8-day flight saw the first use of fuel cells for electrical power production, and showed that human beings could fly in space without ill effect in the time it would take to travel to the Moon and back. Cooper and Conrad evaluated the spacecraft's guidance and navigation systems for future rendezvous missions in their 120 orbits of the Earth. Ironically, incorrect navigation coordinates from ground control resulted in a 144 km (90 mile) undershoot of the planned Atlantic landing zone where U.S.S. *Lake Champlain* was waiting.

GEMINI 6 AND 7

Gemini 6 had been intended to dock with an Agena target vehicle, but this failed to achieve orbit and the count was held at T – 42 minutes. It

The first direct rendezvous between two spacecraft took place in December 1965 between Gemini 6 and 7. The delicate manoeuvres required involved some 35,000 firings of Gemini 6's thrusters, seen here around the circumference of the spacecraft's exterior shell.

was decided to use Gemini 7 as the target instead, so in due course Borman and Lovell were launched on 4 December. A launch attempt for Gemini 6 on 12 December failed when the Titan II engines shut down 1.2 seconds after ignition. But three days later Gemini 6 was finally launched successfully, and headed for a rendezvous with the Gemini 7 crew on their 11th day in space. Both craft came within 2 metres (6 ft 6 in) of each other, and were able to see each other through their cabin windows. After completing 16 orbits, Gemini 6 landed first, within 11 km (7 miles) of the U.S.S. *Wasp* in the Atlantic. Gemini 7 landed on the 18th after spending a fortnight in space – the longest duration flight until 1970. After 206 orbits, the crew landed within 7 km (4.4 miles) of the U.S.S. *Wasp* which had remained 'on station' for their return.

GEMINI 8

The first docking of one space vehicle with another was achieved on this mission. But 27 minutes after docking, the Gemini-Agena combination began to yaw and roll alarmingly. A thruster had jammed open on the Gemini, causing the craft to spin once a second. Armstrong and Scott deactivated the malfunctioning control system of the Gemini, took over manual control, undocked and then activated the re-entry control system. The mission was terminated midway through the 7th orbit and landed in the Western Pacific. Three hours later the crew were picked up by the U.S.S. *Mason*.

GEMINI 9

Problems with the docking targets continued on this mission. First of all, the original target Agena failed to orbit: after a successful launch of the crew two days later, the shroud on the augmented docking adapter sent to replace the Agena failed to separate. The crew described the appearance of the adapter as an angry alligator, and had to make do with making three different types of rendezvous attempts. Cernan attempted a 2 hour 7 minutes EVA during the 44 orbits the spacecraft made of the Earth. The crew landed within 600 m

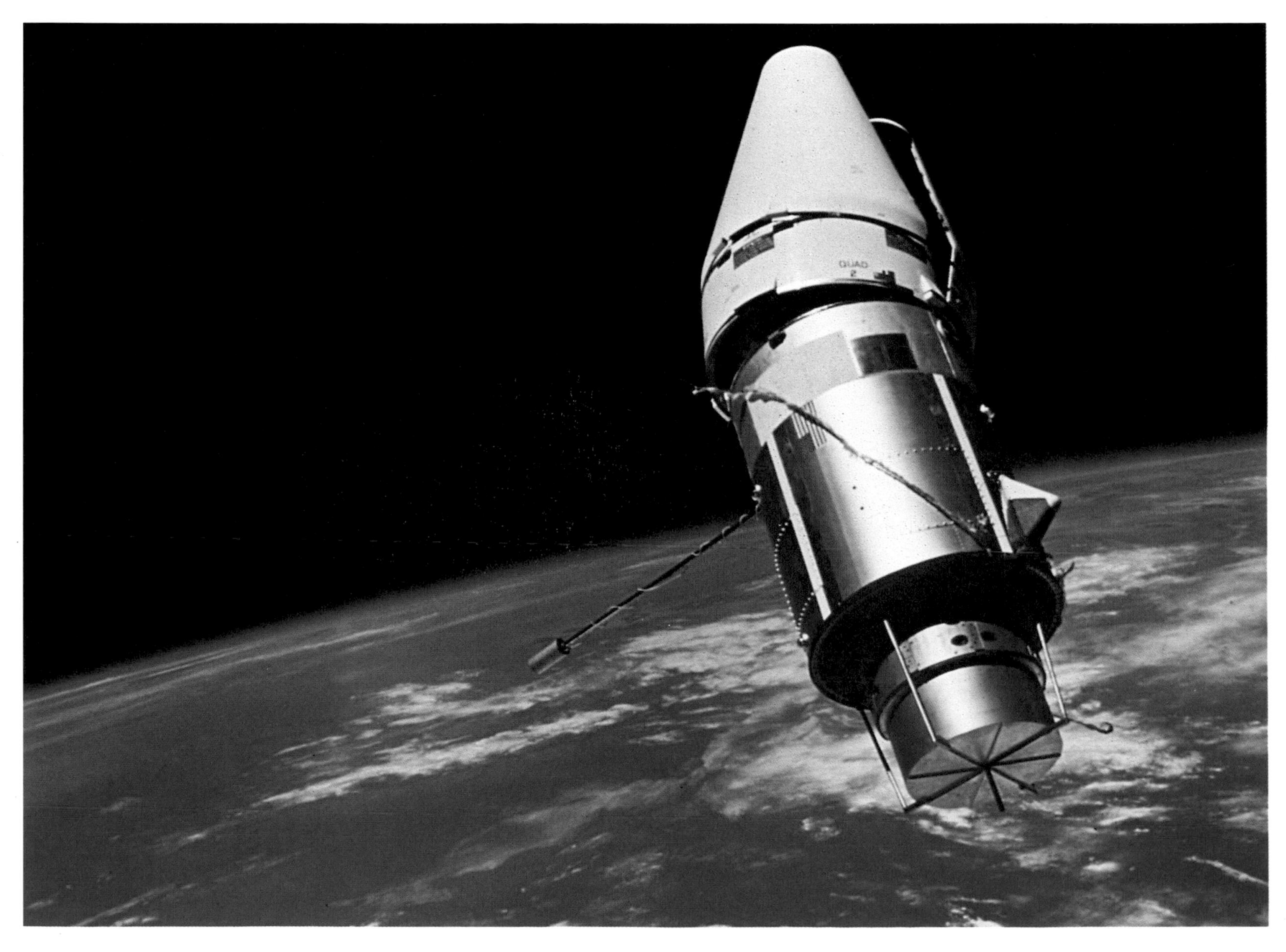

The 'Angry Alligator' that was Gemini 9's Atlas Agena target put paid to ideas of docking with the craft. Gemini 9 was plagued by bad luck from the start: its original crew, Bassett and See, were killed in an air crash. Gene Cernan's two-hour space walk was curtailed because it proved too strenuous.

(1,968 ft) of the planned landing target near to the U.S.S. *Wasp* in the Atlantic.

GEMINI 10

After docking successfully with an Agena target, Young and Collins successfully activated its propulsion systems to reach a higher orbit. They also docked with Gemini 8's target vehicle which had been left in a 'parking orbit' for a later visit. Collins made a 49 minute EVA to retrieve a dust experiment collector on the Gemini 8 Agena target. After 43 orbits, they landed within 8 km (5 miles) of U.S.S. *Guadalcanal* in the Atlantic.

GEMINI 11

Conrad and Gordon successfully docked with their Agena target during their first orbit of the Earth, and used its propulsion engines to reach an altitude of 1,182 km (739 miles) above the Earth. Gordon fastened a 30 m tether to the Agena and the spacecraft made two orbits of the Earth as they separated to extend the tether. After 44 orbits together, an automatic re-entry was achieved for the first time, taking the craft within 2.4 km (1.5 miles) of the U.S.S. *Guam* in the Atlantic.

GEMINI 12

The final Gemini flight saw Buzz Aldrin make three EVAs lasting a record $5\frac{1}{2}$ hours after successful docking with an Agena target. Fifty-nine orbits were made of the Earth, and the computers were used to steer the vehicle for an automatic landing. They landed $2\frac{1}{2}$ miles from the U.S.S. *Wasp* in the Atlantic.

MERCURY

Project Mercury was approved just a week after NASA's formation in October 1958. The White House decided that the military services would provide the potential astronauts who would take America into the manned spage age. NASA stipulated that potential fliers should be under 180 cm (5 ft 11 in) in height, college educated with degrees in engineering or science and be married with families – this latter requirement at the insistence of psychologists who thought marriage would provide them with sufficient mental stability! Over 500 test pilots applied, and this number was whittled down to 32, before the final seven were chosen. On 2 April 1959 NASA announced the names of the seven who had been chosen: they were immediately feted and their life stories exclusively snapped up by *Life* magazine though it was to be two further years before the first of them flew in space. (Only one of the 'Mercury Seven' didn't get to fly: Deke Slayton was grounded due to heart irregularities, but eventually flew in 1975 on the Apollo-Soyuz Test Project.)

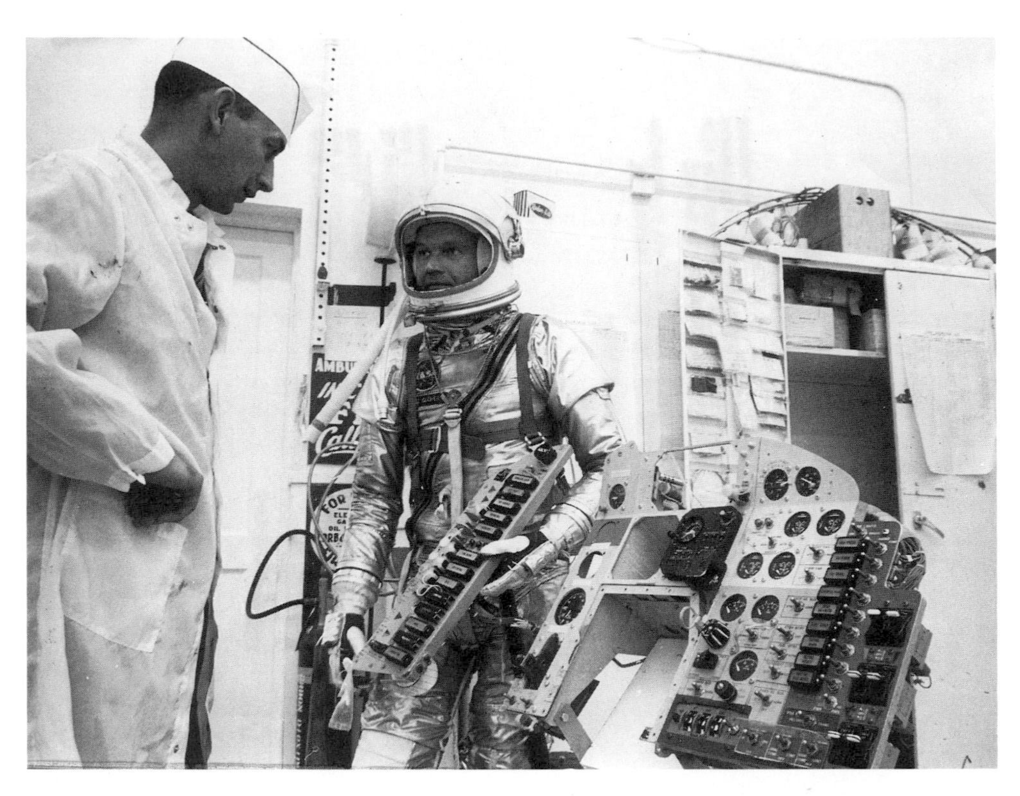

'They don't make' em like they used to!' Gordon Cooper, pilot of the final Mercury mission, is seen with a spacecraft engineer checking the manual override controls for his capsule. This 22-orbit flight was beset by technical troubles, though his only reference to this was typical: 'Things are starting to stack up a little'.

To prepare for manned flights, a number of tests of the Mercury capsule took place on the two boosters that were available. Wernher von Braun's Redstone booster was just about capable of a thrust of 35,380 kg (78,000 lb) and was used to launch the Mercury capsule on a long, ballistic flight path down the Atlantic test range. The Atlas booster – in which development problems caused repeated delays – was used for the four orbital flights. A number of tests with monkeys were flown, perhaps the most famous being Ham in January 1961 – an exact dress rehearsal of Alan Shepard's flight four months later. Originally, the capsules were to be flown automatically from the ground as it was felt that weightlessness might incapacitate the crews. The pilots complained that they would be little more than 'spam in the can' and so they were given the ability to fly the craft themselves. Considering the problems that arose when the automatic systems failed, the decision avoided the first fatalities in space.

THE SPACECRAFT

The first American spacecraft to carry astronauts into space was a bell-shaped capsule, designed with minimizing the heating effects of re-entry in mind. The cabin in which a human being could just about fit included a couch specially moulded for the individual astronaut. Above the crew cabin was a truncated cabin containing the parachutes that opened for re-entry and above that the escape tower, consisting of a small rocket which could pull the capsule to safety in case of a launch pad accident. A safe atmosphere for the astronaut was provided by an environmental control system which maintained a pure oxygen atmosphere at 0.38 kg/cm² (5.5 psi). Two separate control sys-

The orbital flights of the Mercury series required the Atlas booster whose development had been delayed by technical difficulties. Throughout 1961 it was tested with mock-up Mercury capsules, as on this flight of 21 February.

tems were provided: one for the astronaut's suit and the other for the cabin. The spacesuit was only inflated in case of emergencies.

Communications between the astronaut and Mission Control were maintained by a two-way radio system and a radio tracking beacon. Engineering, medical and scientific data from the cabin were relayed automatically back to ground stations. Eighteen thrusters dotted about the capsule provided orientation control by means of firing superheated steam as attitude control gas. Automatic control was provided by the Automatic Stabilization and Control System (ASCS) which could detect instabilities within the capsule's movement. It could also be programmed to orientate the capsule ready for re-entry. However, the pilot could take manual control with the Fly-by-Wire System (FBWS) in case of emergencies. Another addition was a small window to allow the astronaut to view the Earth and take photographs of the surface below.

The heat shield offering protection against the frictional heat of re-entry was provided by a reinforced laminated plastic. Shields were made detachable so that once the main parachute had opened after re-entry, an impact bag of rubberized glass-fibre would cushion the sea landing. Three solid rocket motors were attached to the heat shield as a 'retro-pack' which ensured the capsule was aligned correctly for re-entry.

MERCURY LOG

The designation 'MR' refers to 'Mercury-Redstone', the booster which allowed sub-orbital flights. 'MA' means 'Mercury-Atlas'. The names of the craft were entirely the choice of the astronauts – the '7' referring to the number of pilots originally chosen.

Flight	Spacecraft	Pilot	Date	Duration
MR-3	'Freedom 7'	Alan Shepard	5 May 1961	00:15:22
MR-4	'Liberty Bell 7'	Virgil Grissom	21 July 1961	00:15:37
MA-6	'Friendship 7'	John H. Glenn	20 February 1962	04:55:23
MA-7	'Aurora 7'	Scott Carpenter	24 May 1962	04:56:05
MA-8	'Sigma 7'	Walter Schirra	3 October 1962	09:13:11
MA-9	'Faith 7'	Gordon Cooper	15–16 May 1963	34:19:49

MR-3

Alan Shepard's brief 15-minute foray into space reached a maximum height of 190 km (119 miles) before being recovered by the U.S.S. *Champlain* in the Atlantic. A number of pre-flight delays occurred, so much so that by the time he was actually launched he recalled he was desperate to visit the toilet! Shepard was destined to become the only one of the 'Mercury 7' to walk on the Moon – as Commander of Apollo 14.

MR-4

Though Grissom repeated Shepard's earlier flight, his capsule sank at the end of the flight. To this day nobody knows the reason why, but luckily Grissom was rescued in time by the recovery crew from the U.S.S. *Randolph*. He was weightless for just over five minutes and survived the 11G force of re-entry, when the deceleration caused him to feel eleven times his own weight.

MA-6

NASA's first manned flight in space proved to be more dramatic than intended with a signal that hinted the heatshield was loose. However, John Glenn's three orbit flight was a success, and he landed only 64 km (40 miles) uprange of the U.S.S. *Noah* in the Atlantic. Glenn later became a Senator and made an unsuccessful bid for the Presidency in 1984.

MA-7

Though Scott Carpenter also performed three orbits of the Earth, the excitement during his mission came as he attempted re-entry. By the end of his second orbit, he had used up over half his attitude control fuel as he repeatedly altered the capsule's orientation to get better and better views of the surface. As a result, he had to make a manual retro-fire and overshot the landing site by 400 km (250 miles). Crews from the U.S.S. *Pierce* later found him in his life-raft, but he was effectively grounded. His love of the adventurous life led him to become a deep sea diver and later a breeder of bug-killing bees on the West Coast.

MA-8

Wally Schirra's six-orbit flight went so smoothly that he was able to conserve enough fuel to allow the possibility of a one-day Mercury mission. He landed only 7.2 km (4½ miles) from the U.S.S. *Kearsage* in the Atlantic. Schirra became the only one of the original Mercury astronauts to fly Mercury, Gemini and Apollo spacecraft.

MA-9

Gordon Cooper's 21 orbits of the Earth were designed to evaluate the effect of spending a day in space. Despite continual problems with his automatic stabilization system on the capsule and short-circuiting of the environmental control systems, Cooper was happy with his flight. He had to make a manual re-entry, and came within only 7.2 km (4½ miles) of the U.S.S. *Kearsage* in the Pacific.

MIR

On 20 February 1986, the Soviet Union launched the third generation of its space stations, known simply as Mir ('Peace'). About the same size as the Salyut stations (see page 111), Mir has been occupied since early 1987, and by the end of 1988 had been home to two cosmonauts for exactly a year. Mir allows a greater flexibility because it has six docking ports and larger solar panels which can be used for greater power generation. There is extra room inside the station as scientific equipment is contained in modules which have been added to the station. This has made greater space available for the crews, with the provision of a private sleeping area – a cause of much complaint among earlier crews. Mir's computers are much more advanced, but problems have meant the crews have had to spend as much time in the routine monitoring of essential maintenance tasks. Communications between Mir and mission control are relayed through a satellite called Luch which allows greater data return. Mir is in an orbit inclined at 51.6° to the equator, centred 340 km (212 miles) above the Earth, and the station originally weighed around 20 tonnes.

For the crew, the greater living space has come as a blessing with a greater range of creature comforts. There is an exercise bike and treadmill for 'running on the spot' – important for prolonged stays to counteract the effects of weightlessness on the human frame. A hot plate to heat meals plus a table on which to eat were also provided. According to Yuri Romanenko who stayed on board Mir for most of 1987, 'Mir is 90 per cent different from Salyut. It has more equipment on board and is much more spacious and cosy, and even has a TV and video.'

It is the docking ports on Mir which allow the station its utmost flexibility. There is one docking port at each end of the station, but the docking adapter at the front end also contains four ports around the circumference of the adapter. Addi-

Slightly larger, and definitely more versatile, than the Salyut space stations, *Mir* is seen here being prepared for launch. The size of the station can easily be gauged by the technicians. *Mir* was launched on 20 February 1986 and will probably be superceded by a second station in 1992.

Right: The key to *Mir's* versatility is the fact that modules can be added to the central station. In April 1987, this astrophysical module call '*Kvant*' (or Quantum) was docked with Mir. It carries astrophysical telescopes from Western countries that were used to look at Supernova 1987A among other objects.

Below: This unusual view shows the unfurlable structure 'ERA' built by Aerospatiale and deployed by Jean-Loup Chrétien in December 1988. The gantry-like structure is a prototype to allow the construction of larger scale stations by French industry.

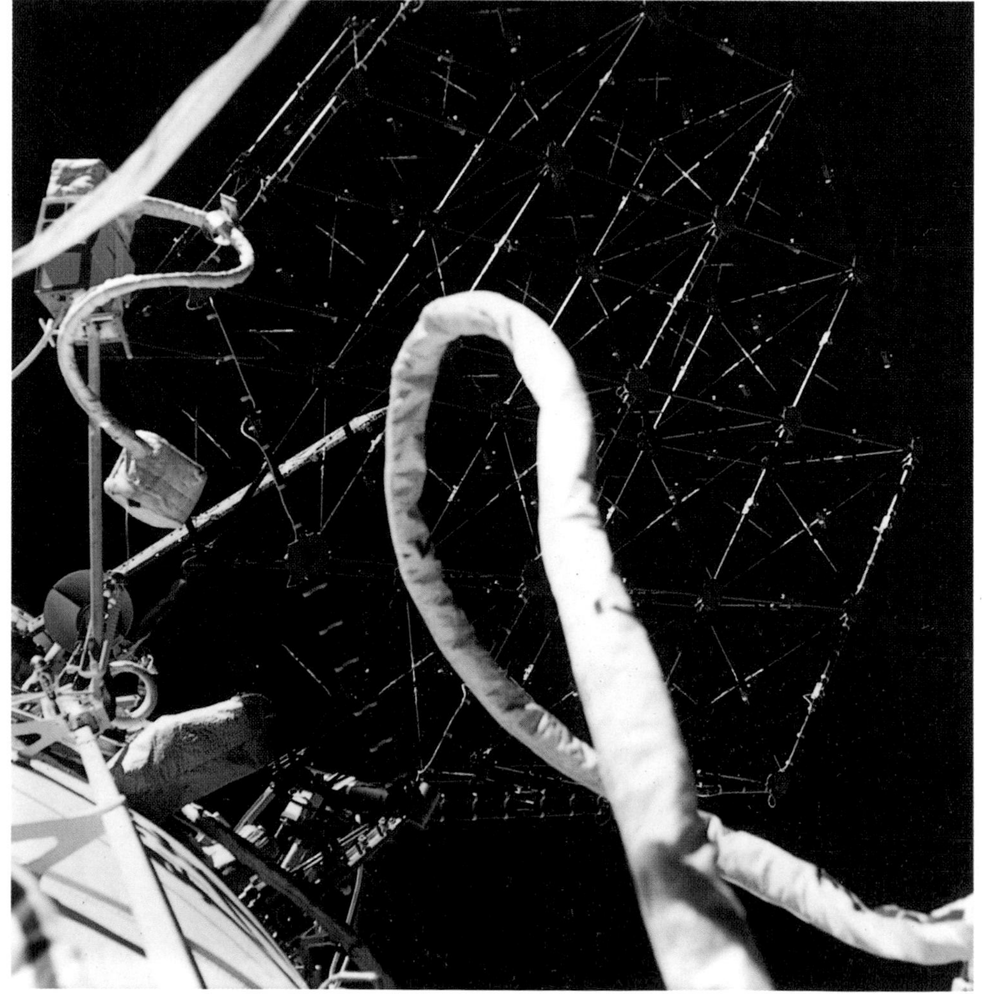

tional modules attached to Mir will be docked at the front end and then later manoeuvred to one of the side ports by use of a remote manipulator arm. The philosophy behind Mir is simple: it is designed as a modular station, onto which a variety of modules are planned to be added.

The first module to dock with Mir was the Kvant ('Quantum') astrophysical module which carries a battery of four X-ray telescopes. With a mass of 20 tonnes, it is nearly as heavy as Mir and has its own onboard computers for control. In late 1989, a further 'equipment' module was expected to be launched, and attached to the uppermost docking module at the front. Kvant is attached to the rear of Mir, and having its own docking module allows Progress or Soyuz ferries to dock.

Scientific research on Mir has covered a range of disciplines which include remote sensing, materials science and high energy astrophysics. Using special cameras, cosmonauts on Mir have already photographed most of the U.S.S.R. in an effort to detect oil and other natural resources. In weightlessness, materials mix more easily and a smelting plant called Korund has been put to good advantage. Kvant has revealed much data to international astronomers, including information on Supernova 1987A. Biomedical, earth resources and materials science modules are expected to be launched during the 1990s, when a further Mir complex will be launched using the Buran Space Shuttle.

SALYUT

The Salyut series of space stations gave the Soviet Union the unparalleled lead in long-duration missions, despite the handicap of many problems along the way. The stations themselves underwent continual development, though the basic modules were of the same design. Essentially, they appeared like short, squat telescopes with as many as three solar panel wings. The whole vehicle was 15 m (49 ft) long, and at its widest 4.2 m (13 ft 9 in) in diameter. The vehicle was split up into separate compartments, including transfer compartments which were used in the transfer of materials from Progress ferry craft. A staggering variety of experiments, including biological and metallurgical ones, were carried out on the Salyuts, all of which were contained in the main living section, giving the station a cluttered appearance.

The first in the series were more or less equally divided between being military or civilian in nature, though the introduction of Salyut 6 seems to have ended that division. An advantage with the Salyuts was their orbital manoeuvring engines, which allowed the craft to boost up to higher orbits, as decay took its natural toll. Both Salyuts 6 and 7 could be re-fuelled which allowed them to be extended beyond their nominal lifetimes. The docking of Progress vehicles allowed the stations to use the supply craft's engines, and by the last in the series, the technique of turning the whole vehicle to face approaching Soyuz craft had been developed.

SALYUT 1

The world's first space station was home only to the Soyuz 11 crew, who died when their Soyuz capsule fatally depressurized after their 24-day stay. The craft was originally launched into a 200 × 222 km (125 × 139 miles) orbit, inclined at an angle of 51.6°, and was manoeuvred for re-entry nearly six months after launch.

SALYUT 2 AND COSMOS 557

Nearly two years after the first Salyut, the next was launched as an improved version, but the craft became unstable and broke up on 14 April, before re-entering on 28 May. Cosmos 557 was launched in haste (believed to be a civilian Salyut) but also decayed 11 days after launch.

SALYUT 3 AND 5

Both Salyuts 3 and 5 were launched into similar orbits, centred on 260 km (162 miles), and very few photographs of either were ever released. The Soyuz crews who attempted to dock with the craft were military in nature, and the fact that these missions used different frequencies and code words suggested their military nature. After the last crews visited the stations, recoverable capsules were returned, believed to carry film taken from orbit of military targets.

SALYUT 4

After 11 days in orbit, Salyut 4 was boosted into a higher orbit than the earlier stations, roughly 345 × 355 km (216 × 222 miles). Unlike Salyut 1, which had two solar panels stretching from either side of its main section, Salyut 4 had three solar panels each separated by 120°. Visited by two crews, it was also the target for the unmanned Soyuz 20 vehicle, a forerunner of the Progress supply craft. It stayed in orbit for 90 days, showing that a Soyuz ferry craft could survive that long in orbit.

SALYUT 6

The 'second generation' Salyut was launched on 29 September 1977 and over the next 4½ years was home to 16 separate Soyuz crews. Though similar in design to Salyut 4, it had a second docking port at the rear of the station which allowed continual resupply of the craft by Progress ferry vehicles and docking by short-staying crews. By the end of the station's life, the crews were able to notch up ever greater stays in space (up to 185 days) and hosted Intercosmos cosmonauts. In June 1981,

On her second visit to Salyut in 1984, Svetlana Savitskaya became the first female to make a space walk. Along with Vladimir Dzhanibekov, they spent a total of 3 hours and 35 minutes outside the station, performing routine maintenence experiments.

SALYUT LOG

Salyut	Launch	Re-Entry	Crews visited
1	19 April 1971	11 October 1971	Soyuz 10 attempted docking, but crew did not enter; Soyuz 11 crew did, staying for 24 days.
2	3 April 1973	28 May 1973	Unmanned due to untimely break-up.
Cosmos 557	11 May 1973	22 May 1973	Launched into orbit similar to Salyut 2, but broke up in orbit.
3	25 June 1974	24 January 1975	Military Salyut: occupied by Soyuz 14 crew, as Soyuz 15 crew failed to dock.
4	26 December 1974	3 February 1977	Civilian Salyut: crews of Soyuz 17 and 18 visited. Soyuz 20 was unmanned, a prototype Progress transporter.
5	22 June 1976	8 August 1977	Military Salyut: visited by crews of Soyuz 21 and 24: Soyuz 23 failed to dock.
6	29 September 1977	28 July 1982	Second generation Salyut: visited by crews of Soyuz 26 to 40. Soyuz 25 and 33 failed to dock. Cosmos 1267, the mysterious Star Module, also docked.
7	19 April 1982	—	Final Salyut: visited by Soyuz T5 to T12, though Soyuz T8 failed to dock. Second Star Module docked, Cosmos 1443.

the last vehicle to dock with Salyut 6 was launched; merely called Cosmos 1267, it was twice the size of a Soyuz. It seems to have been 14 m (46 ft) long and had three solar panels. Its engines were used to fire the Salyut 6 into a higher orbit and higher inclination, before undocking before the station re-entered. The Soviets have never explained what this 'Star Module' was used for.

SALYUT 7

The last in the Salyut series was launched on 19 April 1982, and was home to ten crews and resupplied by no less than 13 Progress vehicles. Similar to Salyut 6, it had an improved navigation and computer system which allowed the craft to spend more time doing scientific work. Salyut 7 was also docked with two further 'Star Modules', Cosmos 1443 and 1686, in March 1983 and September 1985. The latter enabled the crews of Soyuz T-13 and T-14 to augment Salyut's damaged power generating capacity and fuel supplies. Both vehicles – variously described as 'Space Tugs' or 'Cargo Ships' – increased Salyut 7's living quarters to an extra 50 m^3 (1,764 cu ft). After the Soyuz T-15 crew left Salyut 7 in June 1986, Salyut 7 was boosted to a 495 × 475 km (309 × 296 miles) orbit, described as a 'parking' orbit. In 1988, the Soviets announced their intention of returning Salyut 7 section-by-section back to Earth using their Buran Space Shuttle in order to re-launch it at a later date.

The crews of Soyuz 27 (left) and Soyuz 26 (right) prepare to open the first 'post office' in space. Though it appears distinctly cluttered, weightlessness enabled cosmonauts to live quite unhindered aboard Salyut 6. The docking module is seen at the far end of the living quarters.

SKYLAB

To date America's only experience of long duration spaceflight has been the Skylab programme in the early 1970s. It was an unprecedented triumph in the development of space stations, particularly after the station itself was damaged during launch. The station was a converted upper stage of a Saturn V booster, which looked something like a vast butterfly when deployed. The main section of the station known as the Orbital Work Station provided the astronauts with 368 m^3 (13,000 cu ft) in which to live – a 50 times increase in size afforded by the Apollo CSMs. The Orbital Work Station was about the size of a medium sized house and for the first time contained proper washing and toilet facilities, as well as sleeping cubicles. A series of lockers around the central hub of the station contained changes of clothes, domestic items and even a small vacuum cleaner. Perhaps the greatest luxury was a small, 46 cm (18 in) window out of which the astronauts were able to spend hours gazing at the panorama of the Earth below.

Solar panels were attached to the main section which provided power for the station, as well as an airlock module which allowed astronauts to enter space by EVA at the front of the station. In front of that was a docking port which allowed an Apollo CSM to dock with the station. Attached to the top of this was the Apollo Telescope Mount, essentially a converted Lunar Module Ascent Stage containing scientific instruments, from which four large solar panels unfurled. With an Apollo CSM docked, the station 'cluster' was some 36 m (118 ft) in length and weighed a total of 90 tonnes.

The station itself was referred to by NASA as Skylab 1, with the first manned mission referred to as Skylab 2. Skylab 1 was launched as part of a modified Saturn V booster on 14 May 1973, and right from the start things went wrong. A minute into the flight, vibrations broke off part of the station's external shield against heat and meteoroid impacts, as well as a large solar panel. Though Skylab entered the correct orbit measuring 440 × 430 km (275 × 269 miles) at 50° inclination, it also transpired that the second solar panel had not opened. This left Skylab with no power and without protection from the searing heat of the Sun.

The first manned launch was delayed while NASA decided the best plan of action – but it was a case of Skylab 2 to the rescue. So on 25 May, Charles Conrad, Paul Weitz and Dr. Joe Kerwin (the first U.S. medical doctor in space) were launched in a Saturn 1B to rescue the craft with a variety of makeshift tools. After closing in on the stricken station, Weitz stood in the Apollo hatchway and attempted to prod the second solar panel open – but without success. The crew then found difficulty in docking, but did so and remained in the Apollo CM overnight. On the next day, they entered the orbital workshop, finding the metallic surfaces too hot to touch and acrid smells of smouldering plastic. Their first job was to deploy an orange-coloured sheet of nylon, mylar and aluminium foil through an airlock over the main workshop. This acted as an impromptu parasol, and soon conditions within the station were bearable. But still there was the problem of power, and on 7 June, Conrad and Weitz performed a spacewalk, after much effort releasing the undeployed solar panel by cutting through a metallic strip that was in the way. They returned to Earth after 28 days having restored the station to full power and operations.

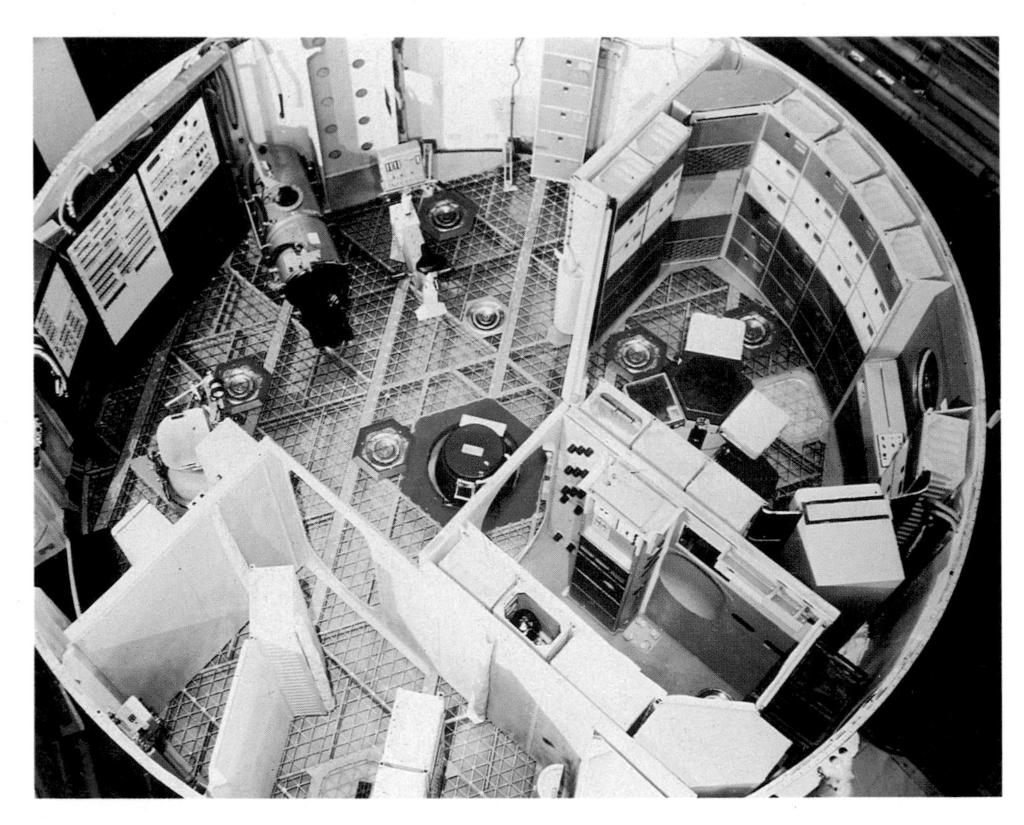

The main living section of the Skylab orbital workshop was compartmentalized like the slicing of a cake. The main eating area is seen to the right of the picture by the 46-cm (18-inch) porthole which provided remarkable views of Earth.

On 28 July, the next crew were launched to the station, destined to spend 59 days in orbit. Conrad's Apollo 12 companion Al Bean was the commander of the mission with civilian engineer Dr. Owen Garriott and Major Jack Lousma his companions. This time, the Apollo CSM was the cause for greatest concern, as it appeared at one point that its attitude control thrusters were leaking fuel. Contingency plans to convert another Apollo to bring the crew home were well in advance when it was realized that the problem wasn't as bad as had been feared. The good-natured crew performed many experiments including one which captured the world's imagination which involved two spiders called Anita and Arabella. Both learned to spin webs, but Anita died before landing, and Arabella did likewise three days after the crew landed, on 25 September 1973.

Battered, but unbowed – Skylab as it appeared to the crew of the first manned mission. After 28 days aboard, the Skylab 2 crew performed a final fly-around of the station before re-entry when this photo was taken. The orange-coloured parasol is immediately noticeable aft of the windmill-like solar panels.

The final crew for Skylab demonstrated that mankind could live in space for nearly three months, but more importantly, pointed out some of the psychological problems associated with long duration flights. Lieutenant Colonels Gerry Carr (Commander) and Bill Pogue (Pilot) were joined by Dr. Edward Gibson, a solar physicist, in the Skylab 4 CSM which was launched on 16 November 1973. After docking, Pogue was overcome with space sickness and vomited: the crew didn't report the incident, and were 'found out' when they were overheard by ground controllers discussing the incident. This led to strained relationships with Mission Control, not helped by a catalogue of complaints. The crew found the levels of noise inside the station impaired their sleep, and the lack of gravity meant that they had to shout because sound didn't travel as well. Internal tannoys squawked and were unreliable. Drinking water was filled with bubbles which led to chronic flatulence. They became bored with their diet and complained that their replacement clothes were all the same colour. Continual changes to – and increases in – their scientific workload caused them to literally go on strike after their sixth week in space. As a result, they spent an inordinate amount of time gazing at the Earth below.

After hasty re-assessment by Mission Control, the crew were sufficiently placated to perform their work more harmoniously. Gibson used the ATM to observe the Sun and on Christmas Day 1973 both he and Pogue performed a seven hour spacewalk. They retrieved film of Comet Kohoutek, a comet which promised to be bright but sadly wasn't. The crew returned back to Earth in February setting the then record of 84 days – to date, the longest time that an American crew has stayed in space.

They left behind a selection of film, unused food, clothing, instrument parts to see how they would fare in zero gravity – but to little avail. Unlike the Soviet Salyuts, Skylab did not have any onboard propulsion motors to lift it to a higher orbit. Very quickly, the expansion of the Earth's atmosphere due to increased solar activity took its toll and Skylab began to descend much more rapidly than had been predicted. By summer 1979, its demise was imminent and on 11 July it re-entered over Western Australia, scattering its remaining 77 tonnes over a large area of the outback.

It was a sad end to a project which had done so much for solar observation, Earth resources as well as biomedical and materials processing. Over 750,000 pictures of the Sun allowed solar physicists to get a clearer, more consistent view of it. Skylab also returned 46,000 photographs of the ground which were invaluable for Earth Resources studies.

SKYLAB 2

First U.S. manned orbiting space station mission. The crew deployed a solar shield and released a stuck solar panel, which allowed manned operations of the orbital workshop after meteoroid shield was damaged and torn off during lift-off. Data was obtained from 46 of the 55 experiments. The crew performed three spacewalks totalling 5 hours 41 minutes in duration. After 433 orbits of the Earth, the crew were recovered by the U.S.S. *Ticonderoga* in the Pacific.

SKYLAB 3

The second Skylab crew performed systems and operational tests, as well as deploying a new solar shield and replacing gyroscopes for the station's orientation. The crew exceeded pre-mission plans for scientific activities and performed three spacewalks totalling 13 hours 44 minutes. After 919 orbits of the Earth, they were recovered by the U.S.S. *New Orleans* in the Pacific.

SKYLAB 4

The final Skylab manned visit still holds the longest flight record for U.S. astronauts in space to date. Crew replenished coolant supplies, repaired a communications antenna and made observations of Comet Kohoutek. They performed four spacewalks totalling 22 hours 21 minutes. Set record for spacewalk duration – 7 hours 1 minute – on 25 December 1973. Recovered by the U.S.S. *New Orleans* in the Pacific after 1,298 orbits of the Earth.

SKYLAB LOG

All the manned Skylab missions were launched by Saturn 1Bs.

Mission	Crew	Date	Duration
Skylab 2	Charles Conrad Dr. Joseph Kerwin Paul Weitz	25 May–22 June 1973	672:49:49
Skylab 3	Al Bean Owen Garriott Jack Lousma	28 July–25 September 1973	1427:09:04
Skylab 4	Gerald Carr Edward Gibson Bill Pogue	16 November 1973– 8 February 1974	2017:15:32

Jack Lousma, atop the main solar telescope, performed one of many routine inspections and replacement of film on Skylab 3. The main telescope mount was adapted from an ascent stage of a Lunar Module. The results from all three manned missions re-drew the portrait of our neighbouring star.

SOYUZ

Twenty years after its introduction, the Soyuz capsule remains the mainstay of the Soviet manned space programme. Though the craft has undergone a series of improvements as technology has improved, the basic design is the same. As far as is known, the Soyuz capsule dates from the Soviet Union's plans to send men to the Moon in the early 1960s. Korolev designed a number of variants which saw service in a number of guises in the late 1960s and 1970s. The basic module could ferry as many as three men into orbit, as was shown from Soyuz 1 to 11. A slightly smaller version, without an upper orbital section, was the Zond craft which could conceivably have launched a lone pilot to the Moon. A heavier version, referred to as the Heavy Zond, would have allowed three men to journey to the Moon. Though the Russians have never admitted as much, calculations show that a lunar mission was just about possible using the Soyuz technology.

After the accident with Soyuz 11, the spacecraft was virtually redesigned, though it retained its outward appearance. The most visible difference was that the solar panels were removed, so that the crew relied on batteries for power. Such a vehicle assumed that dockings would be straightforward, but in reality they weren't. A Soyuz with solar panels was kept for non-Salyut missions, such as the ASTP project and the test of the East German camera on Soyuz 22. The Progress vehicle was introduced in 1978 as a freighter and fuel tanker. In the early 1980s, computer technology had improved so that the control mechanisms in Soyuz could be reduced to allow a third cosmonaut, aided by the introduction of modified spacesuits. Perhaps wisely, solar panels were reintroduced. Further improvements led to the introduction of Soyuz TM in 1987, specifically for docking with Mir.

THE SPACECRAFT

The basic Soyuz craft is made up of three sections: the upper Orbital Module: the middle Descent Module: and the lower Instrument Section. The Descent Module contains the couches in which the cosmonauts are strapped for launch and landing. The Descent Module is bell-shaped and is coated with a thermally protective outer layer against the searing heat of re-entry. The

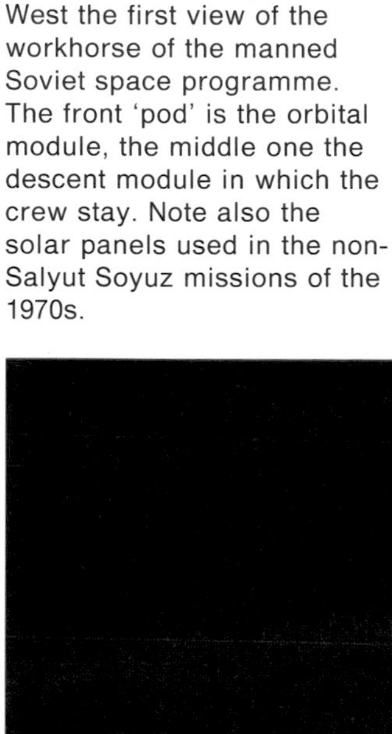

This shot of the Soyuz ferry used in the Apollo Soyuz Test Project (ASTP) gave the West the first view of the workhorse of the manned Soviet space programme. The front 'pod' is the orbital module, the middle one the descent module in which the crew stay. Note also the solar panels used in the non-Salyut Soyuz missions of the 1970s.

atmosphere in Soyuz is a mixture of nitrogen and oxygen maintained at 0.98 kg/cm^2 (14 psi). Water is carried in storage bottles as there is no fuel cell system to produce water as a by-product. There are portholes on either side of the capsule, as well as two TV cameras which observe outside the capsule.

The Orbital Module allows an extra 6 m3 (212 cu ft) of space for the crew when they reach orbit, particularly useful on the earlier Soyuz missions. Lockers within the cabin contain equipment, medical supplies and food. The module also doubles as an airlock when EVAs are required in orbit. A docking probe some 2.7 m (8 ft 9 in) long extends from the Module for use in docking with Salyut stations. At the very top of the module is the escape system, which can lift the crew clear of an emergency and is made up of three separate solid-rocket motors. In the case of the Soyuz 18 and T-10 emergencies, it saved the lives of the crew.

The lower Instrument Module contains the main electronics and orientation equipment, in addition to the communications and main motors. The main engine system is called the KTDU-35 and uses UDMH as a fuel and nitric acid as the oxidizer, carried in four spherical tanks at the rear of the section. There is a main engine, plus a back-up engine which is used to manoeuvre the craft once it is in orbit.

GRADUAL EVOLUTION

The unfortunate deaths of the Soyuz 11 crew were caused by a valve in the Descent Module being jerked open after the Orbital Module was deorbited. The crew were killed almost instantaneously, as they were not wearing spacesuits. Thereafter, spacesuits were to be worn which also saw the need for additional life support systems being added. To save weight, the third seat was removed so that Soyuz 12 to 40 could only accommodate two crew members. The solar panels were removed for the Soyuz station-ferrying role, as it would only fly for two days and could rely on batteries. Fuel margins were less and the tank's capacity was reduced.

For non-Salyut missions, the solar panel variant of Soyuz was continued. Soyuz 19 carried a docking system for ASTP, whereas Soyuz 22 had the MKF-6 camera set up at its front. The Progress transporter, tested in the unmanned Soyuz 20 flight, contained a modified orbital stage attached to the instrument section. All the life support systems were removed, and additional radar antennae were added, enabling the craft to track its quarry. As well as bringing up new supplies, the Progress could bring mail and new foodstuffs. More importantly, by docking at the rear of Salyut stations, the Progress vehicles could refuel the Salyuts with minimal hazards, in close proximity to the fuel lines.

Seen here being mounted on the launchpad at Baikonur is the launcher for the Soyuz 35 mission in April 1980. The multi-fluted appearance of the emergency escape rocket is visible at the front of the rocket. The crews of two launch failures in 1975 and 1983 were ultimately saved by this escape system.

In late 1980 the improved Soyuz T (the T stood for Transport) was introduced. Breakthroughs in electronics meant that the control systems could be miniaturized allowing room for a third cosmonaut, as well as improved digital computers which could fly Soyuz automatically. The Soyuz TM (Modified Transport) contains a new automated docking system, because the Mir space station is not orientated to face the oncoming Soyuz. The 'Kurs' approach system means that less fuel is used, and the Soyuz TM also has more powerful propulsion units and telecommunications systems.

SOYUZ LOG

Because many of the later Soyuz missions involved crews swapping craft after they had rendezvoused with Salyut space stations, it is not possible to record the number of orbits each craft made with reference to their original crews within the confines of this simplified table. The flight duration of each crew is given, though in later flights the crew members returned home at different times.

No.	Crew	Date	Duration
1	Vladimir Komarov	23 April 1967	26:48
2	Unmanned	25 October 1968	–
3	Georgi Beregovoi	26 October 1968	94:51
4	Vladimir Shatalov	14 January 1969	71:21
5	Boris Volynov Alexei Yeliseyev Yevgeni Khrunov	15 January 1969	72:54
6	Georgi Shonin Valeri Kubasov	11 October 1969	118:43
7	Anatoli Filipchenko Viktor Gorbatko Vladislav Volkov	12 October 1969	118:40
8	Vladimir Shatalov Alexei Yeliseyev	13 October 1969	118:51
9	Andrian Nikolayev Vitali Sevastyanov	1 June 1970	424:59
10	Vladimir Shatalov Alexei Yeliseyev Nikolai Rukavishnikov	22 April 1971	47:56
11	Georgi Dobrovolski Vladislav Volkov Viktor Patsayev	6 June 1971	570:22
12	Vasili Lazarev Oleg Makarov	27 September 1973	47:16
13	Pyotr Klimuk Valentin Lebedev	18 December 1973	188:56
14	Pavel Popovich Yuri Artyukhin	3 July 1974	337:30
15	Gennadi Sarafanov Lev Demin	26 August 1974	48:12
16	Anatoli Filipchenko Nikolai Rukavishnikov	2 December 1974	142:24
17	Alexei Gubarev Georgi Grechko	10 January 1975	709:20
'00'	Vasili Lazarev Oleg Makarov	5 April 1975	–
18	Pyotr Klimuk Vitali Sevastyanov	24 May 1975	1,511:20
19	Alexei Leonov Valeri Kubasov	15 July 1975	142:31
20	Unmanned	17 November 1975	–
21	Boris Volynov Vitali Zholobov	6 July 1976	1,182:23
22	Valeri Bykovski Vladimir Aksyonov	15 September 1976	189:52
23	Vyaschlesav Zudov Valeri Rozhdestvenski	14 October 1976	48:07
24	Viktor Gorbatko Yuri Glazhkov	7 February 1977	425:26
25	Vladimir Kovalyonok Valeri Ryumin	9 October 1977	48:45
26	Yuri Romanenko Georgi Grechko	10 December 1977	2,314:00
27	Oleg Makarov Vladimir Dzhanibekov	10 January 1978	142:59
28	Alexei Gubarev Vladimir Remek	2 March 1978	190:16
29	Vladimir Kovalyonok Alexander Ivanchenkov	15 June 1978	3,350:48
30	Pyotr Klimuk Miroslav Hermaszewski	27 June 1978	190:13
31	Valeri Bykovsky Sigmund Jahn	26 August 1978	188:49
32	Vladimir Lyakhov Valeri Ryumin	25 February 1979	4,200:36
33	Nikolai Rukavishnikov Georgi Ivanov	10 April 1979	47:01
T-1	Unmanned	16 December 1979	–
35	Leonid Popov Valeri Ryumin	9 April 1980	4,436:12
36	Valeri Kubasov Bertalan Farkas	26 May 1980	188:40
T-2	Yuri Malashev Vladimir Aksyonov	5 June 1980	94:19
37	Viktor Gorbatko Pham Tuan	23 July 1980	188:42
38	Yuri Romanenko Arnaldo Mendez	18 September 1980	188:43
T-3	Leonid Kizim Oleg Makarov Gennadi Strekalov	27 November 1980	307:08
T-4	Vladimir Kovalyonok Viktor Savinykh	12 March 1981	1,794:38
39	Vladimir Dzhanibekov Jugderdeidyin Gurragcha	22 March 1981	188:40
40	Leonid Popov Dimitru Prunariu	14 May 1981	188:41
T-5	Anatoli Berezovoi Valentin Lebedev	13 May 1982	5,073:05
T-6	Vladimir Dzhanibekov Alexander Ivanchenkov Jean-Loup Chretien	24 June 1982	189:51
T-7	Leonid Popov Alexander Serebov Svetlana Savitskaya	19 August 1982	189:52
T-8	Vladimir Titov Gennadi Strekalov Alexander Serebrov	20 April 1983	48:18
T-9	Vladimir Lyakhov Alexander Alexandrov	27 June 1983	3,585:46

No.	Crew	Date	Duration
T-10-1	Vladimir Titov Gennadi Strekalov	26 September 1983	–
T-10	Leonid Kizim Vladimir Solovyov Oleg Atkov	8 February 1984	5,686:50
T-11	Yuri Malashev Gennadi Strekalov Rakesh Sharma	3 April 1984	189:41
T-12	Vladimir Dzhanibekov Svetlana Savitskaya Igor Volk	17 July 1984	283:14
T-13	Vladimir Dzhanibekov Viktor Savinykh	6 June 1985	Crew returned at different times
T-14	Vladimir Vasyutin Georgi Grechko Alexander Volkov	17 September 1985	Crew returned at different times
T-15	Leonid Kizim Vladimir Solovyov	13 March 1986	3,000:01
TM-1	Unmanned	21 May 1986	–

No.	Crew	Date	Duration
TM-2	Yuri Romanenko Alexander Laveikin	5 February 1987	Crew returned at different times
TM-3	Alexander Viktorenko Alexander Alexandrov Muhammad Faris	22 July 1987	Crew returned at different times
TM-4	Vladimir Titov Musa Manarov Anatoli Levchenko	21 December 1987	Crew returned at different times
TM-5	Anatoli Solovyov Viktor Savinykh Alexander Alexandrov*	7 June 1988	Crew returned at different times
TM-6	Vladimir Lyakhov Valeri Polyakov Abdul Ahad Mohmand	29 August 1988	Crew returned at different times
TM-7	Alexander Volkov Sergei Krikalev Jean-Loup Chretien	26 November 1988	Chretien returned with Titov and Munarov; Volkov, Krikalev and Polyakov on 17 April 1989

*Alexandrov, who flew on Soyuz TM-5, was Bulgarian and no relation to the Russian who flew on TM-3.

Trying to summarize the activities of over 60 Soyuz flights is no easy task. Here the barest outline of the missions is highlighted, an attempt to simplify and make sense of the complicated crew changes that took place on later flights.

SOYUZ 1

Vladimir Komarov's test of the Soyuz capsule ended in disaster when the lines of his parachute tangled and the vehicle crashed.

SOYUZ 2 AND 3

Soyuz 2 was launched as an unmanned docking target for Beregovoi in Soyuz 3, but though they approached within a few hundred metres, they did not dock.

SOYUZ 4 AND 5

The first docking between two manned spacecraft saw Russia's first spacewalk in five years when Yeliseyev and Khrunov transferred to Soyuz 4 and returned with Shatalov back to Earth. Volynov returned home alone in Soyuz 5.

SOYUZ 6, 7 AND 8

Known as the 'Troika', these three Soyuz craft were launched on successive days, and though Soyuz 7 and 8 manoeuvred within a few hundred metres of each other, observed by Soyuz 6 a few kilometres away, the craft did not dock nor were there any crew transfers. Kubasov and Shonin performed the first welding experiments in Soyuz 6.

SOYUZ 9

The main aim of this 18-day mission was to see the effects of weightlessness on the human body. Ironically, though Nikolayev and Sevastyanov surpassed the 14-day record of Gemini 5, they suffered for many weeks afterwards, mainly the result of not being able to exercise. Sevastyanov was already known to the Soviet public as the host of a science programme on TV.

SOYUZ 10

Launched three days after Salyut 1, Soyuz 10 docked with the station but the crew did not enter it. Western observers presume that either illness or inability to open the hatch was the reason for the mission's untimely return home.

SOYUZ 11

After spending a record 24 days in space, having successfully docked with Salyut, the Soyuz 11

The Soyuz launcher rocket is shown here launching Soyuz TM-7 to Mir on 26 November 1988. On board were Alexander Volkov, Sergei Krikalev and Jean-Loup Chrétien. Just before the crew were sealed into the capsule, Nick Mason of the British rock group Pink Floyd handed them an advance version of their new album. The crew's reaction was not recorded!

Seen here making its final descent is the end of the Bulgarian mission to Mir in June 1988. Seconds before touchdown a retro-rocket fires to cushion the spacecraft on landing.

crew perished when a faulty valve led to the fatal depressurization in their descent capsule. They were found dead in their couches, and were later buried in the Kremlin Wall.

SOYUZ 12

After more than a two-year break, Soyuz 12 was launched with Lazarev and Makarov as a two-day test of the modified Soyuz. They wore spacesuits and this was the first Soyuz variant which was not equipped with solar panels.

SOYUZ 13

The second test of the modified Soyuz saw scientific experiments involving plants and astronomical research undertaken. The landing on Boxing Day was fraught because of blizzards in the landing area.

SOYUZ 14

After a night launch, the first crew to dock with Salyut 3 spent 16 days on board the station. They were heard using codewords, indicating the military nature of their flight.

SOYUZ 15

Failed to dock with Salyut 3 and had to make an emergency return home at night.

SOYUZ 16

Six-day dress rehearsal for Apollo–Soyuz Test Project (ASTP) by back-up crew for that mission.

SOYUZ 17

The crew spent nearly a month aboard Salyut 4, the first successful civilian space station after Salyut 1 four years earlier.

SOYUZ '00'

Planned as Soyuz 18, the crew who had flown the Soyuz 12 mission escaped from a booster rocket which malfunctioned at 90 miles, and remained in the capsule which parachuted to safety.

SOYUZ 18

The crew successfully docked with Salyut 4 and spent 61 days aboard the station.

SOYUZ 19

The highly successful Soviet contribution to ASTP saw Leonov and Kubasov land in Kazakhstan televised 'live' for the first time.

SOYUZ 20

Unmanned test of the prototype Progress supply craft which docked with Salyut 4 and remained for just under three months as a test of the ability of a Soyuz craft to remain in space unharmed by the extremes of heat and radiation, etc.

SOYUZ 21

Military mission that docked with Salyut 5, but evacuated the station very quickly after 48 days in space.

SOYUZ 22

A week-long flight, launched into a 65°-inclination orbit and no possibility of docking with Salyut 5. The crew tested an East German camera built by Zeiss, known as the MKF-6 which was later used on Salyut.

SOYUZ 23

The crew failed to dock with Salyut 5, and made an emergency return to Earth. The crew came down in a blizzard and crashed into an icy lake, but were rescued by frogmen from helicopters.

SOYUZ 24

The final crew to visit Salyut 5 spent a fortnight powering down the station, and its automatic return module returned film, presumably from the station's photographic surveys.

SOYUZ 25

The first crew to visit Salyut 6 failed to dock properly after four attempts, and returned home after just over two days in space. There were fears that the front docking port on the station was defective.

SOYUZ 26, 27 AND 28

Yuri Romanenko and Georgi Grechko surpassed the final Skylab crew's long duration record with their 96-day stay on Salyut 6. They docked at the rear of Salyut given the worries with Soyuz 25, which proved to be unfounded. Dzhanibekov and Makarov docked with Soyuz 27 and returned in Soyuz 26, thereby rotating the fresher craft. The first Progress vehicle docked with them as did the crew of Soyuz 28 with the first non-Soviet cosmonaut, Czechoslovakian Vladimir Remek.

SOYUZ 29, 30 AND 31

In mid-1978, Kovalyonok and Ivanchenkov extended the long duration record to 140 days on Salyut 6. They received the first Polish and East German cosmonauts, the latter on Soyuz 31, who returned in the Soyuz 29 vehicle. The long-term crew returned in Soyuz 31.

SOYUZ 32, 33 AND 34

Lyakhov and Ryumin extended the long duration record to 175 days on their own, because of Soyuz 33's failure to dock. Soyuz 33 included Bulgarian cosmonaut Georgi Ivanov – along with Rukavishnikov, they suffered re-entry forces of 15G when their craft's engine malfunctioned. Because of worries about Soyuz 32's technology, an unmanned Soyuz 34 was sent up to return them home.

SOYUZ T-1

This unmanned test of the improved 'transport' Soyuz variant successfully docked with Salyut 6 on 19 December 1979.

SOYUZ 35, 36, 37, 38 AND T-2

Leonid Popov and Valeri Ryumin extended the long-duration record to 185 days, with Ryumin flying in space for a third time, replacing the original flight member so soon after his previous marathon flight. They reached Salyut in Soyuz 35 and returned in Soyuz 37. They were visited by guest cosmonauts from Hungary, North Vietnam and Cuba from Soyuz 36, 37 and 38 respectively. Malashev and Aksyonov flew to Salyut 6 in the first manned flight of an improved Soyuz transporter, Soyuz T-2. Its automatic docking system failed and the crew were forced to make a manual docking.

SOYUZ T-3

The first three-man Soyuz crew since Soyuz 11 in 1971 was launched in the improved Soyuz capsule with improved spacesuits.

SOYUZ T-4, 39 AND 40

The final crews to visit Salyut 6 ensured that it was powered down before re-entry, with visits by a Mongolian cosmonaut (Soyuz 39) and Rumanian (Soyuz 40). On 25 April 1981, Cosmos 1267, the mysterious Star Module, docked with Salyut 6 and boosted it to a higher orbit before the station re-entered in July 1982.

SOYUZ T-5, T-6 AND T-7

Anatoli Berezovoi and Alexander Serebrov extended the long duration record to 211 days, after launch in Soyuz T-5 and return in T-7. They were visited by two crews that included the first French *spationaute*, Jean-Loup Chrétien, and the

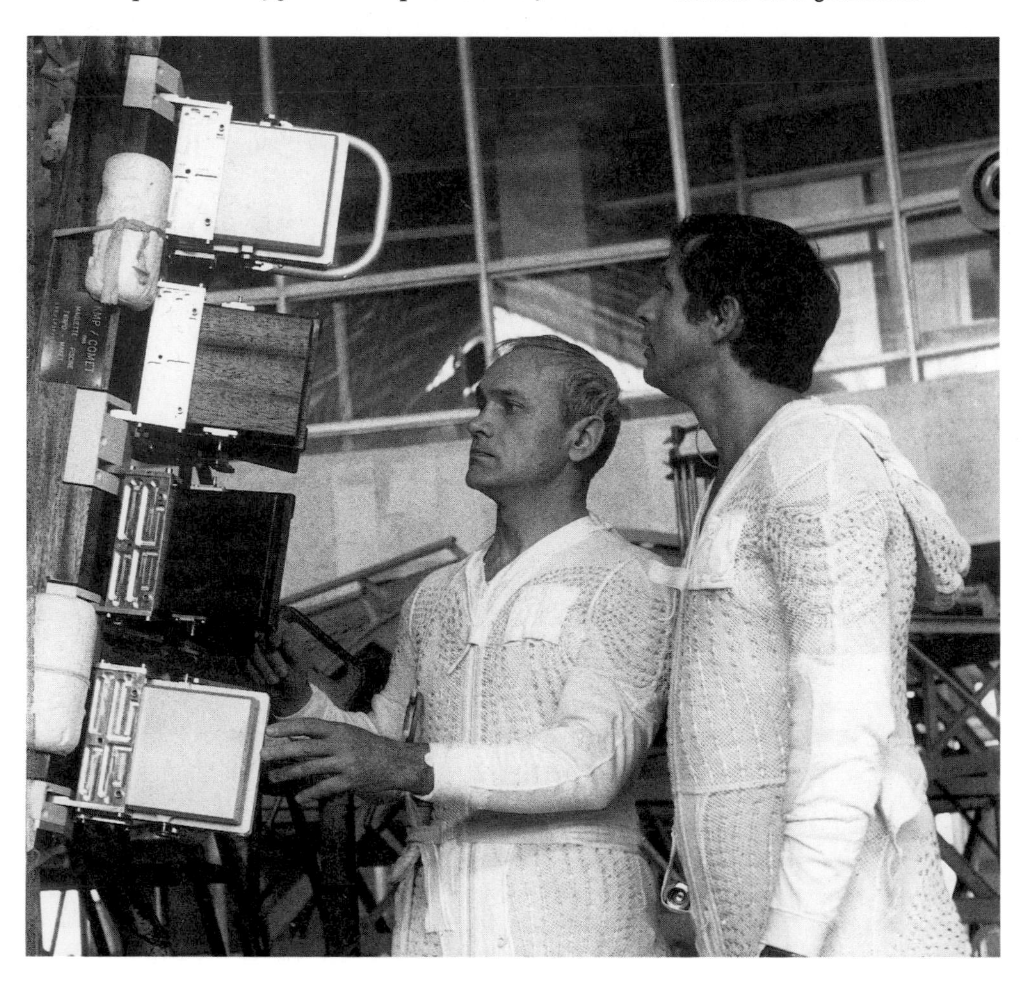

Preparing for their dramatic rescue of Salyut 7 are the crew of Soyuz T-13. Vladimir Dzhanibekov (left) and Viktor Savinykh (right) are seen in their 'thermals', spacesuit undergarments whose temperature is accurately controlled. This was the fifth mission for Dzhanibekov, a former fighter pilot and something of an Anglophile. Savinykh had originally trained as a geodesist.

second Soviet female cosmonaut, Svetlana Savitskaya, on Soyuz T-6 and T-7 respectively.

SOYUZ T-8

The crew failed to dock when the Soyuz radar ranging system failed to deploy properly, and despite attempting a manual docking, returned home when too much fuel had been used up.

SOYUZ T-9

Vladimir Lyakhov and Alexander Alexandrov spent 149 days aboard Salyut 7 and received no visitors due to failure of Soyuz T-10. Performed EVAs to construct additional solar panels for the station. They returned home after 149 days.

SOYUZ T-10-1

Vladimir Titov and Gennadi Strekalov survived the first launch pad explosion when a valve jammed in one of the engines of their SL-4 booster. The crew escaped unharmed by the emergency escape rocket.

SOYUZ T-10, T-11 AND T-12

Leonid Kizim, Vladimir Solovyov and Oleg Atkov – a medical doctor – extended the long duration record to 237 days. They received crews from Soyuz T-11 and T-12 which included the first Indian cosmonaut and Svetlana Savitskaya on her second mission. She entered the history books as the first woman to perform an EVA.

SOYUZ T-13 AND T-14

After rescuing the stricken Salyut 7 station, Dzhanibekov and Savinykh performed the first crew rotation with Soyuz T-14. Dzhanibekov returned with Grechko in Soyuz T-13 on 26 September, leaving Savinykh in the company of Vasyutin and Alexander Volkov. They returned on 21 November because of Vasyutin's illness. On 27 September, Cosmos 1686 docked with Salyut 7 where it remains today.

SOYUZ T-15

Leonid Kizim and Vladimir Solovyov entered the history books as the first cosmonauts to ferry between two space stations, spending a total of 125 days in orbit. After docking with the new Mir space station, they undocked on 5 May and docked with Salyut 7 a day later. On 25 June, they left Salyut 7, re-docked with Mir and then returned home on 16 July. To date, they were the last crews to visit Salyut.

SOYUZ TM-1

The first test of the modified transport Soyuz took place in May 1986, when it automatically docked with Mir on 23 May.

SOYUZ TM-2, TM-3 AND TM-4

Yuri Romanenko and Alexander Laveikin were launched to Mir as the first permanent crew. At the end of July, they were visited by Soyuz TM-3 which included the first Syrian cosmonaut, Muhammad Faris. Because of heart irregularities, Laveikin returned with Viktorenko and Faris in Soyuz TM-2 on 30 July 1987. On 21 December, Soyuz TM-4 docked with Mir bringing up a new long-duration crew, Manarov and Titov: Romanenko, Alexandrov and Levchenko returned on 29 December in the Soyuz TM-3 capsule. Romanenko had spent 326 days aloft, and entered the history books. Levchenko died in August 1988 due to a brain tumour.

SOYUZ TM-5, TM-6 AND TM-7

Whilst Vladimir Titov and Musa Munarov spent a whole year aboard Mir, they were at various times visited by different crews. In June, Bulgaria finally had a cosmonaut aboard a Soviet space station – Alexander Alexandrov. Along with Savinykh and Solovyov, they spent eight days aboard Mir before returning in the TM-4 craft. Afghanistan's first cosmonaut, Mohmand, spent eight days on Mir with Lyakhov and Polyakov – the latter remaining aboard. Lyakhov and Muhmand returned in TM-5 on 7 September, after a delayed landing. On 26 November, Jean-Loup Chrétien made his second journey into space, performing an EVA on 9 December. On 21 December he returned with Titov and Munarov after their record-breaking stay in TM-6. Polyakov, Volkov and Krikalev remained aboard Mir, returning in April with Polyakov.

SPACE SHUTTLE

Any discussion of the U.S. Space Shuttle will almost inevitably focus on the *Challenger* accident and its aftermath. However, the Shuttle has scored some remarkable successes in the face of adversity. The successful return to flight in September 1988 means that the Space Transportation System is back on course for the 1990s in a new era of sober reality in which the limitations of operating a highly complex vehicle have been realized.

The Space Shuttle is made up of three main elements: the Orbiter in which the crew and cargo are carried; the External Tank (ET) in which all the liquid propellants for the Orbiter's main engines (known as the SSMEs) are carried; and two Solid Rocket Boosters (SRB) which provide extra thrust for the whole assembly – christened 'The Stack' by the astronauts – to reach orbit. As a result of the *Challenger* accident, more than a hundred modifications have been made to the Shuttle elements, the most obvious to the field joints of the SRBs which directly led to *Challenger*'s destruction. Others include improving the brakes on the Orbiter's undercarriage, improving

Perhaps the most visible aspect of the post-*Challenger* improvements to the Space Shuttle Orbiter has been the inclusion of an emergency escape system. In the event of an abort, the pole seen here would function similarly to that used by firemen. One of the crew would act as a jump master.

the heat shield and the inclusion of a crew escape system. A pyrotechnically-jettisoned side hatch and a curved, telescoping pole would allow the crew to bale out with survival gear and parachutes in a *Challenger*-like accident. Because they now wear partial pressure suits, the crew could survive such a situation, and would bail-out at the earliest sign of difficulties. An egress slide has been added to make sure that the crew could get out quickly if there was an emergency landing.

A greater emphasis on safety and review procedures before launch has also been instigated. This includes the appointment of astronauts in key management roles as well as an improvement in communications between each of the various NASA centres.

THE ORBITER

The Space Shuttle Orbiter is roughly the same size and weight as a DC-9 airliner and most of its structural elements are made of aluminium. The forward part of the fuselage contains the crew module which is pressurized and looks like the cockpit of an airliner. The flight deck of the Shuttle contains the controls and displays which the pilots use to fly the vehicle. The commander and the pilot fly in the usual pilot and co-pilot arrangement (i.e. left-hand and right-hand seats respectively) with duplicate controls to allow either one to take control. In an emergency, the vehicle could be returned to Earth by one pilot. Seating for two passengers is provided directly behind them, but there are further seats below and behind on the lower flight deck. In this way a maximum of eight crew members can be accommodated.

Beneath the lower deck is the environmental control equipment which can easily be inspected by removing the floor panels. It maintains the nitrogen/oxygen atmosphere inside the crew cabin at sea level pressures around 1.03 kg/cm^2 (14.7 psi), compared to the 0.35 kg/cm^2 (5 psi) on earlier American spacecraft. Temperature can be regulated so that the crew can work in shirtsleeves in orbit. As a result of the *Challenger* accident, partial pressure suits have been re-introduced for lift-off and re-entry, though the bulky pressure suits worn on the first flights are no longer used.

The Shuttle's remote manipulator arm was developed by Canada, and was extensively tested before its first use on STS-7. Technicians are seen here experimenting with the arm's joints to enable fully autonomous operations in space.

On the lower flight deck are the kitchen, toilet and bedroom for the crew. As no refrigerator is carried, most of the foods are dehydrated and are brought to life by a hot or cold water dispenser. There is a small oven which is used to heat the main meals. Toilet facilities are positively luxurious compared to those on earlier missions. The euphemistically-termed 'latrine' is located in a corner of the lower flight deck and a small curtain rail allows privacy. A small cup-like device is used for urine extraction, while a bowl is used for solid wastes which are removed by a strong airflow. In the first flights there was some horror that the airflow blew in the wrong direction; and later the urine outlet on the fuselage became blocked when ice crystals formed. Both these faults have now been rectified.

Sleeping facilities consist of a number of bunks in which sleeping bags are located. Normally, four crew members are allowed to sleep while others work, though activity around them has been the cause of lack of sleep, and the sensation of weightlessness has not been without problems.

AIRLOCK AND EVA FACILITIES

At the rear of the crew cabin is the airlock, through which spacesuited astronauts pass to enter the vacuum of space. It is essentially a cylindrical-shaped pressure vessel which will allow two fully-suited astronauts access to space. Two spacesuits are hung on the walls of the airlock and are much more flexible than earlier spacesuits, essentially a self-contained environment in which the astronauts can be provided with water, oxygen, and warmth for over seven hours. In official NASA parlance, the Shuttle suit is called the 'Extravehicular Mobility Unit' (EMU). The Manned Maneuvering Unit (MMU) which allows Shuttle astronauts to become human satellites is usually stored outside the airlock in the crew cabin. The device can be attached to the normal spacesuit backpack, and has two forward-facing 'arms' on which the controls are located. The MMU has 24 thruster jets which fire nitrogen to manoeuvre, plus a number of power and attachment points which enable it to be used as a work station. Pairs of MMUs were carried on earlier Shuttle flights to allow the rescue of stranded satellites.

PAYLOAD BAY

The main section of the Orbiter fuselage is termed the 'Payload Bay' in which the mission payloads are carried, such as satellites to be launched, or else the Spacelab modules. Satellites are cradled inside the bay by special frameworks tailored for each satellite, and some are contained in protective covers before they are deployed. The payload bay is truly cavernous, some 18.3m (60 ft) in length. The doors are like clam shells, and have to be closed completely for re-entry. Their mechanism is worked electronically by one crew member, but in an emergency, an EVA (Extravehicular Activity) would have to be made to manually crank the doors shut.

One of the most versatile features of the Shuttle as a whole is its 'robot arm', known officially as the Remote Manipulator System. The arm is controlled from the lower flight deck, and allows crews to deploy and/or capture satellites and move equipment around the payload bay. The arm is located on the left-hand side of the payload bay and stretches some 15.3 m (50 ft) in length. The end of the arm consists of a mechanical 'wrist' to which TV cameras are attached to avoid collision with delicate mechanisms.

THERMAL PROTECTION SYSTEM

The Orbiter's external fuselage is covered with the revolutionary heat shield consisting of a variety of thermal tiles which proved troublesome to develop. Part of the problem is that the tiles are very fragile and can break up at the slightest touch. As a result, thousands of them had to be densified before the first flight of *Columbia* to keep them in place. As well as protecting the Shuttle from the heat of re-entry, they are used for protection against solar radiation in orbit. Coated silica tiles and extremely pure silica fibre insulation sheets cover the top and sides of the Orbiter. The tiles offer protection up to 650°C (1,202°F) and the flexible insulation up to 370°C (698°F). On the underside of the Orbiter and the leading edges of the wing – the areas which are heated the most – a high temperature reusable black reflective coating is applied to the tiles as protection up to 1,260°C (2,300°F). A reinforced material made of carbon-carbon is used on the nose cap and the leading edges of the wing where temperatures exceed 1,260°C (2,300°F) on re-entry.

MAIN ENGINES

At the rear of the Shuttle Orbiter are the Space Shuttle Main Engines (SSMEs), the most advanced liquid-fuelled engines ever built. Liquid oxygen and hydrogen enter through a complicated set of valves and turbopumps. Three of the engines are mounted in a triangular pattern and can be pointed directionally (gimballed) to steer the Shuttle as it heads towards orbit. To ensure that the combustion is very efficient, high speed turbopumps are used, which proved difficult to operate. Thrusting supercold propellants into superheated combustion chambers with pressures greater than 200 times that of the Earth's atmosphere caused many headaches. Their total thrust rating is 640,000 kg (1.4 million lb) but this can be altered by throttling the engines, sometimes to 109 per cent of their rated performance. The Shuttle main engines are the first to employ a built-in electronic digital controller which acts on commands from the main computers. In the event of failure, the automatic controller tries to overcome problems or shuts down the engines – as in the Spacelab 3 'Abort to Orbit' incident. At that time, the Orbital Maneuvering System (OMS) was used to augment the Shuttle's main engines. The OMS consists of two smaller engines which are housed in pods either side of the main tail fin and are mainly used to alter the Shuttle's position in orbit.

The main engines have the highest thrust per weight of any rocket engines developed, and though they only operate for at most nine minutes on each flight, they are designed to be re-used on as many as 50 missions. Problems with operating the main engines meant that they had to be heavily serviced and often replaced before the next flight. The electronic engine controller, valve actuators,

THERMAL PROTECTION SYSTEM

REINFORCED CARBON-CARBON (RCC)
HIGH-TEMPERATURE, REUSABLE (HRSI) SURFACE INSULATION
LOW-TEMPERATURE, REUSABLE (LRSI) SURFACE INSULATION
COATED NOMEX FELT (FRSI) REUSABLE SURFACE INSULATION
METAL OR GLASS

COLORING
HRSI – BLACK
LRSI – OFF WHITE
FRSI – WHITE
RCC – LIGHT GRAY

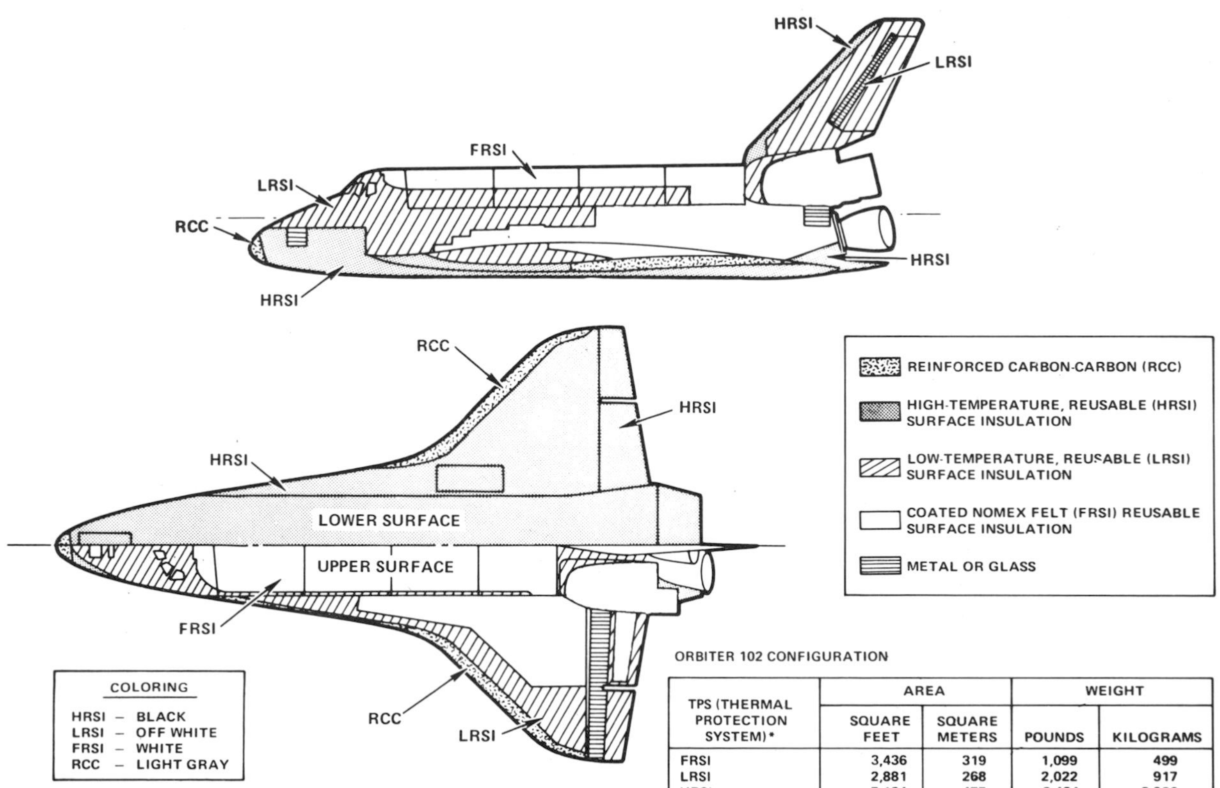

ORBITER 102 CONFIGURATION

TPS (THERMAL PROTECTION SYSTEM)*	AREA		WEIGHT	
	SQUARE FEET	SQUARE METERS	POUNDS	KILOGRAMS
FRSI	3,436	319	1,099	499
LRSI	2,881	268	2,022	917
HRSI	5,134	477	8,434	3,826
RCC	409	38	3,023	1,371
MISCELLANEOUS			1,394	632
TOTAL	11,860	1,102	15,972	7,245

*INCLUDES BULK INSULATION, THERMAL BARRIERS, AND CLOSEOUTS

26

The main elements of the Shuttle's all-important heat shield are shown in this schematic diagram. The most excessive heating reaches the nose-cone, reinforced by carbon-carbon to withstand the friction. With each successive orbiter, more light weight materials were used.

SHUTTLE LOG

In this list 'Cd' stands for Commander, 'Pl' for Pilot – these two astronauts are responsible for flying the Shuttle vehicle. NASA classifies its Shuttle passengers as either Mission Specialists (MS) or Payload Specialists (PS). Mission specialists are full members of the astronaut corps: their job is to monitor the behaviour of equipment aboard the Shuttle and are called upon to make spacewalks. Payload specialists are responsible for particular experiments and though trained by NASA, are not members of the astronaut corps.

No.		Crew	Launch	Duration	Orbits
STS-1	Cd:	John Young	12 April 1981	54:21	36
Columbia	Pl:	Bob Crippen			
STS-2	Cd:	Joe Engle	12 November 1981	54:14	36
Columbia	Pl:	Dick Truly			
STS-3	Cd:	Jack Lousma	22 March 1982	192:05	128
Columbia	Pl:	Gordon Fullerton			
STS-4	Cd:	Ken Mattingly	27 June 1982	169:10	112
Columbia	Pl:	Hank Hartsfield			
STS-5	Cd:	Vance Brand	11 November 1982	122:14	82
Columbia	Pl:	Bob Overmyer			
	MS:	Joe Allen			
	MS:	Bill Lenoir			
STS-6	Cd:	Paul Weitz	4 April 1983	120:24	80
Challenger	Pl:	Karol Bobko			
	MS:	Don Peterson			
	MS:	Story Musgrave			
STS-7	Cd:	Bob Crippen	18 June 1983	146:24	98
Challenger	Pl:	Rick Hauck			
	MS:	John Fabian			
	MS:	Sally Ride			
	MS:	Norman Thagard			
STS-8	Cd:	Dick Truly	30 August 1983	145:09	97
Challenger	Pl:	Dan Brandenstein			
	MS:	Guion Bluford			
	MS:	Dale Gardner			
	MS:	Bill Thornton			
STS-9	Cd:	John Young	28 November 1983	247:47	165
Challenger	Pl:	Brewster Shaw			
	MS:	Owen Garriott			
	MS:	Bob Parker			
	PS:	Byron Lichtenberg			
	PS:	Ulf Merbold			
41-B	Cd:	Vance Brand	3 February 1984	191:16	127
Challenger	Pl:	Robert Gibson			
	MS:	Bruce McCandless			
	MS:	Ron McNair			
	MS:	Bob Stewart			
41-C	Cd:	Bob Crippen	6 April 1984	167:40	112
Challenger	Pl:	Dick Scobee			
	MS:	Terry Hart			
	MS:	George Nelson			
	MS:	James van Hoften			

temperature sensors and main combustion chambers have been improved on the SSMEs as a result of the *Challenger* accident.

EXTERNAL TANK

Propellants for the main engines are carried in the large External Tank (ET), the largest component of the Shuttle system. Inside each tank is a large hydrogen tank and a smaller oxygen tank, with the oxygen tank above the other. Because hydrogen is much less dense, the hydrogen tank is about three times as big. The whole ET is 47 m (154 ft) long and 8.4 m (27 ft 6 in) across. Much of its outer surface is thermally protected to reduce ice or frost formation on the tank during preparations for launch and also to minimize heat leaks. The External Tank is the only part of the system which is not retrieved after launch. Just before the Shuttle reaches orbital velocity, the main engines are cut off. About ten seconds later, a valve at the top of the ET is opened which causes excess liquid oxygen to vent and push the tank away from the Orbiter. It starts to tumble, ensuring that it will break up on re-entry and fall within a specific area of the Indian Ocean.

SOLID ROCKET BOOSTERS

Two Solid Rocket Boosters (SRBs) are used for each Shuttle launch and provide sufficient thrust for the whole vehicle to reach an altitude of about 45 km (28 miles) above the Earth. Each booster contains a solid rocket motor as well as systems for their separation and recovery and electronic instrumentation. The overall length of each SRB is 48.5 m (159 ft 2 in) and the diameter is 3.69 m (12 ft 1 in). At the heart of each booster is the motor, the largest solid rocket motor ever flown and the first designed for re-use. It is made up of 11 segments joined together to make four loading segments which are filled with propellant at the manufacturers' site.

These four steel cases are connected by a field joint made up of a tang and clavis with sealing now provided by three rubber O-rings and asbestos-filled putty that fills the space between them. At ignition, the force of the superhot gas causes the putty to expand and push the O-rings into place to stop the gas from escaping. If the first fails to deploy, then the second and third ensure that the seal holds true. The third O-ring was added after the *Challenger* accident, as well as other amendments. The SRB field joints have also been amended with an extended metal capture feature as well as changes to the insulation surrounding the joint. External heaters with waterproof weather seals now ensure the field joints never fall below 24°C (75°F), as the freezing launch pad conditions on 28 January 1986 significantly degraded the field joints on *Challenger*.

The SRBs burn for a nominal 2 minutes and 4 seconds, after which they are jettisoned by use of pyrotechnic (explosive) devices and eight separation motors. In the nose cap and front sections of each booster is the 'recovery subsystem' which consists of parachutes and location aids for help in the search and retrieval. Three main parachutes are deployed by drogue parachutes to ensure they are not damaged on landing in the Atlantic. On the first Shuttle missions, it was found that the boosters were being damaged and some actually sank, so the size of the parachutes was increased. This now means that the SRBs hit the water at speeds of less than 100 km/h (60 mph), nozzle end first: air inside the boosters is effectively trapped and ensures that the boosters float and can be retrieved by recovery ships in the Atlantic. The motor segments are then returned to the manufacturers where they are cleaned and refurbished. The parachutes are also repaired and used on later missions.

STS-1: *COLUMBIA*

The first flight of *Columbia* was an unqualified success from start to finish. It saw a number of firsts: first landing of an American manned spacecraft on land and the first use of solid rocket boosters in manned flight. *Columbia* landed at the Edwards Air Force Base in California after a total mission of two days and six hours.

STS-2: *COLUMBIA*

The second flight of *Columbia* was the first re-use of a spacecraft. Originally planned for five days, it was cut back to two days when one of the three fuel cells failed. That said, NASA announced 90 per cent of the mission objectives had been met, including the first test of the remote manipulator arm.

STS-3: *COLUMBIA*

The third test of *Columbia* lasted for eight days, a day longer than planned. The dry lakebed at Edwards Air Force Base was waterlogged because of spring rainstorms, so the alternative site at White Sands, New Mexico, had to be used. Only 40 minutes before they were to re-enter, they were 'waived-off' for an extra 24 hours because of high winds at New Mexico.

STS-4: *COLUMBIA*

The final test flight of the Space Shuttle also saw the first classified military payload carried into orbit. The two Solid Rocket Boosters were lost when they sank to the bottom of the Atlantic. After a week in space, the crew landed at the Edwards AFB in California where they were met by President and Mrs Reagan.

STS-5: *COLUMBIA*

A crew of four were carried on this first 'operational' flight of the Shuttle. The Canadian Anik C3 and SBS-3 became the first communications

No.	Crew		Launch	Duration	Orbits
41-D *Discovery*	Cd:	Hank Hartsfield	30 August 1984	144:57	97
	Pl:	Michael Coats			
	MS:	Steve Hawley			
	MS:	Richard Mullane			
	MS:	Judy Resnik			
	PS:	Charles Walker			
41-G *Challenger*	Cd:	Bob Crippen	5 October 1984	197:24	132
	Pl:	Jon McBride			
	MS:	Dave Leestma			
	MS:	Sally Ride			
	MS:	Kathy Sullivan			
	PS:	Marc Garneau			
	PS:	Paul Scully-Power			
51-A *Discovery*	Cd:	Rick Hauck	8 November 1984	191:45	128
	Pl:	David Walker			
	MS:	Anna Fisher			
	MS:	Dale Gardner			
	MS:	Joe Allen			
51-C *Discovery*	Cd:	Ken Mattingly	24 January 1985	73:33	49
	Pl:	Loren Shriver			
	MS:	James Buchli			
	MS:	Ellison Onizuka			
	PS:	Gary Payton			
51-D *Discovery*	Cd:	Karol Bobko	12 April 1985	167:55	112
	Pl:	Don Williams			
	MS:	David Griggs			
	MS:	Jeffrey Hoffman			
	MS:	Rhea Seddon			
	PS:	Jake Garn			
	PS:	Charles Walker			
51-B/ Spacelab 3 *Challenger*	Cd:	Bob Overmyer	29 April 1985	168:09	112
	Pl:	Fred Gregory			
	MS:	Don Lind			
	MS:	Norman Thagard			
	MS:	Bill Thornton			
	PS:	Lodewijk van den Berg			
	PS:	Taylor Wang			
51-G *Discovery*	Cd:	Dan Brandenstein	17 June 1985	169:40	113
	Pl:	John Creighton			
	MS:	John Fabian			
	MS:	Shannon Lucid			
	MS:	Steven Nagel			
	PS:	Sultan Al-Saud			
	PS:	Patrick Baudry			
51-F/ Spacelab 2 *Challenger*	Cd:	Gordon Fullerton	29 July 1985	190:46	127
	Pl:	Roy Bridges			
	MS:	Tony England			
	MS:	Karl Henize			
	MS:	Story Musgrave			
	PS:	Loren Acton			
	PS:	John-David Bartoe			
51-I *Discovery*	Cd:	Joe Engle	27 August 1985	170:18	114
	Pl:	Dick Covey			
	MS:	Bill Fisher			
	MS:	John Lounge			
	MS:	James van Hoften			

SHUTTLE LOG (continued)

No.	Crew		Launch	Duration	Orbits
51-J *Atlantis*	Cd:	Karol Bobko	3 October 1985	97:45	65
	Pl:	Ronald Grabe			
	MS:	Dave Hilmers			
	MS:	Bob Stewart			
	PS:	William Pailes			
61-A/ Spacelab D1 *Challenger*	Cd:	Hank Hartsfield	30 October 1985	168:44	112
	Pl:	Steve Nagel			
	MS:	Guion Bluford			
	MS:	Jim Buchli			
	MS:	Bonnie Dunbar			
	PS:	Reinhold Furrer			
	PS:	Ernst Messerschmid			
	PS:	Wubbo Ockels			
61-B *Atlantis*	Cd:	Brewster Shaw	26 November 1985	165:05	110
	Pl:	Bryan O'Connor			
	MS:	Mary Cleave			
	MS:	Sherwood Spring			
	MS:	Jerry Ross			
	PS:	Rudolfo Neri Vela			
	PS:	Charles Walker			
61-C *Columbia*	Cd:	Robert Gibson	12 January 1986	143:04	95
	Pl:	Charles Bolden			
	MS:	Franklin Chang-Diaz			
	MS:	Steve Hawley			
	MS:	George Nelson			
	PS:	Robert Cenker			
	PS:	Bill Nelson			
51-L *Challenger*	Cd:	Dick Scobee	28 January 1986		
	Pl:	Mike Smith			
	MS:	Judy Resnik			
	MS:	Ellison Onizuka			
	MS:	Ron McNair			
	PS:	Greg Jarvis			
	PS:	Christa McAuliffe			
STS-26 *Discovery*	Cd:	Rick Hauck	29 September 1988	97:00	64
	Pl:	Dick Covey			
	MS:	John Lounge			
	MS:	Dave Hilmers			
	MS:	George Nelson			
STS-27 *Atlantis*	Cd:	Robert Nelson	2 December 1988	105:6	68
	Pl:	Richard Mullane			
	MS:	Mike Mullane			
	MS:	Bill Shepherd			
	MS:	Jerry Ross			
STS-28 *Columbia*	Cd:	Brewster Shaw, Jr.	8 August 1989	121:00	
	Pl:	Richard Richards			
	MS:	David Leestma			
	MS:	James Adamson			
	MS:	Mark Brown			
STS-29 *Discovery*	Cd:	Michael Coats	13 March 1989	119:39	79
	Pl:	John Blaha			
	MS:	James Buchli			
	MS:	Robert Springer			
	MS:	James Bagian			
STS-30 *Atlantis*	Cd:	David Walker	4 May 1989	96:57	64
	Pl:	Ronald Grabe			
	MS:	Norman Thagard			
	MS:	Mary Cleave			
	MS:	Mark Lee			

satellites to be launched from the Shuttle. The first Shuttle EVA was cancelled because of problems with the spacesuit.

STS-6: *CHALLENGER*

The first flight of *Challenger* saw the first Shuttle EVAs by mission specialists Don Peterson and Story Musgrave. However, the deployment of the first Tracking & Data Relay Satellite, a vital part of NASA's communications network, failed to reach its correct orbit when its Inertial Upper Stage 'kick' stage failed to fire properly.

STS-7: *CHALLENGER*

The first crew of five to be carried aboard the Shuttle numbered Dr. Sally Ride, America's first woman in space. Another Canadian Anik-3 satellite and Indonesia's Palipa B were successfully launched into geostationary orbits. A German-built test pallet called SPAS-01 was released and retrieved with the robot arm.

STS-8: *CHALLENGER*

This flight saw the first night launch and landing of a Shuttle mission. Guion Bluford became the first American black to fly in space. The Indian 1B communications satellite was successfully deployed. The robot arm was used to manipulate a dummy payload called the Payload Flight Test Article and intended as practice for later satellite retrievals.

STS-9: *COLUMBIA*

The first crew of six included the European Space Agency payload specialist Dr. Ulf Merbold. *Columbia* carried the first European Spacelab into orbit in which 71 experiments were carried out, which included medical, astronomy and remote sensing studies. Re-entry after ten days was not without incident: two of the main computers failed just before re-entry procedures were undertaken, though luckily they were brought back on-line before the descent began. As *Columbia* headed back through the Earth's atmosphere, a hydrazine leak in one of the Auxiliary Power Units of the main engines caused a small fire, which thankfully extinguished itself minutes before landing.

41-B: *CHALLENGER*

The release of two communications satellites (Westar 6 and Indonesia's Palapa 2) went disastrously wrong when their boosters failed to fire correctly. (They were later rescued on mission 51A.) However, the highlight of the mission was the testing of the Manned Maneuvering Unit (MMU). The crew landed at the Kennedy Space Center for the first time.

41-C: *CHALLENGER*

The primary aim of this mission was to rescue the

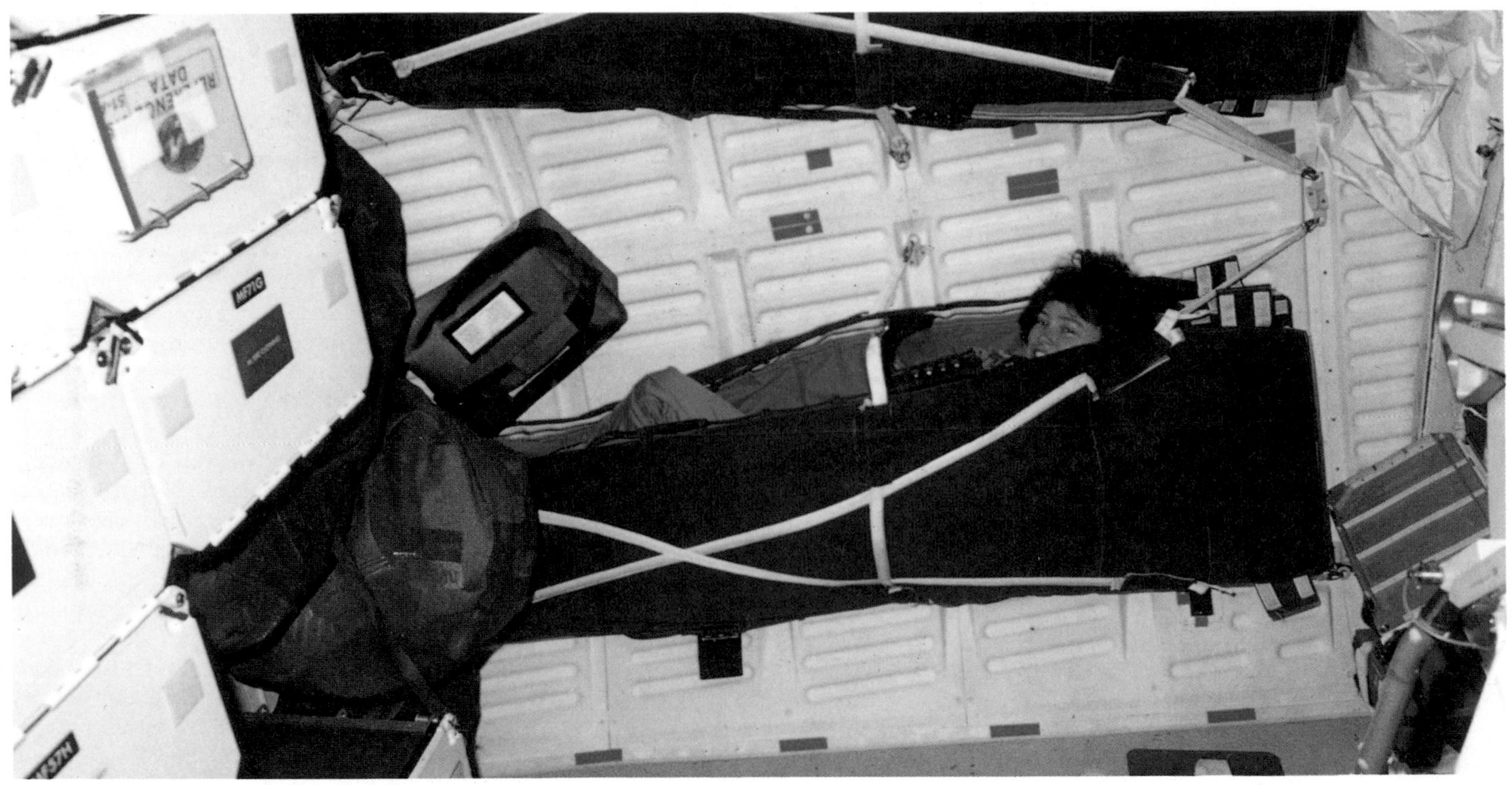

Dr. Anna Fisher is seen in one of the versatile sleeping hammocks in the lower cabin of the Space Shuttle. Normally crews work eight-hour shifts 'on' and 'off', so that there are always crew members awake. Her husband, Bill, also a medical doctor, flew on a later mission.

Solar Maximum Satellite. Using MMUs, Nelson and van Hoften captured the satellite at the second attempt and repaired its faulty electronics in the payload bay. The Long Duration Exposure Facility was also released on this flight, designed to monitor the effects of exposure of materials to weightlessness. Because of bad weather conditions, the crew had to return to the Edwards Air Force Base in California.

41-D: *DISCOVERY*

At the third attempt, *Discovery* was launched on its first flight on 28 August – the second attempt had seen the main engines shut down four seconds before SRB ignition. Three communications satellites (SBS-4, Syncom 4-1 and Telstar 3) were successfully deployed. McDonnell Douglas engineer Charles Walker was able to undertake his electrophoresis experiment for drugs manufacture despite problems with the equipment. After spending six days in space, *Discovery* landed in California.

41-G: *CHALLENGER*

On his fourth Shuttle flight, Bob Crippen finally made his first landing at the Kennedy Space Center after a total of eight days in space. This was the first seven person crew, with two female crew members: Sally Ride was making her second flight, and Kathy Sullivan also made an EVA. The other crew members were Jon McBride and Dave Leestma, and the two payload specialists were Canadian Marc Garneau and Australian-born Paul Scully-Power, who became the first oceanographer in space. After problems with the solar panels, the crew successfully deployed the Earth Radiation Budget Satellite.

51-A: *DISCOVERY*

The ability of Shuttle crews to rescue stranded satellites was demonstrated in this mission, when Dale Gardner and Joe Allen retrieved the Palapa and Westar satellites which had been lost on mission 41-B. Gardner and Allen literally 'man-handled' Palapa into the payload bay and, after a day's rest, they retrieved Westar 6. *Discovery* returned to California after eight days.

51-C: *DISCOVERY*

The first operational Department of Defense mission was shrouded in secrecy, despite NASA's protestations. Even the launch time was not announced until the T–9 minute mark. The crew only spent three days in space, during which time they deployed a communications spy satellite designed to 'listen in' on Soviet military communications channels, before landing at the Edwards Air Force Base.

51-D: *DISCOVERY*

Payload specialist Charlie Walker continued his electropheresis experiments on this flight, which also saw Senator Jake Garn take his place as an observer. The first day saw the successful deployment of another Canadian Anik communications satellite. But when the crew launched the Syncom 4-3 satellite the next day its radio antenna failed to deploy. Despite attempts to revive it, it was clear

Above: Transporting the separate elements of each Solid Rocket Booster is an extremely involved process. The segment shown here weighs 136,080 kg (300,000 lb) and is taken by rail from Utah to Florida. Even the slightest dent would be potentially catastrophic to the integrity of the booster when fully assembled.

Opposite: The most immediate result from the *Challenger* accident was the redesign of the 'O' rings which seal the four separate elements of the SRB. The nozzle of the motors was also redesigned (shown here) to minimize the possibility of catastrophe.

there was a major fault and the crew reluctantly abandoned the satellite. They landed at the Kennedy Space Center where crosswinds resulted in a tyre blowing out due to the harder braking that was required.

51-B: *CHALLENGER*

Because of equipment delays, Spacelab 3 was flown before Spacelab 2, which involved flying the pressurized Spacelab module mainly occupied by experiments investigating biological and materials processing. Lodewijk van den Berg of EG & G Energy Systems and Taylor Wang of the Jet Propulsion Laboratory devoted their time to their crystal growth experiments which were highly successful. After six days in space, the crew landed safely in California.

51-G: *DISCOVERY*

The international flavour of this crew attracted most attention. Prince Sultan Salman Al-Saud was a TV advertising executive who observed the Arabsat satellite being released. A Mexican Morelos and American Telstar were also deployed. French *spationaute* Patrick Baudry performed a number of blood flow and posture experiments. A low-power laser from Hawaii was used to track *Discovery* in connection with SDI or 'Star Wars' research. After just over seven days in space, *Discovery* landed in California.

51-F: *CHALLENGER*

The launch of the Spacelab 2 mission was not without incident at the second attempt: one of the main engines overheated, so the 'Abort to Orbit' command was given, which meant that the fuel in the main engines was channeled to the remaining two, and the Orbital Maneuvering Engines were also fired in support. *Challenger* reached a lower orbit than planned, but the seven-man crew were happy with their scientific experiments located on external pallets in the payload bay. They operated a variety of astrophysical instruments including the Instrument Pointing System which consisted of four telescopes for observing the Sun. *Challenger* landed in California after nine days.

51-I: *DISCOVERY*

The main purpose of this mission was to release three communications satellites, including Aussat-1, Australia's first ever comsat. The crew also retrieved and re-deployed the Syncom satellite that had failed to deploy on the 51-D mission the previous April. After matching the orbit of Syncom on their fifth day in space, van Hoften and Fisher attached a handling bar to the satellite and literally grabbed it into the payload bay for repair. They successfully released it the next day. *Discovery* landed in California after a seven-day mission.

51-J: *ATLANTIS*

The first flight of *Atlantis* was a military mission under the auspices of the Department of Defense. Two military communications satellites were launched to form part of a communications relay network around the globe. Few details were released about the mission because of its classified nature, though it is known that *Atlantis* reached the highest ever orbit for a Shuttle 512 km (320 miles) above the Earth.

61-A: *CHALLENGER*

The 22nd mission of the Shuttle series saw the first eight-person crew and the first foreign-funded Spacelab mission, D-1. The 76 experiments in the pressurized module were under the control of the German Space Operations Centre near Munich, mainly devoted to medicine and materials science experiments. German physicists Reinhold Furrer and Ernst Messerschmid operated the equipment along with Wubbo Ockels, a Dutchman, who flew under the patronage of

The cramped conditions inside Voskhod 2 are immediately apparent in this view of Pavel Belyaev (nearer to camera) and Alexei Leonov. The Soviets have never released pictures of the Voskhod craft without its protective launch shield. This is because they were hastily-modified Vostok capsules.

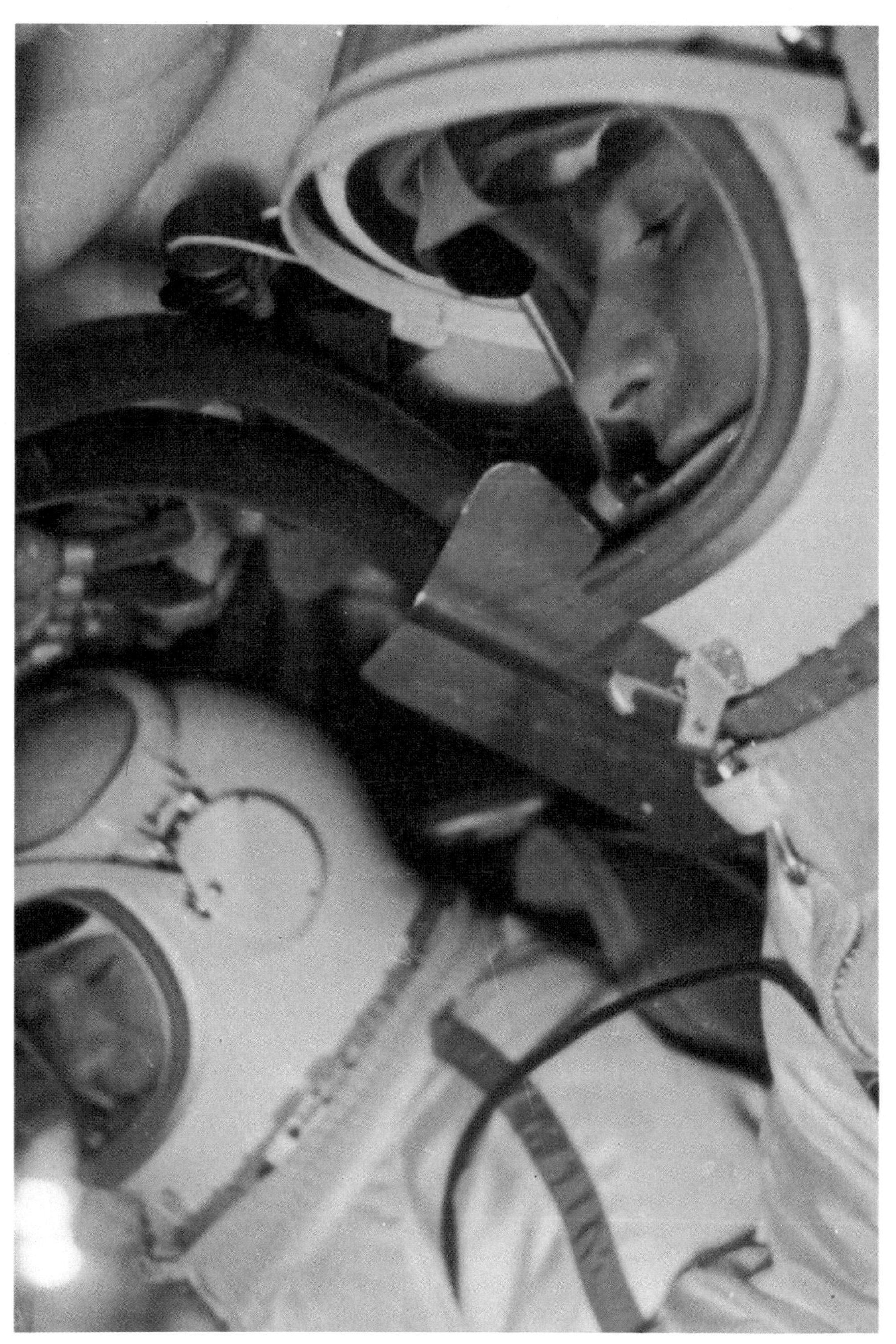

ESA. Because of slight problems with fuel cells, it was not possible to extend the mission beyond the planned seven days after which *Challenger* landed in California.

61-B: *ATLANTIS*

After the second night launch in the series, *Atlantis* successfully deployed three comsats, including the Mexican Morelos 2 satellite for which Rudolfo Neri Vela flew as payload specialist. More importantly, Jerry Ross and 'Woody' Spring built up a tower known as ACCESS (Assembly Concept for Construction of Erectable Space Structures) – essentially a triangular framework made of aluminium struts nearly 14 m (45 ft) long. Another framework was also deployed, proof that astronauts would be able to assemble the Space Station in the 1990s. *Atlantis* successfully landed in California after spending a week in space.

61-A: *COLUMBIA*

The first Shuttle flight of 1986 saw the return of *Columbia* into service after extensive modifications from the first test flights. $9\frac{1}{2}$ hours into the flight, RCA's Syncom K-2 satellite was successfully deployed. Among *Columbia*'s passengers were Congressman Bill Nelson and RCA Engineer Robert Cenker, whose task was to test an infra-red tracking detection system. Astronomer George Nelson (no relation to the Congressman) attempted to photograph Halley's Comet with a telescope called CHAMP which unfortunately failed to operate. After five days, *Columbia* landed in California.

51-L: *CHALLENGER*

The last mission of *Challenger* was intended to deploy the second TDRS satellite and a small telescope called SPARTAN-Halley. Christa McAuliffe, as part of the 'Teacher in Space' programme, was to have conducted two televised lessons from space. Hughes aircraft engineer Greg Jarvis was to have conducted a fluid dynamics experiment of his own design. The loss of *Challenger* 73 seconds after launch led to the Shuttle programme being grounded for over $2\frac{1}{2}$ years.

STS-26: *DISCOVERY*

Announced by NASA as 'The Return To Flight', *Discovery* successfully reached orbit on 29 September 1988, two days short of NASA's 30th birthday – a fitting present. The five man crew led by Commander Rick Hauck held the distinction that they had all flown in space before on earlier Shuttle missions. They successfully deployed a TDRS satellite to replace the one lost in the *Challenger* accident. The other experiments included a protein crystal growth which investigated 60 different enzymes, including reverse transcriptase which plays an important part in the replication of the AIDS virus. After four days, *Discovery* successfully landed in California.

STS-27: *ATLANTIS*

With minimal publicity, *Atlantis* was launched on 2 December 1988 from the Kennedy Space Center on the first classified mission since the *Challenger* accident. The crew deployed an electronic snooper satellite called Lacrosse in a 57° orbit, much higher than the normal Shuttle mission. After returning to California three days later, tile damage was found to be greater than had been expected.

On 18 March 1965 Alexei Leonov became the first human being to walk in space. Only low quality television pictures were recorded as a film camera on the outside of Voskhod 2 was not retrieved because of the drama surrounding his return into the capsule.

STS-28 *COLUMBIA*

Department of Defense mission. Data restricted.

STS-29 *DISCOVERY*

The main objective for STS-29 was the deployment of the third TDRS satellite by an IUS booster. Among the other experiments flown were an infra-red detector involved in USAF ground tests in Hawaii and an IMAX 70 mm film camera used to produce a film to be called *Exploring the Blue Print*. *Discovery* and its crew of five returned to Earth after six days in space.

STS-30 *ATLANTIS*

After a launch ship caused by engine problems and a fractured fuel line, *Atlantis* was launched on Friday, 5 May 1989 and successfully deployed the Magellan spacecraft on its way to Venus. The five-day flight continued the USAF experiment over Hawaii, as well as an investigation into large-scale lightning.

VOSKHOD

Russia's second series of manned spacecraft remains one of its most mysterious and was nothing more than a political 'stop gap'. Very few photographs of the Voskhod spacecraft have ever been released, and the ones that have revealed it to be a modified Vostok capsule, despite earnest Soviet claims that Voskhod was radically different from Vostok. Only by removing the ejector seat and putting in three simple couches could a three-man crew be accommodated on Voskhod 1 without spacesuits: with spacesuits, the two-man crew of Voskhod 2 was launched to perform the first spacewalk. Because the Voskhods were much heavier than Vostok, a more powerful booster had to be used, the 'SL-4' later used on Soyuz missions which launched the craft into higher orbits. Because the Voskhods would not return to Earth within a few days (as was the case with Vostok), an extra retro-rocket was placed at the top of the capsule in case the primary retro-rocket failed. Because a parachute exit with ejector seats was not possible before landing, an additional retro-rocket was used which fired $1\frac{1}{2}$ seconds before landing to soften the impact.

The crew for Voskhod 1 for the first time allowed scientists to fly in space. The pilot was Vladimir Komarov who had shown determination to return to flight status after being grounded because of a heartbeat irregularity. Konstantin Feoktistov was an engineer involved in teaching the pilots the theories of flight, and later went on to design the Salyut space stations. Boris Yegorov was a physician, himself the son of a brain surgeon, who had specialized in studies of the inner ear where the balance mechanism is located which is affected by weightlessness. Voskhod 1 was launched on 12 October 1964 into an orbit which measured 178 km × 409 km (111 × 256 miles) and inclined at 65°. They made 16 orbits of

The clustered propellant tanks around the Vostok capsule are apparent in this colour photograph. The spherical shape of the capsule in which the cosmonaut remained is also obvious.

the Earth in 24 hours 17 minutes, and wanted to continue for another day, but were ordered back by Korolev who quoted Hamlet: "There are more things in heaven and earth than are dreamt of in your philosophy" – taken to be a reference to the deposing of Khruschev.

Voskhod 2 was launched on 18 March 1965, with Alexei Leonov and Pavel Belyayev as the crew. They wore spacesuits and their Voskhod capsule had the addition of an inflatable airlock which would be discarded after Leonov's spacewalk. They were launched into the highest orbit ever attempted – 173 × 495 km (108 × 309 miles), inclined at 65°. After two orbits they were ready for the first spacewalk: Leonov had to be helped into the backpack which contained his life support systems, plus an umbilical chord which kept him attached to the capsule. He spent ten minutes floating in space, and was able to recognize the outline of the Black Sea below. It took him another 12 minutes to get back inside the capsule because his suit had ballooned in zero gravity. On the 16th orbit, the automatic retrofire sequence should have fired, but didn't. Belyayev took over manual control for the first time on a Russian manned spaceflight and fired the retros himself, but the firing was not quite accurate as they landed 2,000 km (1,250 miles) north in the middle of a forest. Total mission duration was 26 hours 2 minutes.

Despite their curious nature, the Voskhods were significant achievements, seeing the first scientific crews and spacewalks. Two further missions were planned, which would have seen a two-week mission and the link-up with an unmanned satellite. However, these missions never flew when it was decided instead to press ahead with Soyuz.

VOSTOK

The spacecraft used to launch the first humans into space was essentially ball-shaped, with a lower instrument panel which contained all the engines and retro-rockets. Designed by Sergei Korolev, it was named Vostok ('East') and was launched into an orbit which would decay within ten days in the event of anything going wrong – that is the capsule would return to Earth within ten days. Even today in the age of *glasnost*, very few details have emerged about the flights, despite their success. The capsules were manufactured by production line means, and 25 years later, the Soviets are offering the basic capsule to Western countries for experimental purposes.

Despite the success with the manned flights, the first Vostok prototype, launched as Sputnik 4 in May 1960, fired its retros into a higher orbit. The next test, Sputnik 5, saw the first return of animals – two dogs called Strelka and Belka – in August 1960. A further failure in December 1960 was followed by two further successful returns of dogs in March 1961. The way was clear for Yuri Gagarin, only 18 days later.

VOSTOK SPACECRAFT

The Vostok capsule in which the cosmonaut was housed was ball-shaped, covered with ablative heat shield material and metallic strips to reflect sunlight and keep the internal temperature reasonably cool. The environmental control systems were located within the crew cabin, as well as the attitude control equipment. In fact, Korolev had realized that a spherical vehicle which was weighted at one end would maintain its correct orientation, and land the correct way up! In front

The extreme effects of re-entry heating are immediately obvious in this view of a Vostok capsule after landing. The spherical shape ensured uniform heating and ablation of the heat shield. Before landing, the cosmonaut parachuted to the ground as the capsule could not support his/her weight.

of the cosmonaut was an instrument panel which contained an Earth globe that moved in synchronization with the spacecraft's motion over the surface and an optical orientation sensor which aligned the craft for re-entry. Vostok operated automatically, as Soviet scientists feared that the effects of zero gravity might incapacitate the cosmonaut. However, in case things went wrong, an emergency key was provided plus a codeword to allow him or her to take over manual control.

A rocket ejection seat was provided within the capsule, inclined horizontally so the cosmonaut would be ejected upwards and away from the craft in case of a launch emergency. Ejection was facilitated by explosive bolts which literally blew open a hatch in the capsule and 'threw' the cosmonaut out of the capsule by means of two powerful rockets. If there was any trouble during launch, the ejection seat would be fired. The ejection seat was actually used before landing, the cause of some controversy, as the Soviets feared that if it be known that Yuri Gagarin had 'bailed out' before landing, it could be construed as an emergency and therefore the honour of completing an orbit would go to somebody else. So it was announced that Gagarin had landed with the capsule, whereas the other Vostok pilots had ejected. It was to be four years before pictures of the spacecraft were released, and not until 1978 that Gagarin's ejection from the capsule was confirmed.

Mission	Pilot	Date	Duration	Orbits
Vostok 1	Yuri Gagarin	12 April 1961	1:48	1
Vostok 2	Gherman Titov	6 August 1961	25:18	17
Vostok 3	Andrian Nikolayev	11 August 1962	94:22	64
Vostok 4	Pavel Popovich	12 August 1962	70:57	48
Vostok 5	Valeri Bykovsky	14 June 1963	119:06	81
Vostok 6	Valentina Tereshkova	16 June 1963	70:50	48

SOLAR SYSTEM EXPLORATION

The following spacecraft have played a significant role in the exploration of the Solar System. Some were more successful than others, yet all have given mankind a great deal of information about our place in the Universe.

EXPLORER

The Explorer series of satellites has been one of the most hardy in the U.S. space programme. The launch of Explorer 1 in January 1958 saw America's entry into the space race: during the 1960s, more in the series went further aloft to study the effect of the atmosphere on satellite motion, and to explore the upper layers of the atmosphere and the interplanetary environment. Further astronomy Explorers followed, and the programme was expanded to include international cooperation.

EXPLORER 1

Three weeks after Sputnik 1, the U.S. Army Ballistic Missile Agency was given the go-ahead to launch two Explorer satellites as part of the U.S. contribution to the International Geophysical Year (IGY). Though much smaller than the Soviet Sputniks, Explorer 1 carried miniaturized experiments to investigate data on cosmic rays, meteoroidal impacts and temperature sensors. A simplified geiger counter built by James van Allen of the University of Iowa discovered the presence of belts of radiation trapped by the Earth's magnetic field. Ironically, three of the first eight Explorers failed to reach orbit, including the second in the series.

IONOSPHERE EXPLORERS

The Earth's ionosphere is the upper part of its atmosphere which stretches from about 40 to 300 km (25 to 185 miles) above the surface. Most of the molecules within this region consist of electrically-charged ions, formed by interaction with the radiation from the Sun (the 'solar wind'). Explorers 8, 20, 22 and 27 measured the density of electrons within this region as part of an international programme from 1960 to 1964.

AIR DENSITY EXPLORERS

Explorers 9, 19 and the dual-launched 24 and 25, and 39 and 40 were essentially inflatable spheres just under 4 m (13 ft) in diameter. By monitoring how they deviated from their 300 × 2,500 km (185 × 1,562 miles) orbits it was possible to examine the seasonal changes in the density of the upper atmosphere.

INTERPLANETARY MONITORING PLATFORMS

During the decade from 1963 to 1973, NASA launched ten Explorers (18, 21, 28, 33, 34, 35, 41, 43, 47 and 50) whose function was to monitor the interplanetary environment by being placed in orbits which swept out as far as the distance of the Moon. They were able to monitor changes in the solar wind over the Sun's 11-year cycle of activity. They also defined the shape of the Earth's magnetotail, the tear-shaped envelope in the solar wind that forms in the wake of the Earth.

Above: The Explorer series of unmanned spacecraft have been among the most venerable in NASA's three decades. Here, one of the Interplanetary Monitoring Platforms is prepared for launch by technicians at NASA's Goddard Space Flight Center in Maryland.

ASTRONOMY EXPLORERS

Explorers 38 and 39 were radio astronomy satellites which were designed to 'listen in' to galactic radio sources. However, Explorer 38's work was hampered by interference from radio broadcasts from the Earth! So Explorer 49 (launched in June 1973) was placed in orbit around the Moon. The Moon effectively shielded the satellite from the Earth's sources of interference when it was orbiting on the far side.

Explorers 42, 43 and 53 were classed as Small Astronomy Satellites (SAS). They were able to map the sky from orbit at gamma and X-ray wavelengths, unseen from the surface of the Earth. Explorer 42 had the distinction of being the first NASA satellite to be launched by a foreign country. The Italian government maintains for launches a modified oil rig off the coast of

Opposite: Mariner 10 is seen here as it would appear in space, its twin TV cameras seen at the top of this view. Because of its Sunward orbit, passing by Mercury and Venus, only two solar panels were required. Engineers were able to 'tack' the panels into the solar wind for additional thrust.

Kenya known as San Marco. It was from here on 12 December 1970 that an Italian crew launched a Scout rocket with Explorer 42 atop. The satellite was soon named Uhuru in honour of Kenya's Independence Day. Uhuru went on to complete the first X-ray survey of the heavens, revealing the existence of 200 X-ray objects in space.

INTERNATIONAL SUN-EARTH EXPLORERS

A further series of Explorers which investigated the interplanetary environment was launched in 1977 and 1978. There were three in the series, the second provided by the European Space Agency. All three were boosted to an orbit around the point in space where gravitational forces of the Sun are balanced with those of the Earth and Moon. Known as Libration Point One, it is some 1.6 million km (1 million miles) from Earth. Their orbits were specially chosen to take them above the plane of the Sun and planets to see if magnetospheric phenomena were caused by the Sun or were within the Earth's magnetosphere. ISEEs 1 and 2 re-entered the Earth's atmosphere in late 1987, but ISEE-3 was jockeyed into a new orbit in late 1982 for the first ever encounter with a comet.

AMPTE

The most recent in the series of Explorers was an international project between NASA, Great Britain and West Germany, launched by a Delta rocket on 16 August 1984. Each country provided one spacecraft each, all of which were launched simultaneously and known collectively as the Active Magnetospheric Particle Tracer Explorer (AMPTE). The German spacecraft released barium and lithium ions into the solar wind, which were then monitored by the U.S. and British satellites after they interacted with the Earth's magnetic field. On 27 December, the German spacecraft released two small canisters of barium to produce an artificial comet. Twelve minutes later a tail of purple-coloured barium ions was seen in the dawn skies on the western seaboard of the U.S. The other spacecraft observed that the barium atoms were ionized almost instantaneously by the ultraviolet portion of light from the Sun.

GIOTTO

As the European Space Agency's first deep space probe, Giotto was an unqualified success. This was all the more significant, given the almost-suicidal nature of its mission: to head towards Halley's Comet with a speed of 70 km/sec (44 miles/sec) and take scientific readings very near to the central core of the comet (known as the 'nucleus'). A double 'bumper shield' had to be developed to protect the spacecraft from the

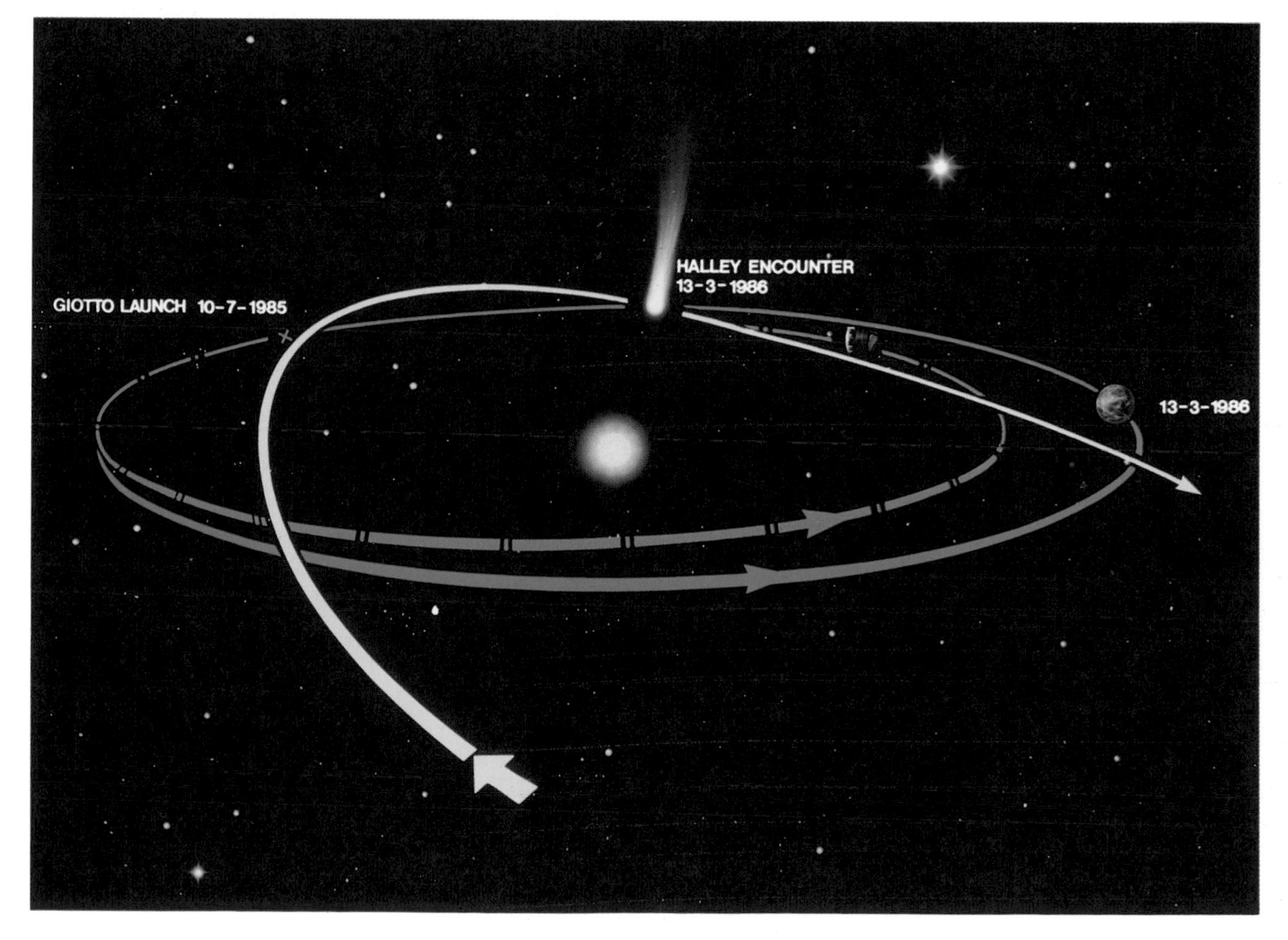

This schematic diagram shows the relative paths of Giotto and its celestial quarry, Halley's Comet. The elliptical path of the comet is seen (yellow), intersected by Giotto (blue), launched in July 1985. The position of the Earth is also shown at the time of close approach. Spacecraft and comet approached each other with a velocity greater than 50km/sec.

potentially lethal dust impacts. At such speeds, dust particles have the energy of hand-grenade explosions, so an aluminium front sheet was designed to take the 'sting' out of the impacts, and the particles were completely vaporized. A Kevlar sheet located 23 cm (9 in) behind the first ensured that the shock of the explosion and its energy were absorbed.

The spacecraft was named after the Italian painter Giotto Di Bondone whose 'Adoration of the Magi' from 1304 depicts a comet which is believed to be Halley's. The comet returns to the inner part of the solar system every 76 years – Edmond Halley was the first to recognize this in the 17th century. To minimize development costs, Giotto was based on an earlier design for a spin-stabilized spacecraft called Geos. Giotto was essentially a 2.75 m (9 ft) high cylinder, 1.8 m (6 ft) in diameter, which spun once every four seconds to stabilize itself. Power restrictions meant that data could not be stored, so it was transmitted 'live' back to Earth with a high-gain antenna transmitting at two frequencies. The craft carried a full complement of scientific experiments: a TV camera; a photopolarimeter to measure the brightness of the nucleus; three mass spectrometers to analyse the chemical composition of the dust and gas within the comet's tail; a number of dust impact detectors and a number of experiments used to investigate the interaction between the comet and the solar wind.

Giotto's TV system – referred to as the Halley Multicolour Camera – was a 998 mm focal length f/7.68 telescope which did not view the comet directly. A protective, rotatable turret carried a forward-looking mirror at 45° which relayed the light to the telescope. As Giotto rotated, the comet was 'scanned' by the camera which used a sensitive Charged Couple Device to build up a picture. A filter wheel was used to produce a colour picture, though the pictures were shown in false colour on the ground at the European Operations Centre in Darmstadt.

Giotto was launched on 2 July 1985 by the 14th Ariane flight and the spacecraft began its journey to meet the comet the following March. From the Earth, the head of the comet (known as 'the coma') is at best seen as a fuzzy white area: the central nucleus is obscured by dust and gas. As the last of the international flotilla of spacecraft to reach the comet, Giotto's flight path was targeted using the results from the other spacecraft – particularly the Soviet VeGas. This allowed Giotto to be targeted to within 605 km (378 miles) of the nucleus, the closest of all the craft. The moment of closest approach was set for a few minutes after midnight (GMT) on 14 March.

In all, the spacecraft returned 2,112 images of the nucleus and the dust impacts were lower than expected until a few minutes before closest approach. The number of impacts increased dramatically to over 200 per second as Giotto passed through a stream of dust thrown off by the nucleus. Fourteen seconds before closest approach, a dust particle only 1 gram (0.035 oz) in mass impacted the various shields with sufficient force to start the spacecraft to wobble off its axis: the high gain antenna was no longer pointing towards the Earth, and contact was lost for about 30 minutes. After that, Giotto's wobble was sufficiently reduced to allow the spacecraft to continue transmitting data.

The nucleus of the comet was revealed to be an irregular object about 15 km (9 miles) long and between 7 and 10 km (4.3–6.25 miles) across. Two large 'jets' of dust were observed on the sunward side which obscured the full extent of the width of the nucleus from Giotto's TV camera. The nucleus was very dark: it reflected only 2–4 per cent of the sunlight reaching it, and image enhancement showed valleys and hills on the nucleus surface. The nucleus seems to be made of dirty ice, and the dark outer coating is made of dust which acts to 'insulate' the whole of the surface. Cracks in the dust layer revealing ice underneath were responsible for the jets that were seen. Dust measurements show that about 3 tonnes of material are ejected per second.

After the encounter – which many predicted the spacecraft would not survive – over half the instruments were damaged, and although the camera optics seem to have survived, it is not known whether the mirror survived. A series of manoeuvres will bring Giotto within the vicinity of the Earth in July 1990, and thereafter it may be possible to re-direct the spacecraft to a comet called Grigg-Sjkellerup in July 1992.

This ESA artwork shows the Giotto spacecraft approaching the head of Halley. Despite the rigours of encountering the comet's swarm of dust particles, Giotto survived, though at least two of its scientific instruments were destroyed. Of the remaining eight, at least four are known to be in working order.

INTERNATIONAL COMETARY EXPLORER (ICE)

Financial cutbacks meant that NASA did not send a probe to Halley's Comet. However, the agency was able to score a remarkable 'first' in cometary exploration, by re-routing a spacecraft designed to study fields and particles around the Earth towards comet Giacobini-Zinner. The International Sun–Earth Explorer 3 had already reached the end of its operational life in 1981, when NASA realized its orbit could be altered to send it off towards the comet whose peculiar name comes from the astronomers who discovered it. The spacecraft was in orbit around a point known as the L1 Libration Point where gravitational forces of the Sun are balanced by those of the Earth and Moon. By a complex set of manoeuvres in 1982, the spacecraft headed within 120 km (75 miles) of the Moon's surface to head off towards the comet which would be in the vicinity of the Sun in late 1985.

The craft was re-named the International Cometary Explorer, and it literally chased the comet as it headed towards the Sun, passing within 8,000 km (5,000 miles) of the nucleus of Giacobini-Zinner on 11 September 1985. The spacecraft's energetic particle and plasma detectors observed the comet's interaction with the solar wind, and dust impacts were noted from changes in the plasma density. Though the craft was travelling at 21 km/sec (13 miles/scc) with dust impacts every second, the craft emerged unscathed. The ICE data was later used by the planners of the International Halley missions.

The third of the International Sun-Earth Explorers, ICE is seen here prepared for launch in August 1978. The ISEE project was planned as a five-year-long endeavour to investigate the interplanetary environment.

LUNA

The Soviet Union's series of Luna spacecraft achieved some notable firsts in mankind's exploration of the Moon. Those achievements include the first man-made objects to hit the Moon, return pictures from its far side, make a soft landing, orbit it, automatically return samples and despatch an automated rover on the surface.

THE FIRST LUNAS

Soviet exploration of the Moon began in earnest in 1959 with the launch of Luna 1 (also known as Lunik 1 or Mechta). It was only the ninth ever Soviet launch, and is believed to have been preceded by four failed attempts. A third stage had been added to the same 'Semyorka' rocket which had launched Sputnik 1, and though it failed to fly past the Moon it became the first man-made object to orbit the Sun. The next attempt with Luna 2 hit the Moon, but it was left to Luna 3 to pass behind the moon returning pictures. Launched two years to the day after Sputnik 1, the probe flew only 6,000 km (3,750 miles) above the Moon's surface and recorded views onto film that was processed inside the capsule, scanned and then transmitted as it headed towards the Earth.

LANDERS AND ORBITERS

A second series of Lunas (4–8) attempted to make soft landings on the Moon, but it was not until 31 January 1966 that Luna 9 successfully touched down on the eastern edge of the Sea of Storms. As the whole spacecraft landed, it ejected a 100 kg (220 lb) ball that was weighted to land the correct way up. Four petal-shaped covers sprang out to stabilize the lander before a simple facsimile TV camera returned nine panoramas of the lunar surface. Its success was repeated the following December with Luna 13. At the same time as the first Luna landers, a series of lunar orbiters was attempted (Luna 10, 11, 12 and 14) all of which were successful. It is believed that Luna 11 carried a TV system to return pictures of the surface which failed; it was left to the next in the series to return pictures of the Moon. Again, a photographic system was used in which the film was developed, fixed and dried, and then scanned by a TV system for transmission.

SAMPLE RETURNS

The final generation of Luna spacecraft involved automatic sample return techniques and roving capabilities. Whereas the first Luna landers headed straight for the surface without entering a parking orbit, Luna 15 heralded a new approach. Just two hours before the Apollo 11 crew were scheduled to leave the Moon's surface, Luna 15 was making its final descent – and sadly, it crashed. It was left to Luna 16 in September 1970 to return the first samples. After landing in the Sea of Fertility, a 90 cm (35 in) long probe with a drill attached at its end took seven minutes to scoop a sample of rock. It then swung back and dropped the drill bit into a return capsule which successfully blasted off and returned to Earth with a total of 100 kg (220 lb) of samples. Four more attempts to return samples were made; two failed (Lunas 18 and 23) and two were successful (Lunas 20 and 24).

ORBITERS

Lunas 19 and 22 were placed into orbits around the Moon in October 1971 and June 1974 respectively. They carried a wide range of scientific instruments similar to those flown on Soviet Mars and Venera probes in the early 1970s. A TV system took highly detailed images of the Moon. Shortly after arrival, Luna 22 was lowered into an orbit that took it as close as 25 km (16 miles) above the surface. A radar altimeter was used to measure how the orbit of the probes altered as they flew across the lunar surface in an attempt to look for mass concentrations (mascons) of material. Gamma ray spectrometers were able to conduct an analysis of the chemical composition of the surface.

LUNOKHODS

The first automated rovers to travel across the lunar surface were carried by Lunas 17 and 21 in November 1970 and January 1973. Known as Lunokhods 1 and 2, they were eight-wheeled, tub-shaped vehicles that were 0.9 m (3 ft) high and 1.2 m (4 ft) in length. The central 'tub' was pressurized and its control electronics were kept at a pressure of 1 Earth atmosphere. A heat exchanger kept internal temperatures within the range of 0–40°C, and excess heat was radiated by a lid that could be removed. The lid was closed during the lunar night when temperatures reach as low as −150°C (−238°F), and a radioactive power source kept the vehicle from freezing. Two TV cameras returned panoramic images and high resolution views to enable a five-man team in Mission Control in Moscow to 'drive' the vehicle by remote control. A variety of instruments were

LUNA LOG

Mission	Launch Date	Remarks
Luna 1	2 January 1959	Intended to hit the Moon, it missed by 6,000 km (3,700 miles). Continued to transmit for 62 hours after launch.
Luna 2	12 September 1959	First spacecraft to hit the Moon, on September 13, near to the crater Archimedes.
Luna 3	4 October 1959	First spacecraft to pass behind the Moon and photograph its far side, on October 7.
Luna 4	2 April 1963	First of new generation lander craft which attempted to soft land. Missed Moon by 8,500 km (5,300 miles).
Luna 5	9 May 1965	Attempted soft landing, but retro-motor failed and vehicle crashed.
Luna 6	8 June 1965	Attempted soft landing, but main motor failed to cut off and vehicle missed the Moon by 160,000 km (100,000 miles).
Luna 7	4 October 1965	Attempted soft landing, but retro-motor fired too quickly and crashed near to crater Kepler.
Luna 8	3 December 1965	Attempted soft landing, but retro-motor fired too late and vehicle crashed into surface.
Luna 9	31 January 1966	First successful soft landing in Ocean of Storms on 3 February. Lander returned 27 panoramic TV pictures of surface.
Luna 10	31 March 1966	Successfully reached lunar orbit on 3 April and returned data for almost 2 months.
Luna 11	24 August 1966	Successfully reached lunar orbit on 27 August and returned data until 1 October.
Luna 12	22 October 1966	Successfully reached lunar orbit on 25 October and was first to return high resolution TV pictures from orbit. Last images received on 19 January 1967.

In January 1966, the Soviet Union's Luna 9 became the first human artefact to land on the surface of the moon. Seen here at a Soviet exhibit, the petal-shaped lander and the carrier spacecraft (upper right) are pictured against a panorama of the surface returned by the lander.

LUNA LOG (continued)

Mission	Launch Date	Remarks
Luna 13	21 December 1966	Successfully landed 400 km (150 miles) away from Luna 9 in Ocean of Storms on 22 December. Returned data until 30 December, including penetrometer to investigate surface soil density.
Luna 14	7 April 1968	After reaching lunar orbit on 10 April, returned gravity measurements but no TV system apparently carried.
Luna 15	13 July 1969	Failed attempt to return lunar samples back to Earth. Vehicle entered lunar orbit on 17 July, but crashed in Sea of Crises on 21 July.
Luna 16	12 September 1970	First successful unmanned sample return mission. 100 kg (220 lb) of soil returned from Sea of Fertility after landing on 20 September.
Luna 17	10 November 1970	First successful deployment of automated rover Lunokhod 1 after landing on 17 November in Sea of Rains.
Luna 18	2 September 1971	Sample return attempt ended in failure when vehicle crashed in Sea of Fertility on 11 September after losing radio contact.
Luna 19	28 September 1971	Successful lunar orbiter arrived on 3 October and returned TV pictures until the end of the month.
Luna 20	14 February 1972	Second successful sample return mission landed in Sea of Fertility on 18 February. About 50 kg (110 lb) soil sample returned.
Luna 21	8 January 1973	Deployed Lunokhod 2 after successfully landing in Sea of Serenity on 15 January.
Luna 22	29 May 1974	After reaching lunar orbit on 2 June, returned TV pictures and physical data until September 1975.
Luna 23	28 October 1974	Though it landed in the Sea of Crises on 2 November with the intention of returning a sample, it seems that the roughness of the terrain damaged the vehicle and it ceased transmission on 9 November.
Luna 24	9 August 1976	Successfully landed in Sea of Crises on 18 August and returned a 170 kg (374 lb) soil sample to Earth on 23 August in Siberia.

carried including an X-ray soil analyser and an array of X-ray telescopes which observed the most remote parts of the galaxy.

After being deployed on 17 November 1970, Lunokhod 1 operated for just under a year; on the anniversary of Sputnik 1 (4 October) the next year, official announcements said that the vehicle was no longer operating. It had travelled over 10 km (6.25 miles) and returned 20,000 pictures. Though Lunokhod 2 carried improved electronic motors and an extra TV camera, it only operated from 16 January to 4 June 1973. It had travelled 37 km (23 miles) and returned 80,000 TV pictures.

LUNAR ORBITER

Starting in August 1966, a total of five Lunar Orbiters were placed into orbit around the Moon at roughly three-monthly intervals. As the second part of NASA's unmanned lunar exploration programme, their task was to return detailed TV pictures of the whole of the lunar surface – a task which they performed brilliantly. The pictures were then used to make high resolution maps of the Moon from which Apollo landing sites were chosen.

All five spacecraft were identical and consisted of a central truncated structure protected by a thermal blanket inside which the attitude control and computer systems were located. Four solar panels which were folded at launch and later deployed gave the spacecraft a windmill-like appearance. Inside the central equipment was the TV imaging system which would allow detailed mapping of the Moon – virtually a miniaturized photographic laboratory. It had been developed

Looking for all the world like a celestial wash tub on wheels, the first Lunokhod belies its complex workings in this mock-up photograph. The 'lid' at the top was a protective covering to keep heat in during the cold lunar night and to let it out during the day.

by the Eastman Kodak company and is believed to have been based on that used for a series of USAF spy satellites called SAMOS. Essentially, photographs were taken that were recorded onto finely-grained film which were then scanned by a high resolution TV imaging device and returned to Earth. Up to 194 dual exposure images could be captured on one roll of 70 mm film which was 61 m (200 ft) in length. The onboard photographic sub-system developed the film and prepared it for transmission back to Earth.

Two lenses were available for photography: one had a 80 mm focal length for wide angle views; the other, focal length 610 mm, was narrow angle. The resolution on the surface depended on the orbital altitude, but was around 2 m by the narrow angle camera at a nominal altitude of 112 km (70 miles) above the surface. The spacecraft were placed in orbits so that the nine potential Apollo landing sites were viewed at optimum lighting conditions – normally when the Sun was between 10 and 30 degrees above the horizon.

The first three Orbiters were placed in equato-

LUNAR ORBITER LOG

Mission	Launch Date	Mission Ended (Lunar Impact)	Comments
Orbiter 1	10 August 1966	29 October 1966	Equatorial Orbit (12° inclination). Photographed Landing Sites north of equator.
Orbiter 2	6 November 1966	11 October 1967	Equatorial orbit, with plane change to 17.5°. Sites south of equator detailed.
Orbiter 3	5 February 1967	9 October 1967	Completed Photography for choice of Apollo landing sites – eight preliminary sites found.
Orbiter 4	8 May 1967	October 1967	First polar Orbiter, 86° inclination. Observed south pole for first time. Contact lost on 17 July.
Orbiter 5	1 August 1967	31 January 1968	Final polar Orbiter. Mapped most of far side. Spacecraft observed around Moon by University of Arizona astronomers.

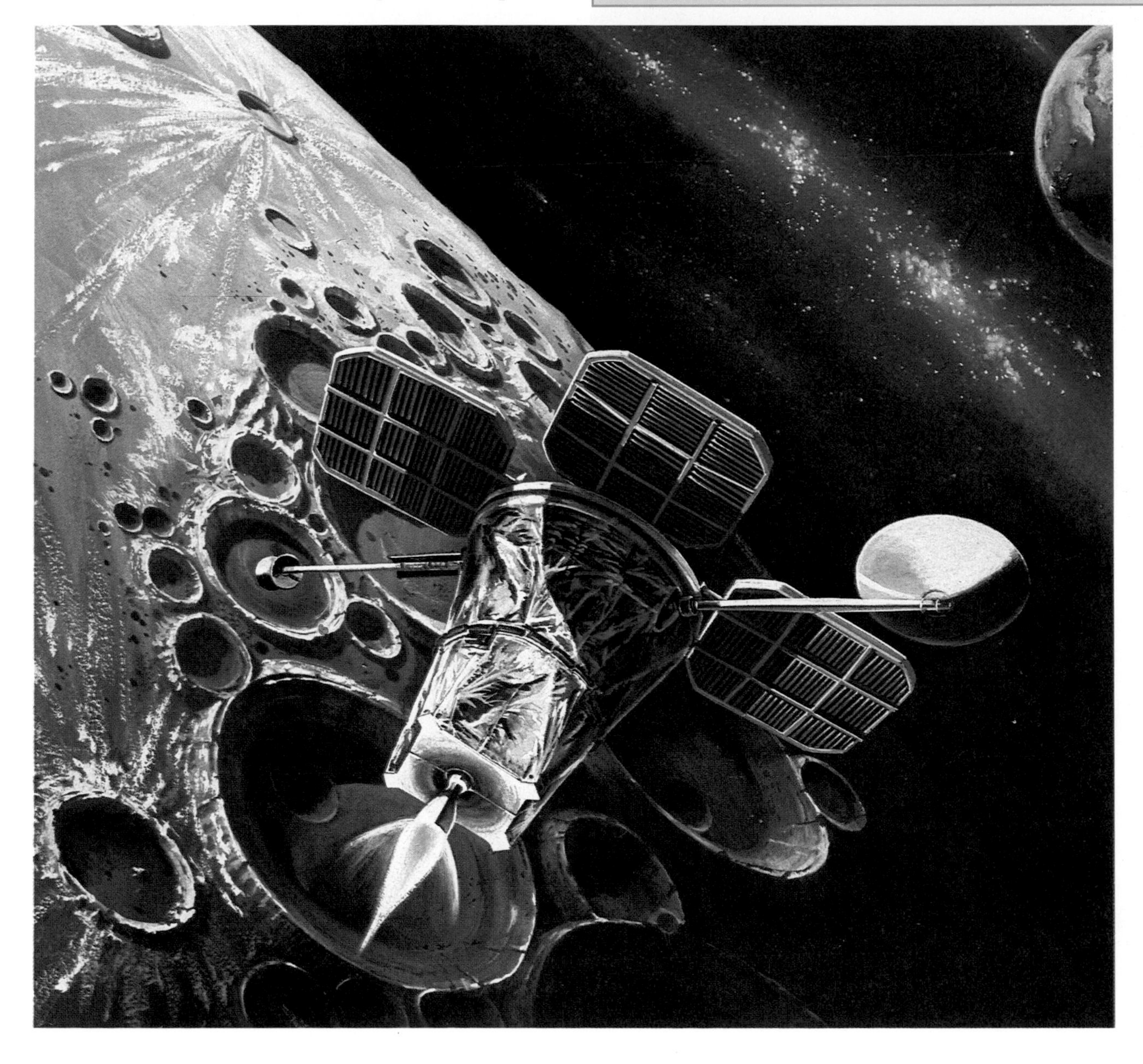

This NASA artist's impression shows a Lunar Orbiter being manoeuvred into lunar orbit. The orbiter's cameras were located in a housing, not seen here, beneath the solar panels. The retro-motor is seen here in action, later used for orbital change manoeuvres.

rial orbits, as most of the Apollo landing sites were located within 30 degrees latitude of the equator. However, the final pair were placed into polar orbits so that the whole of the lunar surface could be mapped in time. Lunar Orbiter 4 viewed the Moon's south polar regions for the first time. At the end of the missions, all five spacecraft were deliberately crashed into the Moon's surface to avoid interference with later Apollo telecommunications.

MARINER

NASA's Mariner series of spacecraft gave mankind the first detailed information about our planetary neighbours – Mars, Venus and Mercury. All were built by the Jet Propulsion Laboratory (JPL) in Pasadena, California, and the first in the series had much in common with the Ranger spacecraft that were sent to the Moon, though the Mariners were rather more successful. As technology improved, each successive spacecraft in the series became more sophisticated.

Technicians at the Jet Propulsion Laboratory in Pasadena ready Mariner 5 for its flight to Venus. The octagonal arrangement of its solar panels is obvious. with a Magnetometer boom pointing at right angles to the solar panels. Mariner 5 returned valuable data about the way in which Venus interacts with the solar wind.

The last probe to date was Mariner 10 which proved that the technique of 'gravity assist' was possible. By using the gravitational influence of a particular planet it was possible to visit others without need of extra rocket stages. Theorists in the 1920s had shown that it would be possible to play a game of 'interplanetary billiards' and visit one planet after another in quick succession. But it wasn't until the early 1960s that computers with sufficient 'number crunching' capability became available to planners at JPL.

It was soon realized that a probe could be launched to Mercury after passing Venus using the Atlas-Centaur booster which did not have enough power to send probes to Mercury directly. By aiming at a point within a 400 km (250 mile) slot of Venus, it was possible to boost the spacecraft on towards Mercury. Though sometimes described as getting a 'free' ride, it should be pointed out that the energy 'taken' from Venus meant that its rotation was effectively slowed down by only a billionth of a second! Two further probes in the series originally called Mariners 11 and 12 were re-named Voyager (see page 161) shortly before launch.

SPACECRAFT

Mariners 1 and 2 were similar to the Ranger spacecraft in that they consisted of a tubular tower attached to a hexagonal base. Two solar panels were attached to the base, providing power for the spacecraft as it headed towards Venus. A variety of instruments were located around the tubular structure, but no TV cameras were carried. Because Mariners 3 and 4 were heading *away* from the Sun towards Mars, they were required to carry four solar panels to generate the same power.

These Mars-bound Mariners consisted of an octagonal central body inside of which the main computers and rocket engine were located. As well as a simple TV system, the spacecraft carried a variety of instruments to monitor the interplanetary environment and how Mars interacted with it. Mariner 5 was similar in design to Mariner 3 and 4 but was modified to fly to Venus: only two solar panels were required plus a thermal protective shield. Mariners 6 and 7 were twice as heavy as their earlier Mars-bound predecessors, again based on an octagonal central body with four solar panels attached. A new feature was a 'scan platform' on which remote sensing instruments such as TV cameras and spectrometers could be moved to observe the planet as it sped past. Mariners 8 and 9 were similar to Mariners 6 and 7 with one notable difference: they carried large retro motors to brake them into orbit, making them twice as heavy again. Because Mariner 10 was using gravity-assist techniques, its total mass was halved and was of a simple octagonal design.

MARINERS 1 AND 2

NASA's first attempt to fly a spacecraft towards a planet failed because of a computer error in the Atlas-Agena B launch vehicle's navigation software: a hyphen had been typed instead of a minus sign, which resulted in the Atlas going wildly off course. It was blown up by the Range Safety Officer at the Cape. The second attempt with Mariner 2 was much more successful. After a 109-day journey, it flew past Venus on 14 December 1962 and obtained just over half an hour's worth of data on the planet's clouds and magnetic field. Flying past at a range of 35,000 km (22,000 miles), its radiometers showed the surface temperature to be 425°C (797°F) – hotter than the melting point of lead. There seemed to be no difference between the night and day hemispheres. The magnetometers and radiation instruments detected no sign of a magnetic field, nor any interaction with the solar wind.

MARINERS 3 AND 4

NASA's first attempt to launch a probe to Mars failed when the fibreglass shield which housed the Mariner 3 spacecraft in the nose cone of the Atlas-Agena booster did not jettison correctly. A more reliable metallic shroud was found for Mariner 4 which successfully headed towards Mars in late November, and eight months later flew within 10,000 km (6,250 miles) of the red planet's surface. Its instruments included a variety of radiation detectors which monitored the solar wind and revealed that Mars had no radiation belts and a weak magnetic field. However, most attention was paid to the TV camera, a simple slow-scanning vidicon tube which could take only 22 frames of the surface as the probe passed by. Mariner 4's simple computers could only return data at the rate of 8.3 'bits' per second. As each frame was made up of 240,000 'bits', it took ten days for the pictures to be returned to Earth. Eleven of the frames showed the surface of Mars, and most revealed craters similar to those seen on the Moon.

MARINER 5

Because of the success of Mariner 4, it was decided to send its back-up vehicle to Venus, with suitable modifications for the increase in solar radiation. The aim was to fly ten times nearer to the Venusian surface, and on 19 October 1967, Mariner 5 flew within 6,400 km (4,000 miles) of Venus. A magnetometer built by James van Allen revealed that the Venusian magnetic field was at most 1 per cent the strength of the Earth's. Instead of a TV camera, Mariner 5 flew an ultraviolet photometer which measured atomic hydrogen and oxygen emissions in the upper atmosphere, revealing much about its chemical processes. A strong hydrogen corona was detected, but there appeared to be no sign of oxygen emission.

MARINER LOG

Craft	Destination	Launch Date	Arrival Date	Comment
Mariner 1	Venus	22 July 1962	–	Launch failure.
Mariner 2	Venus	27 August 1962	14 December 1962	First successful planetary fly-by.
Mariner 3	Mars	5 November 1964	–	Launch failure.
Mariner 4	Mars	28 November 1964	14 July 1965	First successful Mars fly-by.
Mariner 5	Venus	14 June 1967	19 October 1967	More advanced study of Venus.
Mariner 6	Mars	24 February 1969	31 July 1969	First twin mission to fly past Mars.
Mariner 7	Mars	27 March 1969	5 August 1969	Returned much new data about planet.
Mariner 8	Mars	8 May 1971	–	Launch failure.
Mariner 9	Mars	30 May 1971	13 November 1971	First successful Mars orbiter.
Mariner 10	Venus/ Mercury:	3 November 1973	Venus: 5 February 1974 Mercury: 29 March 1974 (1st fly-by) 21 September 1974 (2nd fly-by) 16 March 1975 (3rd fly-by)	

MARINERS 6 AND 7

Both craft were launched within a month of each other in early 1969 using the Atlas Centaur booster heading for a fly-by of the planet in late summer. Mariner 6 flew within 3,410 km (2,131 miles) of the surface, and Mariner 7 within 3,530 km (2,206 miles). Both spacecraft returned over 200 photographs of the southern hemisphere of Mars accounting for about 20 per cent of the total surface area of the planet. Infra-red measurements of the polar cap indicated that its temperature was in the order of −125°C (−193°F), suggesting that it was made of dry ice. Indeed, carbon dioxide seemed to be the major constituant of the atmosphere. A few days before arrival, Mariner 7 suffered a slight 'hiccup' when signals from the craft were lost. It seems that the craft was hit by a micrometeoroid impact, but automatically regained contact within a few hours. Because of Mariner 6's observations, Mariner 7 was reprogrammed to take a better look at the southern pole.

MARINERS 8 AND 9

The 1971 launch opportunity to Mars was particularly favourable, so NASA uprated the

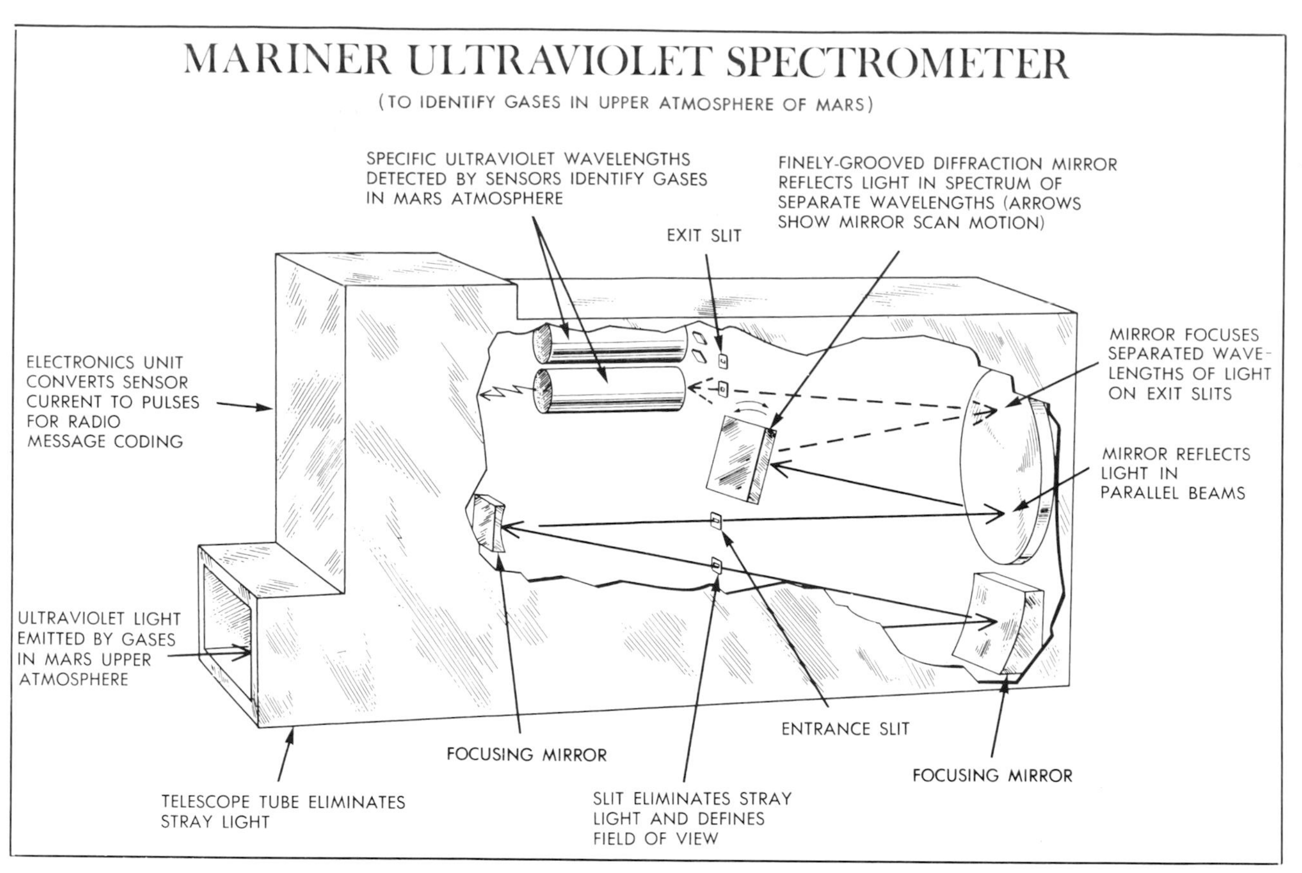

Both Mariners 6 and 7 carried ultraviolet spectrometers to look at the upper atmosphere of the Red Planet. The overall way in which a spectrometer works is shown in this schematic diagram. Through the Mariner series, NASA was able to extend spectroscopy to the atmospheres of both Venus and Mars.

Mariner 6 and 7 craft to enter Mars orbit to return data in tandem in an attempt to map the whole of the planet. On 8 May, however, Mariner 8's Atlas-Centaur launcher suffered a failure in the Centaur second stage, causing it to crash into the Atlantic. Though it was the last failure in the American planetary programme to date, it was no comfort to JPL scientists. Mariner 9 was reprogrammed to do Mariner 8's work and was successfully launched on 30 May. It reached Mars on 13 November 1971 to find the planet engulfed by dust with no features visible. So Mariner 9 was reprogrammed to look at Phobos and Deimos, the tiny Martian moons. Both were revealed as dark, irregular objects that were most probably captured asteroids. Mariner 9 continued to operate for nearly a year, and by the time it was shut down on 27 October, after it started to tumble because its attitude control gas was depleted, a total of 7,329 pictures had been received. Mariner 9 mapped virtually all the surface, and its high resolution cameras had revealed 2 per cent of the surface area down to 100 m (328 ft) on the surface. Its more staggering discoveries included the largest volcano in the Solar System, named *Olympus Mons* after the home of the Greek gods. It is three times higher than Everest and cliffs at its base are as deep as 4.8 km (3 miles) in places. Scarring the equator was a vast canyon nearly 4,800 km (3,000 miles) long, named *Valles Marineris* in honour of the spacecraft itself. In places it is 400 km (250 miles) across, as deep as 6.4 km (4 miles), with enormous landslides and hanging valleys in its walls. Also enormously long dried-up river channels were found that stretch for many hundreds of miles, estimated to have been caused by catastrophic flooding billions of years ago. These features were left to be studied in greater detail by the Viking spacecraft in 1976.

MARINER 10

After launch on 3 November 1973, Mariner 10's TV cameras were tested out on the Moon and its north polar regions as the spacecraft headed towards Venus. On 5 February 1974, Mariner 10 passed Venus at a distance of 5,760 km (3,600 miles) and returned 3,500 pictures. Using an ultraviolet filter, the first details in the cloud tops were revealed for the first time. A Y-shaped feature that had sometimes been seen from Earth was revealed, and its four-day motion around the planet was tracked. Just over a month later, Mariner 10 flew within 300 km (187 miles) of Mercury and returning the first pictures of its surface. The innermost planet was shown to be a heavily cratered body similar in appearance to the Moon. A vast multi-ringed basin which was later named Caloris was seen, appearing like a vast bullseye some 1,280 km (800 miles) across. Lava-filled craters hinted at the existence of past volcanism, and the spacecraft's interaction with the planet's gravitational field led to density measurements similar to those for the Earth. It is believed that Mercury has a metallic core, most probably iron in content. Because Mariner 10 was in orbit around the Sun, it was able to intersect the orbit of Mercury on two further occasions revealing over half the planet's surface area.

MARS

Soviet attempts to send spacecraft to Mars were dogged with failures in the 1960s and 1970s, despite the success of their Venus-bound spacecraft. The main problems concerned the relatively unsophisticated technology which often failed at critical moments. After the final series of disasters in 1974, the Soviets left Mars well alone. But a look at the Mars series shows the dogged persistence of Soviet scientists, which taught them a great deal about interplanetary travel.

Compared to Venus, Mars is a much more difficult object to 'aim' for. Rockets are simply not powerful enough to launch spacecraft to the planets at will: they have to be despatched only when the Earth is relatively close. The Earth and Mars are at their closest – within 64 million km (40 million miles) – every 17 years, but every 25 months they are so aligned as to allow spacecraft to travel between them. However, these 'launch windows' are variable and some are more favourable than others. With each opportunity, the Soviets readied three spacecraft, but quantity did not count for much as many of them failed.

In 1960, a launch window for Mars opened in October and it is known that three attempts were made, all ending in failure. The third resulted in a terrific explosion at Baikonur which destroyed much of the facilities and killed key personnel (see page 19). Three more attempts were made at the next launch window, the first at the end of October 1962. On 24 October, the first Mars vehicle reached Earth orbit but its third-stage rocket exploded: it was soon named Sputnik 22. On 1 November, the next vehicle was attempted, following an orbit much like its predecessor: it worked, and was soon heading towards Mars. It was named Mars 1. On 4 November, the third vehicle was launched: it suffered a similar fate to Sputnik 22, and was named Sputnik 24.

The elation over Mars 1 was short-lived: three months before it should have reached Mars, its antenna failed to point towards the Earth and contact was lost on 21 March 1963. When the next launch window occurred in November 1964, Zond 2 was launched only to suffer a similar fate: contact was lost less than six months later. It was left to Mariner 4 to enter the history books as the first successful Mars probe. Zond 5 was launched in July 1965 and travelled as far as the orbit of Mars while still in radio contact, but this success was also short-lived when two further Mars probes failed in 1969.

The 1971 launch window in May was one of the best this century, and the Soviets attempted to make the first landings on Mars. Unlike NASA's later Viking craft, the Russian approach was to 'drop' the lander into the atmosphere before the carrier craft went into orbit around the planet. The Soviet computer technology was limited, so that the lander could not be reprogrammed: they *had* to land where planners had decided before the mission began. As a result, the Mars 2 and 3

Left: The solar panels and central section on Mars 1 are seen in this Novosti photograph. The intricate web-like pattern behind the left-hand panel is the high-gain antenna with which the craft communicated with Earth.

Below: The structure of Mars 2 and 3 is seen in this pre-launch photograph. The lander is housed at the top, with the central pressurized compartment and radiator seen in the centre. The radiator (with its 'U'-bend) dissipated excess heat from inside the spacecraft. Its scientific instruments are seen at the bottom, pointing vertically downwards.

MARS LOG

The missions mentioned in this table are known to be Mars-related from their times of launch, orbital inclinations, etc. The different names given to each were assigned by Soviet authorities at the time.

Mission	Launch Date	Arrival Date	Comments
1960A	10 October 1960	–	Failed to reach Earth orbit.
1960B	14 October 1960	–	Failed to reach Earth orbit.
Sputnik 22	24 October 1962	–	Failed to leave Earth orbit.
Mars 1	1 November 1962	June 1963	Contact lost before fly-by.
Sputnik 24	4 November 1962	–	Failed to leave Earth orbit.
Zond 2	30 November 1962	August 1965	Contact lost before fly-by.
Zond 3	18 July 1965	–	Successful communications test out to Mars orbit.
1969A	27 March 1969	–	Failed to reach Earth orbit.
1969B	14 April 1969	–	Failed to reach Earth orbit.
Cosmos 419	10 May 1971	–	Failed to leave Earth orbit.
Mars 2	19 May 1971	27 November 1971	First Soviet Mars orbiter. Deployed Lander which failed.
Mars 3	28 May 1971	2 December 1971	Second Soviet Mars orbiter; lander also unsuccessful.
Mars 4	21 July 1973	10 February 1974	Intended orbiter; flew by when retro-rocket failed to fire.
Mars 5	25 July 1973	12 February 1974	Successful orbiter.
Mars 6	5 August 1973	12 March 1974	Intended Lander; contact lost before touchdown.
Mars 7	9 August 1973	9 March 1974	Intended Lander: vehicle missed planet altogether.

landers came to grief in ferocious dust storms. However, the Mars 3 lander transmitted for 20 seconds on the surface with the Soviets claiming that the vehicle had flown a hammer and sickle pennant. The 1973 launch window was not as favourable as the previous one, and so it was not possible to send combined Mars orbiter/lander craft. Mars 4 and 5 were orbiters which carried TV equipment in place of the landers, while Mars 6 and 7 just carried landers. Only Mars 5 was successful.

PHOBOS

What should have been the start of a new chapter in planetary exploration went disastrously wrong for the Soviet Union when both of its Phobos spacecraft failed. Their failure, the recriminations that have followed and the lack of confidence in the Soviet unmanned space programme have cast a blight over future missions to Mars and international collaboration.

The Soviets had heralded the craft as the first in a new generation of planetary spacecraft, involving unprecedented levels of international cooperation. The stated objects of the Phobos missions were to orbit the Red Planet, make a close pass of Phobos, its larger moon, and continue to investigate the planet and the solar wind environment.

Each Phobos spacecraft consisted of a central cylinder housing propellant tanks, inside of which was a pressurized compartment containing the main control and communications systems. The Soviets had made great truck of the fact that control was being facilitated via a computer which would allow greater flexibility of operation than earlier Mars probes.

The craft's main rocket engines fired through a lower hole in this central section. Two large solar panels extended from the central section which stretched for a total width of over 3.3 m (11 ft).

The Phobos craft were launched on 7 July and 12 July 1988 respectively by Proton boosters from Tyuratam. Both craft returned a steady stream of data about the solar wind and its interaction with the 'geotail', the region downwind of the Earth's magnetosphere. At the end of August, however, it was reported that contact had been lost with Phobos 1.

A hasty re-assessment of procedures led Western observers to believe that all would be well for Phobos 2. At the end of January 1989 it arrived in Mars orbit, and over the next three months proceeded to home in on its quarry. Phobos is

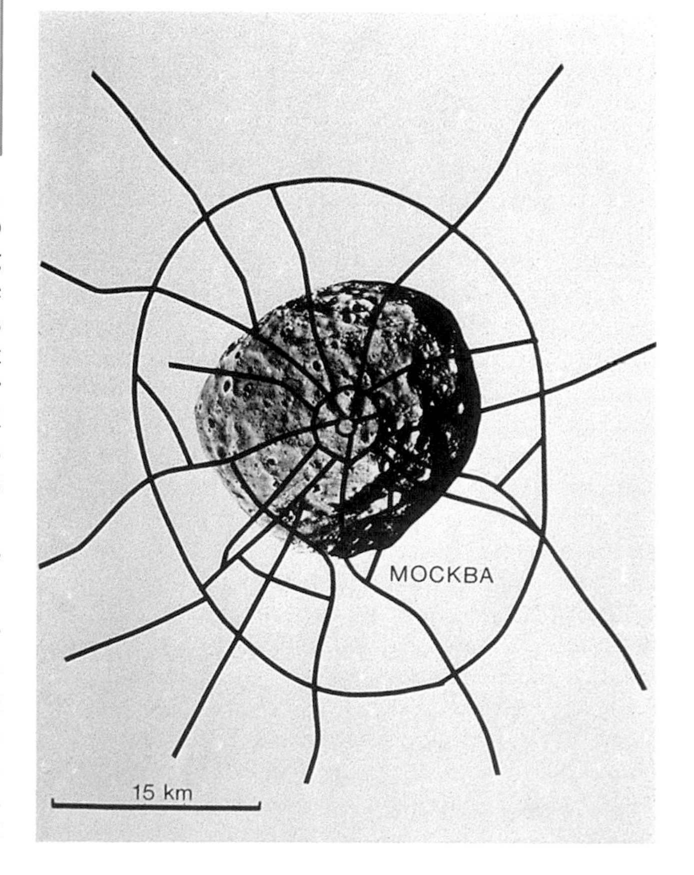

Right: The smallness of Phobos may be gleaned from this comparison of the major roads around Moscow (Mockba). Sadly, though the second Phobos craft was within days of making a close pass of its quarry, it failed.

minute, measuring only 20 × 23 × 28 km ($12\frac{1}{2} \times 14 \times 17\frac{1}{2}$ miles), so required unprecedented levels of navigational accuracy. Five sets of TV transmissions from mid-February onwards showed the target in site, and all seemed to be set for a close approach in the first week of April.

Then suddenly, on 27 March, shock news came from Moscow. Contact had been lost with Phobos 2. A final sequence of TV pictures, to enable the last manoeuvres to be performed, was being taken, which involved the craft (and communications aerial) pointing away from the Earth. Instead of automatically 'locking on' to Earth again, Phobos 2 failed to make contact.

A stream of data was then returned for 13 minutes, which stopped as suddenly as it began. Engineers tried to communicate with the craft in vain; contact had been well and truly lost. In mid-April the Soviets announced that from those last signals it appeared that the craft was rotating and had lost its bearings. Sadly, this was the ignominious end to the Soviets' most ambitious mission.

The bad luck which has dogged Soviet attempts to send probes to Mars has continued, and ambitious plans for roving vehicles, balloons and the eventual return of samples from Mars during the 1990s have now been put on hold.

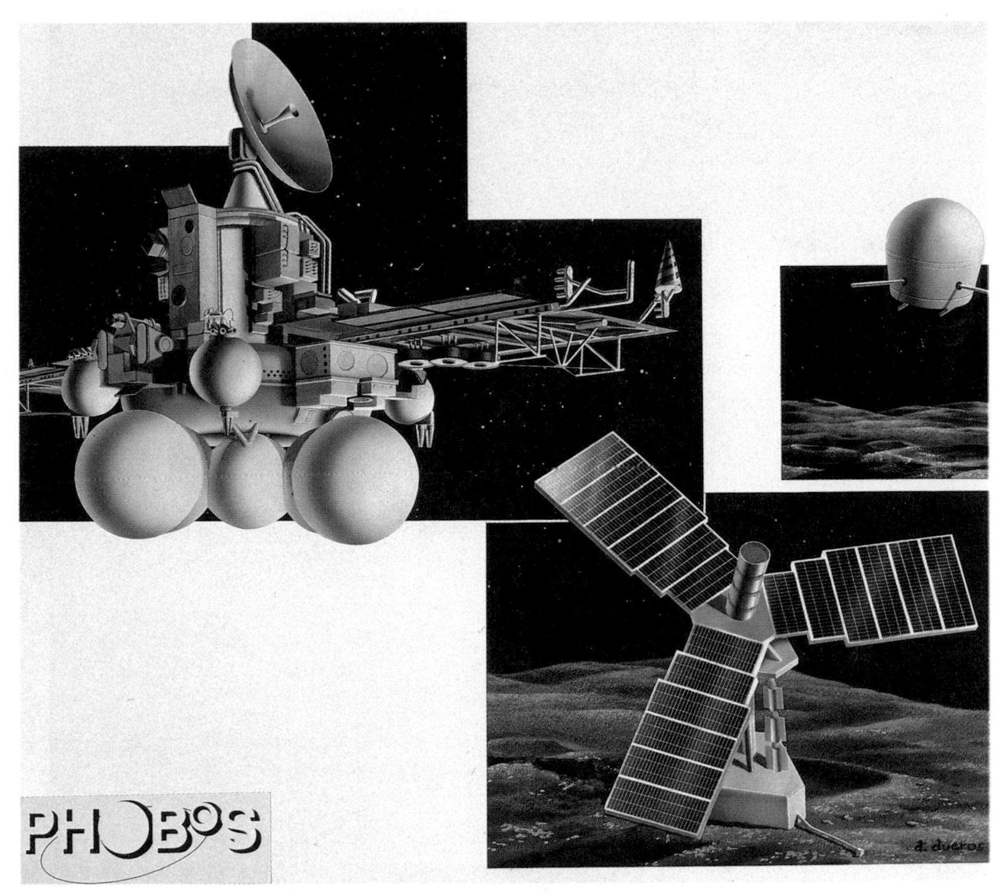

This artwork from the Centre National d'Etudes Spatiales, the French space agency, shows the three elements of the Phobos spacecraft. The main spacecraft is seen at left, with its windmill-like long-term lander beneath it. The football-sized 'hopper' is seen in mid-hop, its spring horizontal.

THE MOONS OF MARS

Phobos and Deimos were discovered as recently as 1877 by the American astronomer Asaph Hall. Subsequent Earth-based observations indicated that both moons were of low density, and in 1959 a Soviet theorist called Iosef Shklovskii suggested, with his tongue planted firmly in his cheek, that they might by hollow! However, Mariner 9 showed both to be heavily cratered objects in late 1971, corroborated by the Viking orbiters, one of which was manoeuvred to within 120 km (75 miles) of Phobos. By measuring its gravitational interaction with the spacecraft, it was concluded that Phobos had a density of around half that of the basaltic rock commonly found on the surfaces of the terrestrial planets. One of the Viking scientists aptly described the moon as 'more marshmallow than rock'.

The Vikings showed that the Martian satellites are also among the darkest objects in the solar system, reflecting only 6 per cent of the light incident upon them. Spectroscopic measurements revealed that their chemical compositions are similar to carbonaceous chondrites, a rock type found within the asteroid belt beyond Mars. This suggested that they were captured asteroids, a plausible hypothesis, but one with a number of problems. For a start, both moons are in circular, equatorial orbits: if they had been captured asteroids, they would be more likely to end up in eccentric orbits, inclined to the equator. Both moons are in orbits near to the theoretical limits at which they should become dynamically unstable. In the case of Phobos, which is only 5,980 km (3,737 miles) distant from the planet's surface, it is close to the point at which it should start to break-up due to the gravitational pull of Mars. Deimos, on the other hand, is near to the point at which it should escape the gravitational pull of Mars altogether.

Scientific experiments on the Soviet Phobos craft will address these problems. A laser (called LIMA-D) was fired from the main spacecraft, vaporizing soil particles, and allowing spectroscopic measurements of their composition. Because it is so small – 20 × 23 × 28 km ($12\frac{1}{2} \times 14 \times 17\frac{1}{2}$ miles) – the gravitational pull of Phobos is tiny: to escape the moon, a velocity of only around 13 m (43 ft) per second is needed. So there was a very real danger that the two probes would literally bounce off the surface! To prevent this, the larger of the two carried a harpoon-device to anchor it onto the surface. The smaller probe, quite aptly known in Russian as 'the frog', is a ball-shaped device that 'hopped' over the surface by the use of powerful springs. The full scientific results from the probes are expected to be announced in Paris in October 1989.

PIONEER

NASA's Pioneer spacecraft complemented the Mariner series in the first years of the space age and very quickly lived up to their name. The first four in the series were launched towards the

PIONEER LOG

Mission	Launch Date	Remarks
Pioneer 1	11 October 1958	All four were NASA's first probes to the Moon – sadly, none of them reached their target.
Pioneer 2	8 November 1958	
Pioneer 3	6 December 1958	
Pioneer 4	3 March 1959	
Pioneer 5	11 March 1960	Prototype 'interplanetary' spacecraft which returned data on the Sun and the solar wind.
Pioneer 6	16 December 1985	All four spacecraft together provided data on the solar wind and the interplanetary environment. All still operating twenty years later.
Pioneer 7	17 August 1986	
Pioneer 8	13 December 1967	
Pioneer 9	8 November 1968	
Pioneer 10	3 March 1972	Mankind's first probe to Jupiter. Successful fly-by on 5 December 1973. Left Solar System in June 1983.
Pioneer 11	5 April 1973	Reached Jupiter 3 December 1974, before becoming first probe to reach Saturn on 1 September 1979.
Pioneer Venus 1	20 May 1978	Entered orbit around Venus on 4 December 1978 and celebrated ten years of continuous operation. Returned first radar maps.
Pioneer Venus 2	8 August 1978	Four individual atmospheric probes carried by spacecraft 'bus' entered Venusian atmosphere on 9 December 1978.

Housed safely atop its Atlas-Centaur, Pioneer 10 is launched at 8.40 pm local time on 2 March 1972. Twenty-one months later it reached Jupiter, and just under a decade later left the Solar System altogether.

Moon, but all sadly failed. However, Pioneer 3 did discover the Earth's second radiation belt. Pioneers 5 to 9 were intended to investigate the interplanetary environment, and were launched in large orbits around the Sun. Pioneers 10 and 11 headed towards Jupiter, and Saturn was later added to Pioneer 11's itinerary. Both are now heading towards the stars and Pioneer 10 has already left the Solar System. The Pioneer Venus craft added immeasurably to our knowledge of our sister planet, and the Pioneer Venus Orbiter was still operating ten years after it had been launched.

PIONEERS 1 TO 4

NASA 'inherited' the first four lunar Pioneer spacecraft from the USAF after its formation on 1 October 1958. Just 11 days later Pioneer 1 was launched from Cape Canaveral. Though it failed to reach the Moon, it did show the extent of the Earth's vast radiation belts. Pioneer 2's Thor-Able third stage failed to fire after launch on 8 November 1958. And though Pioneers 3 and 4 missed the Moon altogether after launches on 6 December 1958 and 3 March 1959, they passed within 102,000 km (64,000 miles) and 60,000 km (38,000 miles) of the surface respectively. Pioneer 4 entered solar orbit and returned much useful information.

PIONEER 5

Pioneer 5 was launched on 11 March 1960 and returned information about the solar wind and solar flares until 26 June 1960. It stopped transmitting on that date at a distance of 36.8 million km (23 million miles) away from the Earth.

PIONEERS 6 TO 9

The next spacecraft in the Pioneer series were launched into orbits around the Sun to investigate the solar wind and its interaction with the Earth. Pioneers 6 to 9 were destined to become 'solar weather stations' and though they were designed to last a minimum of six months, all were still operating two decades later (Pioneer 9 was declared 'dead' in early 1986). An instrument aboard Pioneer 8 which had been used to measure electric fields in space was sucessfully switched on after it had been out of operation for 13 years in 1984. On 26 November 1988, NASA reported that Pioneer 6 had returned within 1.8 million km (1.16 million miles) of the Earth and was still operating.

All the spacecraft were drum-shaped and spin-stabilized. Solar cells covered their exteriors and three instrument-bearing booms stuck outwards from the spacecraft's body, each separated by 120°. During the late 1960s they provided data to allow scientists to predict solar flares that would affect the Apollo missions en route to the Moon.

'There is a light that never goes out.' Pioneer 10 as it now appears, travelling through interstellar space, away from the centre of our galaxy. The relatively bright star is the Sun, with the spectacular backdrop of the Milky Way in the mid-distance.

In 1973, Pioneer 6 became the first spacecraft to investigate a comet's tail when it measured the effects of Comet Kohoutek on the solar wind. The Interplanetary Pioneers showed that the solar wind is far from the gentle, steady flow it had been thought before the Space Age. The stream of ionized gases which spiral outwards from the Sun is much more violent: faster moving regions of particles pass through slower moving regions, causing shock waves where they slam into each other. The Pioneers were also able to monitor solar storms and their interaction with the Earth's magnetic field which give rise to 'geomagnetic' storms, ultimately responsible for long-term changes in the Earth's climate.

PIONEERS 10 AND 11

Launched a year apart in the early 1970s, Pioneers 10 and 11 are identical, spin-stabilized craft that were sent to explore the outer Solar System. Both probes carried radio-isotope thermonuclear generators, small canisters of Plutonium 238, for the provision of power, located on booms well away from the scientific instruments. A 2.7 m (9 ft) diameter radio dish provides each probe with the capability to receive and transmit signals to and from Earth.

Interest in Pioneers 10 and 11 focussed on the curious fact that they will ultimately become mankind's first envoys to the stars. Both craft carry a 15 × 23 cm (6 × 9 in) gold-plated aluminium plaque onto which a drawing of a naked man and woman standing next to an outline of the spacecraft (to give an idea of scale) have been etched. A schematic diagram shows the planets of the Solar System and the spacecraft's trajectory. To show where the probes have come from, the Solar System's position is shown relative to 14 nearby pulsars. These are the rapidly-spinning remnants of stars which give off high-frequency radio waves and act as unmistakable celestial 'navigation beacons'. Though most people approved the idea, NASA did receive a number of complaints from people about 'sending smut into space'.

Pioneer 10 passed within 131,200 km (82,000 miles) of Jupiter on 3 December 1973, and a year later, Pioneer 11 flew within 41,600 km (26,000 miles) of the cloud tops. Because it travelling faster than its twin, it spent less time inside Jupiter's vast radiation belts. The identical spacecraft carried some 11 scientific instruments destined to investigate the giant planet's magnetic and radiation environments. An infra-red radiometer measured the temperature of the cloudtops, whilst an ultraviolet photometer determined the amounts of hydrogen and helium in the atmosphere. The spacecraft's eyes were the Imaging Photopolarimeter which scanned the planet through blue and red filters in 'strips' as the spacecraft spinned. In this way it was possible to build up TV pictures of the planet.

Pioneer 11 was re-routed to visit Saturn and passed within 20,800 km (13,000 miles) of its cloud tops on 1 September 1979. Though many scientists wanted to fly the spacecraft within the planet's ring system, a more modest trajectory outside the rings was chosen. Ironically, Pioneer 11 passed within a few thousand kilometres of a previously-unknown moon, and also discovered a new outer ring, soon named the 'F' ring.

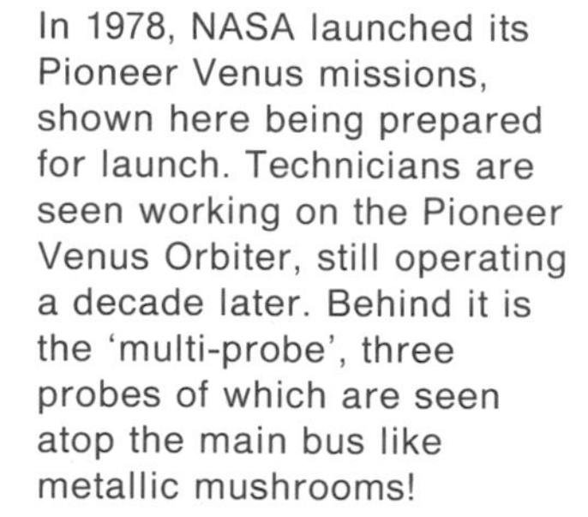
In 1978, NASA launched its Pioneer Venus missions, shown here being prepared for launch. Technicians are seen working on the Pioneer Venus Orbiter, still operating a decade later. Behind it is the 'multi-probe', three probes of which are seen atop the main bus like metallic mushrooms!

Both craft are travelling faster than 40,000 km/h (25,000 mph), the velocity needed to escape from the Sun's gravitational influence altogether. Pioneer 10 passed the orbit of Neptune in June 1983, so has effectively 'left' the Solar System. By monitoring the spacecraft's trajectory and spin, NASA scientists concluded in mid-1987 that Pioneer 10 showed no signs of being influenced by a tenth planet beyond Pluto, usually referred to as Planet X. If such a planet exists, Pioneer has relegated its existence to a highly inclined orbit virtually at right angles to the plane of the Solar System. Pioneer 10 is heading in the general direction of the constellation of Taurus and in 33,000 years time will pass near to an obscure star called Ross 248.

Pioneer 11 is heading in the opposite direction, and analysis of its motion will refine the possibilities of finding Planet X. As both spacecraft are still in good working order, their instruments are continually monitoring the solar wind in the outermost reaches of the Solar System. NASA hopes that the Pioneers will reach the 'heliopause' – the point where the solar wind interacts with the interstellar wind – before they pass out of communications range which should be some time in 1994.

PIONEER VENUS

The last spacecraft in the highly successful series of Pioneer probes were targeted to Venus and arrived there in December 1978. Pioneer Venus 1, more often referred to as the Pioneer Venus Orbiter (PVO), was first to arrive on 4 December. Pioneer Venus 2 was usually referred to as 'the Multiprobe' as it consisted of a main spacecraft 'bus' which carried four small landers as 'passengers'. The carrier 'bus' burned up in the dense Venusian atmosphere on 9 December 1978, before which the four probes had been successfully despatched.

The highlight of the Pioneer Venus Orbiter's scientific investigations was a radar mapping instrument, that curiously enough malfunctioned during the first few months of operation. The spin-stabilized craft entered into a highly-inclined elliptical orbit around the planet, so that in the meantime its other instruments (including spectrometers and polarimeters) could observe the Venusian poles for the first time. Ultraviolet cloud images were also taken, allowing meteorologists to learn more about the structure of the Venusian atmosphere. Within two years, 93 per cent of the surface had been mapped by radar so that topographic maps accurate to 100 m (328 ft) vertically could be assembled. In 1986, the UV spectrometer was used to look at Halley's Comet on its return to the inner Solar System. The spacecraft's orbit is slowly spiralling downwards, and in August 1992, the craft will burn up in the Venusian atmosphere.

The second craft's Multiprobe bus and individual entry probes carried a variety of instruments to measure the planet's atmospheric structure directly. Though none were designed to transmit once they reached the surface, one (called the Sounder Probe) survived for over an hour. Data from the probes revealed that there were three major layers within the cloud deck and a 14.4 km (9 mile) thick smog above the clouds. The clouds themselves seem to be made of sulphuric acid, and evidence suggested that billions of years previously there may have been oceans on Venus. It is theorised that the hydrogen escaped into space while the oxygen was absorbed by the rocks, leaving the planet to be the boiling, dry world seen today.

RANGER

NASA's Ranger series of spacecraft represented its first attempts to investigate the lunar surface in support of the Apollo programme. The Rangers crashed onto the Moon with the intention of returning high resolution TV pictures before impact. The first Rangers launched from 1961 to 1963 were unsuccessful for two main reasons. Firstly, the spacecraft had been designed to fly on a more powerful booster than the Atlas-Agena which was used, so no back-up systems could be carried. NASA also insisted that the spacecraft be 'sterilized' against contaminating the lunar surface, which involved the spacecraft components being heated to 125°C (257°F) for over a day. It was later shown that many of the spacecraft subsystems failed as a result of this sterilization process, the temperature being too high for many of the delicate components.

By 1964, an improved version of the Ranger had been developed which included more onboard batteries and, most importantly, the central computerized elements were not sterilized before launch. These later Rangers consisted of a hexagonal central body to which two solar panels were hinged. A conical tower contained the TV cameras which looked down onto the Moon via a cut-away, consisting of two wide-angle and four narrow angle cameras which could operate for about 15 minutes. However, the first of these new versions (Ranger 6) failed when its TV system shorted out after separation from its booster en route to the Moon.

On 31 July 1964, Ranger 7 finally succeeded: it had been aimed at a point in the Sea of Clouds where the surface appeared to be smooth and uncratered. It returned 4,308 pictures of the area which was then christened Mare Cognitum – 'the sea which is now known'. Ranger 8 returned over 7,000 TV pictures of the Sea of Tranquillity in February 1965, Both Ranger 7 and 8 showed that the lunar seas were smooth enough for moonlandings So Ranger 9 was targeted towards the crater Alphonsus in the lunar highlands, an area in which volcanic activity had been suspected. After launch on 21 March 1965, Ranger 9 hit the floor of the crater three days later. Despite the earlier problems, the final trio of Ranger spacecraft operated perfectly and returned much valuable information about our planetary neighbour.

As it sped towards the Moon, Ranger 8 returned this view of the Sea of Tranquillity from an altitude of 250 km (155 miles). Though they have long been superseded, the Ranger views gave astronomers their first really detailed picture of the Moon.

SAGIKAKE AND SUISEI

Japan's first deep space probes were part of the international armada of spacecraft sent up to investigate Halley's Comet in 1986. Though both were relatively unsophisticated, they should not be overlooked: they returned much useful information about the processes that occur in cometary tails. Both spacecraft were identical in construction, although they carried different payloads. They were spin-stabilized and cylindrical in shape: 137 cm (4 ft 6 in) in diameter and 96.5 cm (2 ft 4 in) in height. Sagikake's aim was to investigate the interaction between the solar wind and the comet at large distances from the nucleus. Suisei carried an ultraviolet imaging experiment to investigate the growth and decay of the comet's hydrogen corona and the interaction between the solar wind and the comet itself. Sagikake was aptly named – it means 'Pioneer' in Japanese – and was launched on 7 January 1985 from the Kagoshima Space Centre. Its main result was to find that the solar wind was disturbed by the comet some 7 million km (4.4 million miles) away from the nucleus. Suisei ('Comet') was launched on 18 August and headed within 150,000 km (94,000 miles) of the nucleus. Both craft are still orbiting the Sun and will reach the vicinity of the Earth in 1992, after which they may be re-directed to other objects in the Solar System.

SURVEYOR

In support of the Apollo Moonlandings, NASA's Jet Propulsion Laboratory developed the Surveyor programme to return detailed information about the Moon by landing on the lunar surface. After the problems with the Ranger programme Surveyor came as something of a blessing, as Surveyor 1 made a successful landing at the first attempt. All seven in the series were built by Hughes Aircraft, and directly descended to the surface without entering lunar orbit. The second crashed because its engines failed to fire for a soft landing and radio contact was lost with the fourth just before landing.

The Surveyors were just over 3 m (9 ft 10 in) high and had three landing legs, the theory being that a tripod would stay upright in a rugged area. Attached to the main, triangular-shaped body of the craft was the main antenna and a single solar panel. Scientific instruments included a steerable TV camera (which used mirrors that could rotate through 360°) which was carried on all seven of the spacecraft, used to return panoramic views across the landing site. Surveyor 3 had a sampling arm, while Surveyors 5 and 6 had a magnet attached to one footpad, and an alpha-scattering instrument to return data about the physical properties of the lunar surface. Surveyor 7 had all these plus the surface sampler. Such analytical instruments allowed scientists to determine the surface strength of the soil, as well as its major elemental composition and magnetic properties.

The Surveyors showed beyond a doubt that a manned lunar landing was possible. Indeed, Surveyor 3 – which made the first sampling of the lunar surface – was later visited by the Apollo 12 astronauts who landed nearby in the Sea of Storms. After landing, Surveyor 6 fired its soft-landing engines and 'hopped' a couple of metres away to test the strength of the soil. Whereas the previous landings were achieved in the equatorial regions of the Moon, Surveyor 7 landed near to the crater Tycho at 41°S, a remote area too difficult for the Apollo astronauts to reach.

JPL engineers test the landing system of a Surveyor mock-up on the Californian coast. Its descent was slowed by three tiny vernier engines and on landing, the shock was absorbed by the crushable aluminium footpads.

VEGA

The Soviet Union's contribution to the series of spacecraft which investigated Halley's Comet was launched in December 1984 and dropped balloons into the atmosphere of Venus in June 1985 before reaching the Comet in March 1986. Though based on earlier Venera technology the VeGas employed more sophisticated sub-systems allowing them to continue operating until early 1987. The VeGa spacecraft also saw unprecedented international cooperation in the planning, building and execution of the mission.

VeGa 1 was launched on 15 December 1984 and VeGa 2 on 23 December from Baikonur by Proton boosters. The spacecraft were named after the Russian words 'Venera' and 'Gallei' – the former meaning 'Venus', of course, and the latter, 'Halley' (there is no letter 'h' in the cyrillic

On 19 April 1967, Surveyor 3 landed on the Moon. It later 'hopped' a few centimetres by firing its vernier engines. At the end of 1969, Surveyor was visited by the Apollo 12 astronauts, who were amazed at the clarity of the footprints left by the probe.

alphabet). For the first time, the Soviets released TV pictures of the launches which revealed the Proton vehicle in operation some 15 years after its introduction into service. By June 1985, both were approaching Venus, and VeGa 1 passed within 40,000 km (25,000 miles) of the cloudtops on 9 June, and its twin within 25,000 km (15,600 miles) on 13 June. Both spacecraft 'buses' dropped landers with balloons into the atmosphere. Both landers were based on earlier Venera vehicles and carried a variety of instruments to investigate the atmosphere on the way down and the surface chemistry. It seems that the VeGa 1 lander malfunctioned and its surface drill began to operate before it reached the surface, but VeGa 2 was rather more successful. It landed in the equatorial mountain range Aphrodite and returned data for 57 minutes.

However, most attention on the VeGas at Venus focussed on the balloons that detached

SURVEYOR LOG

Mission	Launch Date	Landing Date and Site	End of Mission
Surveyor 1	30 May 1966	2 June 1966 Ocean Of Storms	7 January 1967
Surveyor 2	September 20 1966	Crashed 22 September 1966 near to crater Copernicus	
Surveyor 3	17 April 1967	20 April 1967 Ocean of Storms	4 May 1967
Surveyor 4	14 July 1967	Crashed 17 July 1967 in Sinus Medii	
Surveyor 5	8 September 1967	11 September 1967 Sea of Tranquillity	17 December 1967
Surveyor 6	7 November 1967	10 November 1967 Sinus Medii	14 December 1967
Surveyor 7	7 January 1968	10 January 1968 Crater Tycho	21 February 1968

from the landers at about 60 km ($37\frac{1}{2}$ miles) above the surface. They were originally proposed by the French, but weight restrictions meant that they were eventually built by the Soviets. An instrument 'gondola' was hung beneath a 3.54 m (11 ft 7 in) diameter plastic balloon filled with helium. The gondola was made up of three sections which contained temperature and pressure sensors as well as instruments to measure the cloud composition. Each balloon floated in the middle cloud deck of the planet, and neither found any conclusive evidence for lightning. Both survived for 46 hours and did not 'pop' as a result of increases in sunlight as they passed to the day side of Venus. Both balloons travelled a third of the way around the planet's circumference. Radio transmissions from the balloons allowed them to be tracked by a network of Earth-based radio telescopes, including NASA's Deep Space Network, 11 other dishes organized by the French space agency, CNES, and six Soviet antennas. Though the first balloon encountered more turbulence, the average wind speed seemed to be around 250 km/h (156 mph) – many times greater than had been expected.

Meanwhile the main spacecraft buses of the VeGas continued in their journeys towards the Comet. Each identical craft was about 11 m (36 ft) in length and had four large solar panels to provide the power – they were the only craft to investigate Halley which were not spin-stabilized. VeGa 1 made its closest approach to within 8,900 km (5,562 miles) of the cometary nucleus on 6 March 1986 and VeGa 2 within 8,100 km (5,062 miles) on 9 March, both on the sunward side. Each VeGa was equipped with two TV cameras and instruments to analyse the dust, such as spectrometers and impact detectors. The TV cameras used narrow angle and wide angle lenses as well as a highly-sensitive CCD (Charge Couple Device) to observe the cometary nucleus. The cameras were the result of work by Soviet, Hungarian and French scientists. VeGa 1 took around 300 individual images of the nucleus, which appeared to show a double nucleus – later found to be a dust jet, as the camera was slightly out of focus. VeGa 2 returned 700 pictures of greater clarity and there seemed to be less dust. Both VeGas' pictures were used by the European Space Agency to pinpoint exactly where the

This artist's impression shows the second VeGa spacecraft passing through the dust and gas surrounding Halley's Comet. The outline of the craft shows its central hub with solar panels. Both VeGas were in good shape after their encounters.

nucleus was for the close approach of the Giotto spacecraft on 13 March. After the Halley encounters, the VeGa spacecraft continued to orbit the Sun and were shut down in early 1987.

VENERA

Compared to its Mars series of spacecraft which were continually beset by disaster, the Soviet Venera series was rather more successful. As with the Mars spacecraft, early disasters were covered up with 'Cosmos' designations. Some Western analysts suggest there may have been as many as eight failures up until the launch of Venera 8 in 1972. Nevertheless, the Venera series scored a significant number of firsts, including the first penetration of the Venusian atmosphere, the first successful landings, the first Venus orbiters, the first TV pictures and the first detailed mapping. Considering that the Venusian atmosphere is 90 times as dense as the Earth's and the surface temperature is over 450°C (842°F), Soviet engineers had to use their ingenuity to ensure the Venera spacecraft could withstand the very great pressures and temperatures.

The first generation of spacecraft in the series (Veneras 2 to 8) were designed to land on the surface. Each consisted of a cylindrical main bus beneath which was located a spherical descent capsule. On reaching Venus, the lander separated and headed through the atmosphere; with each successive mission, improvements ensured that the vehicle could penetrate further downwards. The main parachute of the lander could withstand temperatures above 500°C (932°F) and opened in the upper reaches of the atmosphere. Venera 7 returned signals for 23 minutes after landing but the signal strength was very weak. So Venera 8 employed a second radio antenna which came into use on landing, and a refrigeration unit allowed it to operate for a total of 50 minutes after touchdown.

Veneras 9 and 10 were the next generation of Soviet Venus landers, and although they looked similar in design to their predecessors, they were much more sophisticated. For Venera 9 and 10, the main spacecraft bus entered orbit around Venus and acted as a relay for the lander's communications. This meant the landers required a less heavy communications system to allow a greater range of instruments. Another new feature occurred during the landing: three main parachutes were jettisoned 50 km (32 miles) above the surface to ensure the landers reached the surface as quickly as possible, where they were cushioned by shock absorbers. Almost immediately, the scanning TV camera began to record a panoramic view around the spacecraft, and within 15 minutes was returned to Earth. Scientists were amazed at the clarity. Venera 9 seems to have landed on a steep slope of a mountain, whilst Venera 10, by contrast, landed in a much lower, stonier region.

The Venera 11 and 12 missions seem to have been a repeat of the earlier missions, but because the launch window was not as good, the carrier bus did not enter orbit around the planet. Data had to be returned from the lander as the orbiters were above the horizon. It seems that although the landers returned data on the atmospheric composition – including electrical activity consis-

VENERA LOG

Mission	Launch Date	Arrival	Comments
Venera 1	12 February 1961	–	First Soviet Planetary Mission: Radio contact lost at 7.6 million km (4.75 million miles) from Earth.
Venera 2	12 November 1965	27 February 1966	Failed to return data as it passed within 24,000 km (15,000 miles) of cloud tops.
Venera 3	16 November 1965	1 March 1966	Contact lost before it reached planet; became first probe to impact planetary surface.
Venera 4	12 June 1967	18 October 1967	First probe to return direct data on Venusian atmosphere as it descended. Transmitted for 94 minutes during descent.
Venera 5	5 January 1969	16 March 1969	Descent probe returned data for 53 minutes before being crushed by atmospheric pressure.
Venera 6	10 January 1969	17 March 1969	Descent probe transmitted for 51 minutes.
Venera 7	17 August 1970	15 December 1970	First successful landing on surface of Venus; continued to transmit for 23 minutes afterwards.
Venera 8	27 March 1972	22 July 1972	Second successful landing; found levels of lighting much less than on Earth.
Venera 9	8 June 1975	22 October 1975	First Venus orbiter used to relay data from lander, including the first TV picture from the surface of Venus.
Venera 10	14 June 1975	23 October 1975	Second Venus orbiter and TV picture on landing. Landing site much lower and noticeably different to first.
Venera 11	9 September 1978	25 December 1978	Lander recorded lightning in Venusian atmosphere: no TV pictures returned.
Venera 12	14 September 1978	21 December 1978	Identical mission with Venera 11: recorded 15 minute thunderclap.
Venera 13	30 October 1981	1 March 1982	First colour TV pictures returned from surface, as well as surface soil analysis.
Venera 14	4 November 1981	5 March 1982	Identical mission with Venera 13; landing site on small hill.
Venera 15	2 June 1983	10 October 1983	
Venera 16	7 June 1983	14 October 1983	First high resolution radar mapping by Soviet Venus orbiters.

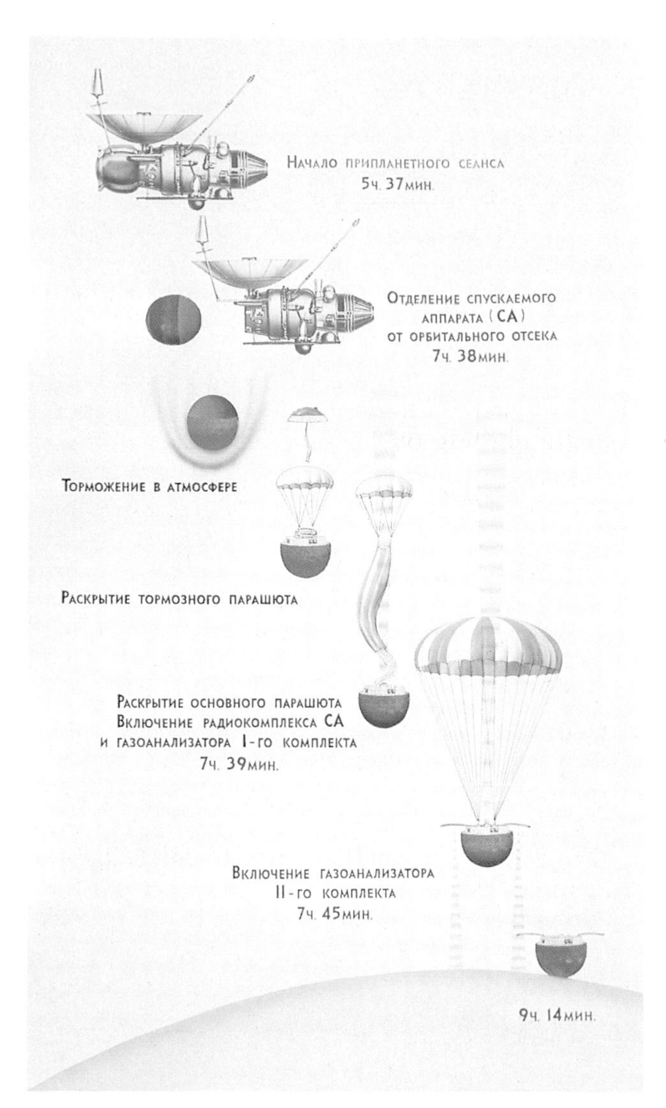

In 1967, Venera 4 successfully descended onto the surface of Venus – no mean feat, considering the awful conditions there. This Novosti illustration shows how the landing was achieved, with re-entry, heating and parachute deployment in the dense atmosphere.

tent with lightning – many of the instruments failed. Certainly no TV pictures were returned. The next in the series, Veneras 13 and 14, more than made up for these deficiencies. Again, both landers returned their data through the spacecraft carriers which had dropped them into the atmosphere and which then later 'played it back' for transmission back to Earth. Venera 13 returned eight colour pictures of the surface revealing a yellow/orange landscape strewn with rocks and stones. A sample arm was quickly extended, which used a drill and bit mechanism to remove surface material for analysis. Venera 14 landed 1,000 km (625 miles) to the south-east of its twin, and seems to have landed on a small hill where stones were noticeable by their absence. Chemical analysis of the soil revealed that at both sites it was basaltic in composition, similar to that found in Hawaii. Veneras 15 and 16 were placed into polar orbits around Venus with a low point of 1,000 km (625 miles) and a high point of 65,000 km (40,600 miles). Where there had been landers in the earlier craft, a radar mapper that looked sideways took their place. Both craft continued to function for over a year and mapped the Venusian poles for the first time.

VIKING

NASA's Viking mission brought its exploration of Mars in the 1970s to a fitting climax with the first successful landings on the red planet. The

The most recent Veneras were the 15th and 16th in the series, shown here being readied for launch in 1983. The propellant tanks of the central 'bus' are visible at the bottom as a technician works on a solar panel. The large radio dish at the other side was used in the radar mapping of the surface.

agency had hoped for a mission of at most three months' duration, but all the Viking spacecraft operated for many years. Each Viking spacecraft consisted of an orbiter based on the Mariner 9 design, with a lander safely ensconced in a 'protective aeroshell'. The main aim of the landers was to investigate the possibilities of life on Mars, so it was imperative that the spacecraft would not contaminate the planet with terrestrial microbes which might be mistaken for Martian ones! So the whole lander was 'sterilized' at a temperature of 112°C (44°F) for 40 hours before launch.

After reaching Mars orbit, the Viking orbiters found that the proposed landing sites for the landers were far rougher than had been expected, so its TV cameras were used to find alternatives. The Orbiter contained two other instruments on a scanning platform in addition to the TV cameras: a spectrometer called the Infra-Red Thermal Mapper measured the temperature profile of the atmosphere and surface; another instrument called the Mars Atmospheric Water Detector searched for water vapour in the thin atmosphere.

The Viking landers were perhaps the most complex machines ever built. Each was about 2 m (6 ft 6 in) tall and 3 m (9 ft 10 in) across, with a central hexagonal body attached to which were three landing legs. Two radio aerials allowed the lander to return data via the orbiter or directly back to Earth. The landers carried 11 experiments, including two TV cameras, an automatic weather station and a variety of spectrometers and analytical experiments. Surface samples were obtained by a sampling arm that dropped soil into a hopper device for analysis by three biological experiments – sadly, the results were deemed inconclusive. The cameras were located within 'turrets' above the main body of the lander which were able to rotate and produce panoramas of the Martian surface.

The Viking orbiters were shut down after their attitude control gas supplies were exhausted. Radioisotope Thermonuclear Generators were used to power the landers, and though Viking 1 had enough power to continue transmissions into the 1990s, an erroneous command accidentally switched its communications system 'off'.

VIKING LOG

	Launch	Arrival	Landing	End of Operations Orbiter	End of Operations Lander
V–1	20 August 1975	19 June 1976	20 July 1976	7 August 1980	November 1982
V–2	9 September 1975	7 August 1976	3 September 1976	24 July 1978	11 April 1980

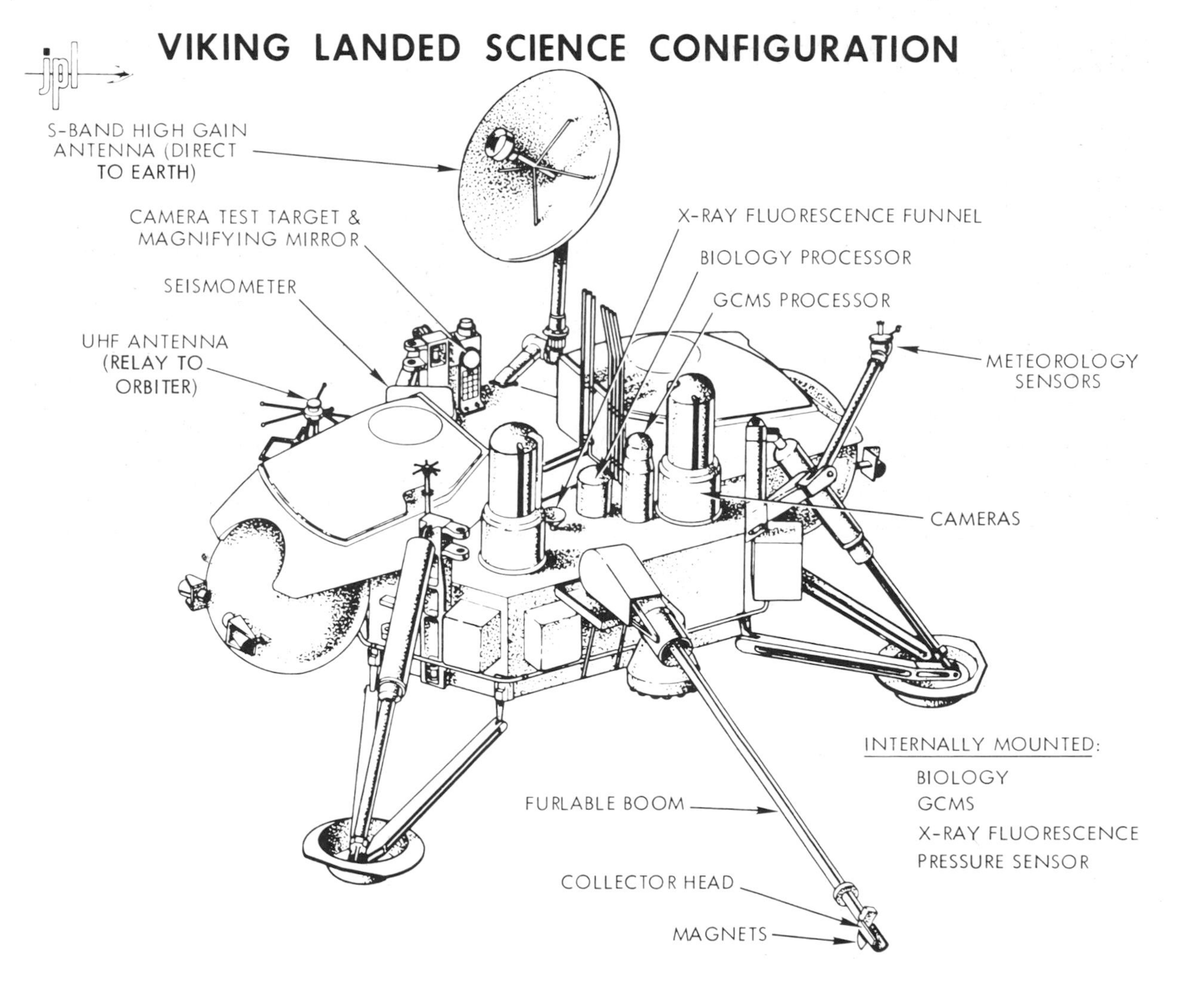

This schematic diagram shows the main instruments of the Viking lander. The furlable sample arm could be retracted and dropped through a funnel on the main body of the craft. The two TV cameras appear as turrets, the delicate optics contained inside. Because of a failure with the seismometer on Viking 1, it was not possible to pinpoint Marsquakes – the only failure in the mission.

After arriving in Mars' orbit, the Viking lander was despatched towards the surface of the Red Planet. The orbiter was an improved version of Mariner 9, while the lander was contained in a biologically-clean aeroshell. This protected the craft from the heat of entering the Martian atmosphere.

RESULTS

Both Viking landers have given us the most detailed insight into the surface of another planet. Both landed in the northern hemisphere, separated by a few thousand miles, and returned staggering TV pictures of a rock-strewn desertscape that was reminiscent of the Arizona desert. Viking 1 was very lucky: it landed only a few feet away from a large rock (later named Big Joe) which could have destroyed the spacecraft. It also saw sand dunes, which remained static, unlike the dunes on Earth that are blown around by the wind. One of Viking 2's footpads rested on a small rock which tilted the body of the spacecraft: as a result the horizon appears tilted by about 8*! Many of the rocks it observed appeared to be pitted with holes like Swiss cheese suggesting they were volcanic in origin: bubbles of gas would have escaped when they were molten.

The Vikings showed that the Martian skies are pink, because of dust suspended in the atmosphere which scatters red light more effectively. During the global dust storms, sunlight levels decrease considerably though the surface can still be seen. Because the Martian day is about the same length as our own, the Sun moves through the sky at roughly the same rate, though it appears two-thirds its size as viewed from Earth. The Vikings recorded that the sunrises and sunsets on Mars are white. There is an eerie red afterglow on the horizon after the Sun has set.

The Vikings also carried meteorological instruments that showed Mars experiences some fascinating weather. After its first day on the surface, the weather forecast from Viking 1 was: 'Light winds from the east in late afternoon, changing to light winds from the south-east after midnight. Maximum winds around 24 km/h (15 mph). Temperatures ranging from −83°C (−117°F) to −33°C (−27°F). Pressure steady at 7.7 millibars.' This pattern was repeated for many weeks afterwards, typical of the northern hemisphere summer. Viking 2 landed further north and measured temperatures that were cooler by about 10°C (50°F). As autumn approached, the wind patterns changed and it became more blustery and cooler.

During the winter time a strange weather phenomenon occurred: the atmospheric pressure dropped by as much as 20 per cent. Most (96 per cent) of the Martian atmosphere is made of carbon dioxide: at the poles temperatures around −150°C (−238°F) meant that it started to freeze out onto the surface, which lowered the atmospheric pressure. Such extremes of temperature and pressure mean that water cannot exist as a liquid on the surface – it would explosively evaporate! But clouds of water vapour are common, and the Viking 2 lander often returned pictures of frost, believed to be a mixture of water and carbon dioxide ice.

VOYAGER

The Voyager project has been the most successful planetary mission ever attempted: trying to summarize the scientific results from the mission is near impossible, but both spacecraft have shown that the outer planets are far more puzzling than had ever been thought. Astronomers have had to totally alter their perception of the outer Solar System.

The idea for a grand survey of the outer planets originates from the late 1960s. Planners at the Jet Propulsion Laboratory (JPL) in Pasadena, California, knew that in the mid-1970s the positions of the outer planets were so aligned that it would be possible to visit each in turn by the technique of gravity assist. (By using the gravitational field of each planet, a spacecraft could head to the next planet out without need for an extra rocket motor.) Originally it was intended that a 'Grand Tour' would be possible, in which two spacecraft would visit all the outer planets, including Pluto. Cutbacks meant that the mission was scaled down to visit just Jupiter and Saturn. Approved in 1972, it was known as Mariner Jupiter-Saturn before its name was changed to Voyager in 1977. By that

This enhanced view from Viking 1 shows the surface of Mars as it would appear to the naked eye. The instruments on the body of the spacecraft – seen in the foreground – are obvious.

time, JPL engineers realized that the route of the second spacecraft could be altered to include Uranus and Neptune in its path, but only if the first were successful. Thankfully, despite occasional mishaps, both Voyagers have performed far beyond their original specifications.

THE SPACECRAFT

Both Voyager spacecraft are identical and were built by the Jet Propulsion Laboratory and consist of a central compartment with propulsion module. At the top of this main compartment is a large radio antenna some 3.7 m (12 ft) in diameter. On one side of the central module is its power source or 'Radioisotope Thermonuclear Generator' – essentially three containers of uranium that undergoes radioactive decay to provide power. In the outer Solar System, the Sun becomes progressively dimmer so that solar panels are unthinkable. Because of the Uranus and Neptune option, Voyager 2's RTG provides a higher power output. To avoid interference from their radiation, most of the scientific instruments are located on the opposite side of the spacecraft from the RTG. This science boom contains most of the scientific instruments, including a scanning platform on which the remote sensing instruments are located. The notable exceptions are the magnetometer located on a 13 m (42 ft 7 in) long boom which was unfurled after launch, and two whip aerials for radio astronomy purposes. Both spacecraft are stabilized on three axes, and the central bus contains the fuel for flight manoeuvres (four thrusters) and 12 attitude control thrusters as well as housing the spacecraft's main computers and back-up systems.

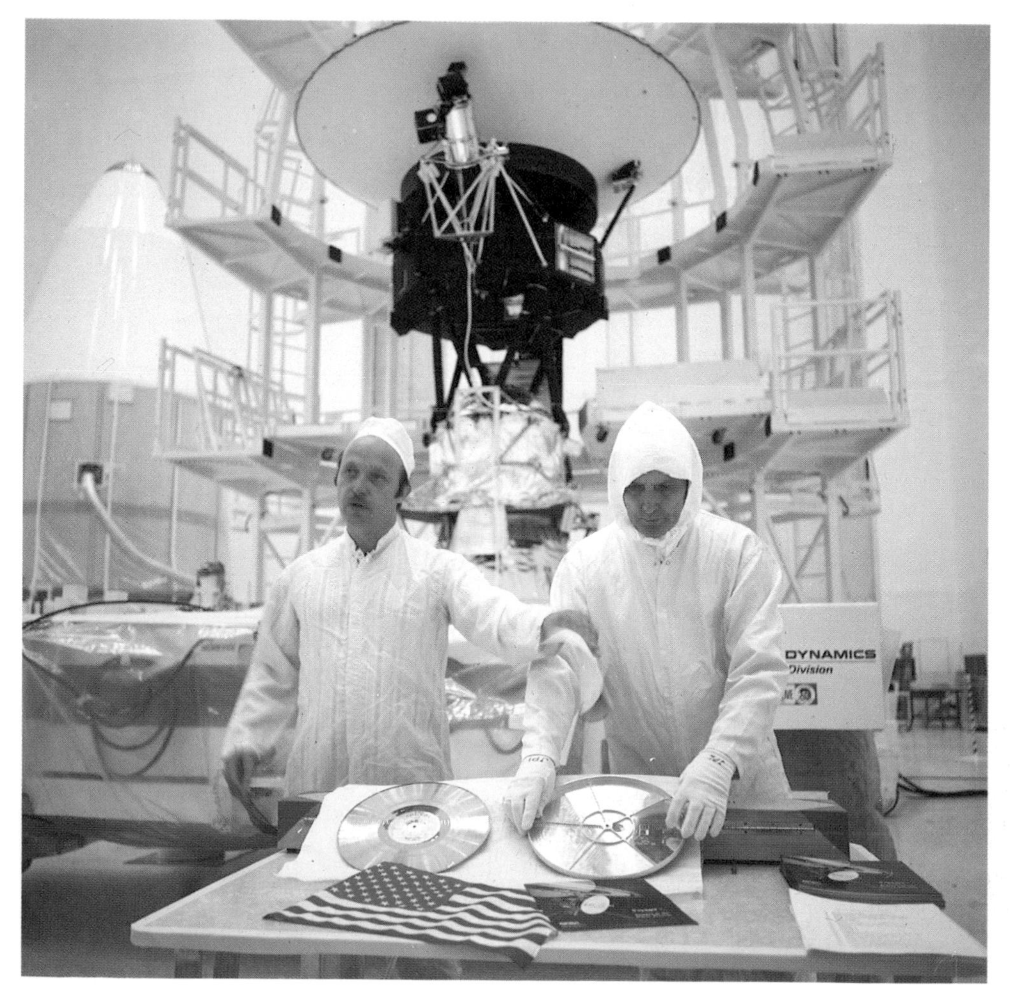

Technicians prepare the final items of Voyager's payload in the 'clean room' at Cape Canaveral. Project Manager John Casari (left) is seen explaining the purpose of the gold-plated 'interstellar record'. An aluminium cover, also gold-plated, is being held by a technician.

The full tally of scientific instruments is listed below: the first four are located on the spacecraft's scan-platform which allowed the instruments to 'target' their observations.

IMAGING SCIENCE

Two TV cameras enable high resolution reconnaissance of planets, moons and rings in the outer Solar System. Weather systems on the planets, geological structure of the moons and structure of rings were all scrutinized by Voyager's imaging experiment. Both wide- and narrow-angle cameras are carried to allow the optimum study of objects. The wide-angle camera has an f/3 20 cm focal length lens, whilst that for the narrow-angle camera is 150 cm at f/8.5. A rotating filter wheel allows different colours to be selected which allow specially enhanced views to be produced. The light from the lenses reach a slow-scanning selenium-sulphur television tube which takes 48 seconds to 'take' one picture.

INFRA-RED INTERFEROMETER SPECTROMETER (IRIS)

The infra-red spectrometer known as IRIS uses a 51 cm diameter telescope to observe spectra over 2,000 wavelength bands in the 4–50 micron radiation band. The 'interferometer' comes from the fact that light of different wavelengths interfere to produce a pattern which is transmitted back to Earth where it is analysed by computer. The spectra allow atmospheric composition, thermal structure and dynamics to be determined. The surface composition of satellites and their thermal structure can also be inferred from the data.

ULTRAVIOLET SPECTROMETER (UVS)

A straightforward spectrometer with a diffraction grating enables 128 wavelengths to be analysed in the 50 to 170 nanometres range. Its aim is to observe the structure and composition of the upper atmospheres of the outer planets, as well as aurora and electrically-charged particles trapped in their vast magnetospheres.

PHOTOPOLARIMETER (PPS)

A 15 cm aperture Cassegrain telescope is able to accurately measure the brightnesses and polarization of the light observed. Unlike the TV imaging system, it can only observe one particular point at any time. Sadly, the instruments on both spacecraft were plagued with troubles, particularly with the filter wheels which allowed different coloured light to be observed. Voyager 1's instrument failed completely, but that on Voyager

This 'raw' image of Saturn's moon Titan has been enhanced to show the extent of its dense, nitrogenous atmosphere. The moon is eclipsing the Sun, though light is being scattered by its atmosphere.

2 was brought to life for its encounter with Saturn when it observed the planet's rings were made of thousands of individual ringlets.

PLANETARY RADIO ASTRONOMY (PRA)

To 'listen' in to the radio emissions from the outer planets, Voyager has two 10 m (32 ft) long metal poles as receivers which are tuned to between 1.2 kilohertz and 40.5 megahertz – a wide range of frequencies. As a result of interactions between charged particles in the magnetic fields of the outer planets, a variety of emissions are given off. The PRA enables scientists to determine their origin, though to the 'untrained' ear, when they are converted to sound signals and played through a speaker they sound like hiss and static!

MAGNETOMETER

The instrument consists of two different components which monitor the intensity and direction of the magnetic fields surrounding the spacecraft as it passes through the magnetospheres of the planets it is encountering. The two instruments allow the magnetic interaction of the spacecraft to be eliminated, and one of them is carried on a 13 m (42 ft 7 in) boom which slowly unfurled after launch. When analysed on Earth, the magnetometer measurements allow scientists to determine the intensity of the planets' magnetic fields and the processes occurring within them.

PLASMA PARTICLES

The interaction of the solar wind and the magnetospheres of the planets and moons is observed by this instrument. A 'plasma' is a gas made of electrically-charged particles where there are equal numbers of negatively-charged electrons and positively-charged protons, resulting in zero net charge. If the electrons and protons have velocities less than 0.1 per cent of the speed of light (in the energy range 10 – 60,000 electron Volts, where an 'electron Volt' is the amount of charge one electron possesses) then this instrument will be able to detect them.

PLASMA WAVES

Unlike an ordinary gas, plasmas are easily affected by the influence of the solar wind – because both are electrically-charged. As a result, the plasma oscillates in density and produces electric fields which cover the audio range of frequencies (0.01 to 56 kHz). From such measurements, it is possible to determine the oscillations within the plasmas surrounding the spacecraft and determine their temperatures. The plasma wave instrument uses one of the PRA antennas.

LOW ENERGY CHARGED PARTICLES

Charged particles with speeds up to a few per cent of the speed of light cannot be detected by the plasma particles instrument. However, they are detected by this instrument in the energy range 10 kilo electron Volts (keV) to 11 million electron Volts (MeV) for electrons and 15 keV to 150 MeV for protons and ions. The instrument is thus highly sensitive over a wide range of energies

VOYAGER LOG

Craft	Launch Date	Dates of Encounter			
		Jupiter	Saturn	Uranus	Neptune
Vgr 1	5 September 1977	5 March 1979	12 November 1981		
Vgr 2	20 August 1977	9 July 1979	25 August 1981	24 January 1986	24 August 1989

which allows scientists to determine the distribution, composition and flow of energetic particles surrounding each planet.

COSMIC RAY TELESCOPE

The highest energies measured by Voyager are cosmic rays, which are highly-charged particles which originate from beyond the Solar System. The description of a 'ray' comes from their interaction with the Earth's upper atmosphere, when X- or Gamma-Ray showers are produced. In a sense, the solar wind protects the planets from this constant bombardment, but as Voyager heads towards the interstellar medium it will reveal more about their mysterious origins. As these particles range in speed from 10 per cent to 99 per cent of the speed of light, it requires a detector sensitive to energy regions from 0.5 to 500 MeV.

RADIO SCIENCE

The Voyager spacecraft's radio signals are used to reveal much about the outer Solar System. To study many of the physical properties of the outer planets, such as their masses and densities, requires accurately measuring their interaction with the spacecraft's motion. By careful monitoring of its radio transmissions, such interactions can be measured. As the spacecraft passes behind the rings and atmosphere of planets (as seen from the Earth), the way in which the signal is altered reveals much about their structure.

THE MISSIONS

Voyager 2 was launched first, but on a longer flight path, so that Voyager 1 actually reached Jupiter first. The mission timeline is shown below. There was some concern with Voyager 2's telecommunications systems, particularly after the primary radio receiver failed six months after launch and then it was realized that the secondary system could not 'lock onto' signals from the Earth. This is because of the Doppler effect – the frequency of the radio signals sent from Earth is altered because the spacecraft is heading away at tremendous speed. The only solution was to calculate the frequency at which radio signals would be 'heard' by Voyager and then change the frequency of broadcast to compensate for the Doppler shift. Despite the complex hindrance that this entails, the system has worked well and Voyager 2's telecommunications system has behaved itself ever since.

After its Saturn encounter in late 1980, Voyager 1 was directed upwards and away from the plane of the Solar System. Its fields and particles instruments are still operating, though its TV cameras were switched off a month after passing Saturn. It had been suggested that the spacecraft be rotated to allow a TV picture of the whole of the Solar System be taken. However, it was soon realized that the cameras would reveal very little, and would waste valuable fuel required to keep the spacecraft aligned correctly.

As Voyager 2 reached Saturn, it passed underneath the planet and through the plane of the rings revealing much information about their structure. However, 45 minutes after passing through the ring plane, the scan platform jammed and though the scanning instruments continued to operate they were pointing in completely the wrong direction because the gearing mechanism of the platform malfunctioned. Within a few days, it was possible to unjam the platform, and it was later determined that lubricant had been lost as a result of repeated high-speed scanning motions. As luck would have it, the platform had been jammed in a position which would allow scientists to observe Uranus and Neptune with minimal scanning motions of the platform.

By the time Voyager 2 began to return data from Uranus in November 1985 all seemed to be well with the spacecraft. For Voyager engineers, the most important aspect of the encounter was to ensure that the quality of data transmitted from the spacecraft was not lost because of its increasing distance from Earth. So they devised some ingenious ways of communicating with the spacecraft. Firstly, they lowered the rate at which data was transmitted back to Earth by increasing the time taken to transmit each 'bit' of information. This is rather like speaking slower to enunciate more clearly, and though less is said, there is a greater chance of being heard.

Computer techniques were developed to 'squeeze' the data returned, performed by transmitting new software up to Voyager. Each TV picture, for example, is made up of 800 lines with each of those lines in turn containing 800 picture elements or 'pixels'. Because 256 exposure levels are possible within each pixel, an individual picture requires over 5 million bits of information for transmission. The only way to return high-quality pictures from Uranus was to use less data bits, in other words, describing the picture in fewer words. Instead of returning the actual exposure of every pixel, only that of the first of each line is transmitted. Thereafter the difference in exposure between each successive pixel and the first is returned, so less data bits are needed. Another technique was to allow the scanning instruments to track their targets while taking long exposures – noon on Uranus is dimmer than dusk on Earth! So the entire spacecraft was rotated while the shutter was open for exposures up to 15 seconds in length.

By the time Voyager reached Neptune, the spacecraft was nearing 5,000 million km (3,200 million miles) from Earth, so there is a further handicap to receiving data clearly on Earth. For many years now, NASA has operated the Deep Space Network (DSN) of three 64 m (209 ft) radio dishes in Goldstone, California, Madrid and Tidbinbilla, Australia. They are positioned so that at least one of them will be in communication with the spacecraft when they are above the horizon. For Uranus, DSN engineers electronically linked two or more of the dishes together to reinforce the signal strength received as though a single, larger dish were operating. For the Neptune encounter, the Goldstone dish was linked with the National Radio Astronomy Observatory (NRAO) in New Mexico, a series of 27 movable radio dishes. This will allow data to be received at the same rate as the Uranus encounter.

At Neptune, Voyager 2 will fly over the planet's north pole within 5,000 km (3,125 miles) then head downwards, passing within 40,000 km (25,000 miles) of Triton, before leaving the plane of the Solar System by 48°.

THE VOYAGER INTERSTELLAR RECORD

Like Pioneers 10 and 11, the Voyagers are heading out of the Solar System and towards the stars. The spacecraft will pass out of communications range by 2010, so Voyager scientists are confident that the location of the heliopause will be found. In 40,000 years time, Voyager 1 will reach an otherwise obscure dwarf star called AC +79 3888 in the constellation of Camelopardus. Similarly, Voyager 2 is heading towards the star Sirius, and will pass close by it 360,000 years hence.

Being much more sophisticated than the Pioneers, the Voyagers also carry much more sophisticated messages from its builders. A 0.3 m (12 in) diameter gold-plated copper record has been encased in a protective aluminium jacket and carried on the main spacecraft bus. A variation of the Pioneer plaque has been etched on it with instructions how to play the record: just in case any intercepting lifeforms do not possess a hi-fi system, NASA has thoughtfully provided a cartridge and needle. The 'Voyager Interstellar Record' doubles as a videodisc and contains 115 images of life on Earth as well as a choice selection of sounds, languages and music. The 60 languages range from Sumerian used in biblical times, to Wu, a modern Chinese dialect. The music ranges from Melanesian pan pipes to Chuck Berry, though sadly, it wasn't possible to include the Beatles' 'Here Comes The Sun'!

The choice of music and images was overseen by the noted American astronomer, Carl Sagan. Responding to criticisms of the futility of such gestures, Sagan has suggested that launching such 'bottles into the cosmic ocean' says something very hopeful about life on Earth.

Voyager's original flight plan called for encounters with Jupiter and then Saturn. While Voyager 1 has headed upwards and outwards after its Saturn encounter in late 1980, Voyager 2 has since reached Uranus and Neptune.

EARTH SATELLITES

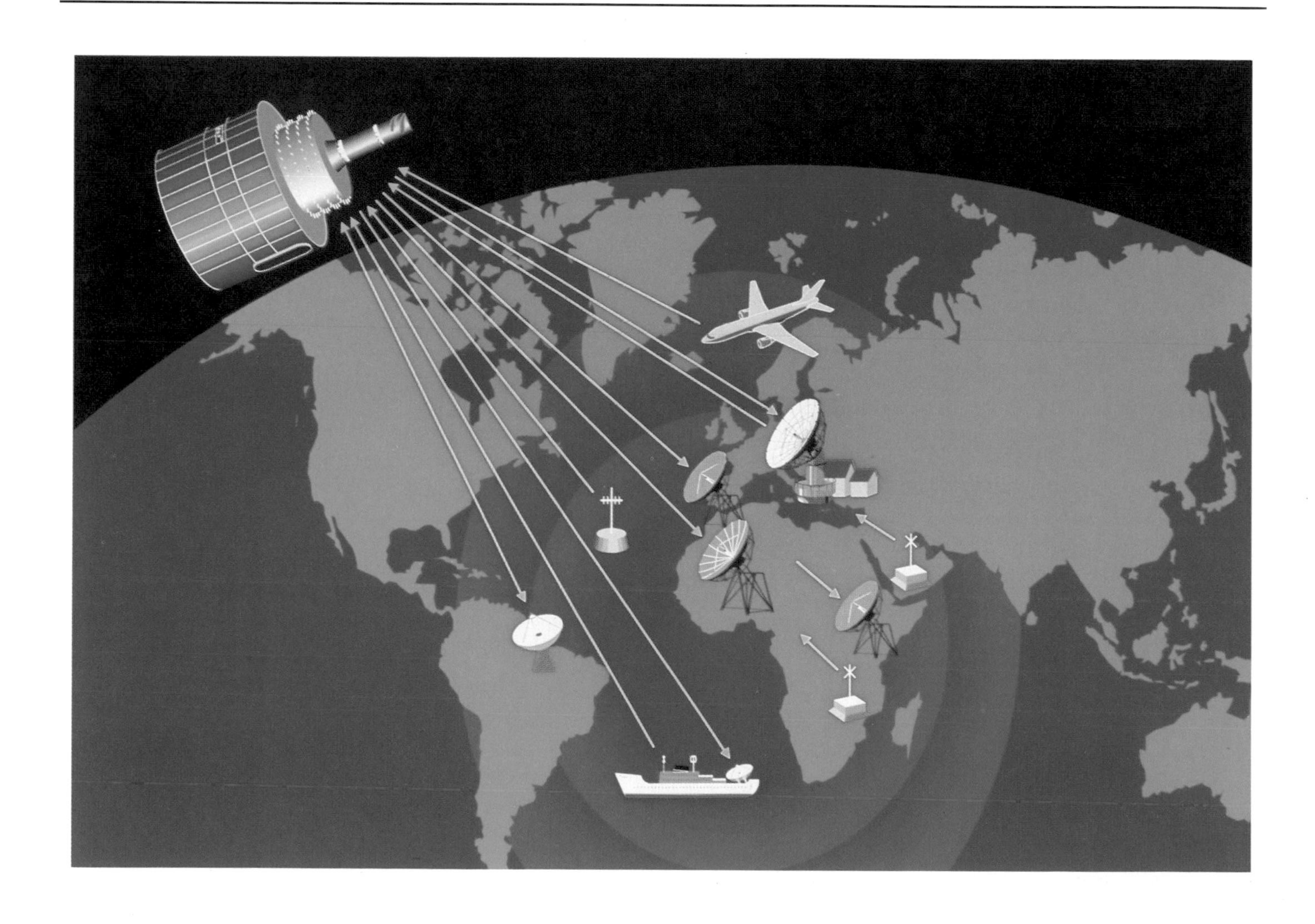

By the end of 1988, over 3,000 satellites had been launched by the spacefaring nations of the world. In the few pages available, it would be nigh impossible to cover every single one without reverting to a vast, boring list of launch dates and satellite names that would tell the reader very little. So, here are summarized the details of a couple of dozen of the more interesting satellites which have been launched since 1957 – many of which were mentioned in the Developments in Space section at the start of this book.

ATS (ADVANCED TECHNOLOGY SATELLITE)

Perhaps the finest demonstration of the uses of geostationary orbit were NASA's Advanced Technology Satellites (ATS). The series of six were envisaged to perform the first in-orbit tests of communications, weather and navigation techniques for later, more advanced satellites. Despite some failures (ATS-2 in April 1967 and ATS-4 in August 1969 – both suffered launcher failures) the series far exceeded expectations. ATS-1 was launched in December 1966 and entered orbit over the Pacific where it was used to return the first full-globe weather pictures of the Earth, and relay live TV pictures in the Pacific regions. ATS-3, launched in November 1967, returned the first colour TV weather pictures and was later used to monitor storm systems to provide flood warnings for farmers. It was still operating in 1985, stationed at 105°W where it covered most of the North American continent. In March 1985, it was used by the emergency services for communications as a result of the Mexican earthquake.

By far the most famous of the series was undoubtedly ATS-6 which provided the first educational lessons from space. ATS-6 was as tall as a two-storey house, essentially a large, pointable radio dish with an Earth-resources package underneath. In mid-1975 it was used to re-transmit TV pictures from an Indian ground station to more than 2,000 villages equipped with only 3 m (9 ft 10 in) dishes and TV sets for educational purposes. For the next year, ATS-6 broadcast basic education lessons for around four hours per day across the Indian sub-continent. Arthur C. Clarke described ATS-6 as 'the most ambitious and important educational experiment in history'.

COSMOS

By the end of 1988, the Soviet Union had launched nearly 2,000 in its Cosmos series, perhaps the most wide-ranging of any series of satellites. Though many analysts are quick to point out that over half those launched were military in nature, the rest cover a remarkable range of uses. The Cosmos series numbers scientific, communications, and meteorological satellites as well as failed lunar and planetary missions and unmanned tests of manned spacecraft. As a result, the spacecraft have varied in size from relatively small spheres to vast laboratories like the Cosmos 1686, the so-called 'Star Module' which weighed over 50 tonnes and carried astronomical, biological and technological workshops. An average of 90 satellites have been launched per year: the maximum was 101 in 1976, out of which 90 were directly related to military systems. Most of the Cosmos launches have been from the Northern Cosmodrome at Plesetsk, but others

Above: The most obvious feature of ATS-6 was its 9m (30 ft) diameter parabolic mesh dish. The main 'bus' of the craft contained the Earth Viewing Module, including electronics and control systems. Experiments were located above the antenna.

Opposite: ESA's Meteosat system disseminates different information to 'end-users' on Earth. The network is two-way, as weather readings from aircraft, ships and ground stations are sent up to the satellite and processed though central ground stations.

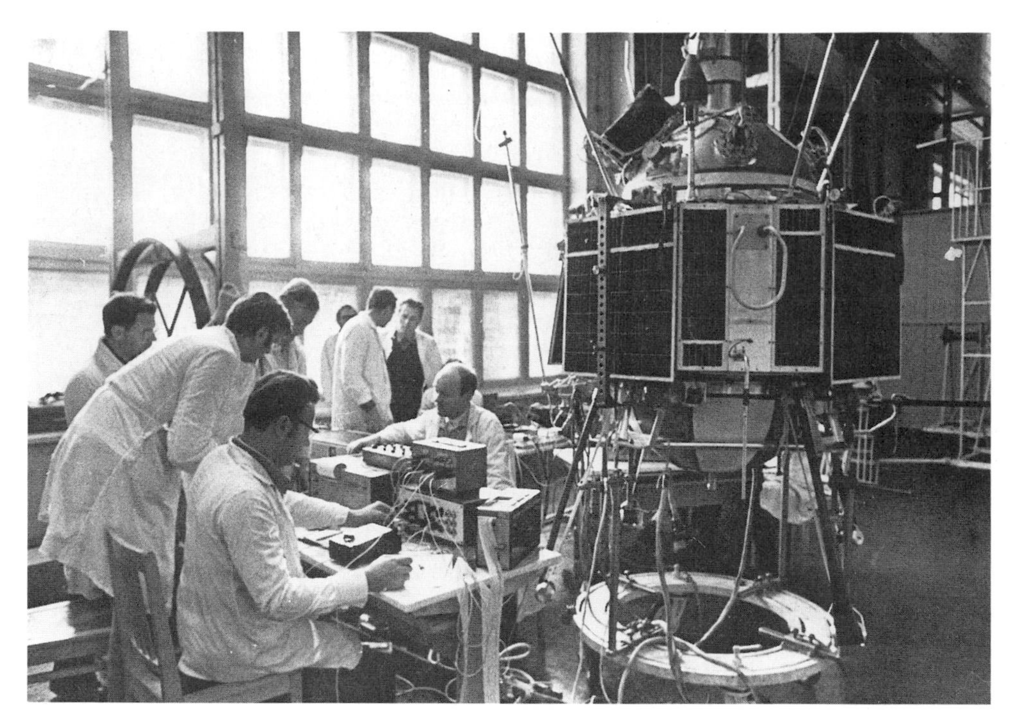

Above: The USSR's Cosmos programme notched up its 2,000th launch in 1989. Among its most notable were the Intercosmos series, the 10th of which is seen here. Launched in 1973, it investigated the Earth's ionosphere and magnetosphere.

have used Kapustin Yar and Baikonur.

The Military Cosmos flights run the full gamut of military applications satellites, such as electronic 'ferrets' and reconnaissance roles as well as the rather more dangerous hunter-killer satellites and RORSATs (Radar Ocean Reconnaissance Satellites). Since the mid-1960s the Soviet Union has launched nearly 30 nuclear-powered RORSATs to snoop on foreign naval powers. They carry powerful radar dishes to monitor maritime traffic from only 270 km (169 miles) above the sea. Nuclear fission provides a great deal more power than solar panels, but has led to very grave worries when they have re-entered the Earth's atmosphere far sooner than expected. Although they were designed so that their reactors (containing 50 kg (110 lb) of enriched uranium) would separate from the main body of the spacecraft on re-entry, this failsafe mechanism failed in January 1978. Cosmos 954 fell foul of atmospheric drag and re-entered before the reactor was ejected to a higher orbit. Low-level radioactive debris was strewn over the Great Lakes in Canada. Soviet RORSATs were then redesigned so that the uranium core could be separated from the reactor itself. But there were worries when Cosmos 1402 re-entered in 1983 and Cosmos 1900 in 1988 with no indication of separation – but thankfully separation had occurred. However, an estimated tonne of fissile products remains in Earth orbit for the next generation to clear up.

The scientific Cosmos satellites, however, have been rather more of a blessing to the Soviet Union. Cosmos 1 in March 1962 was a simple scientific satellite which monitored the ionosphere. Subsequently, the Cosmos series has included satellites to investigate the Aurora Borealis, carried ultraviolet telescopes, measured ocean temperatures, and mapped polar icefields.

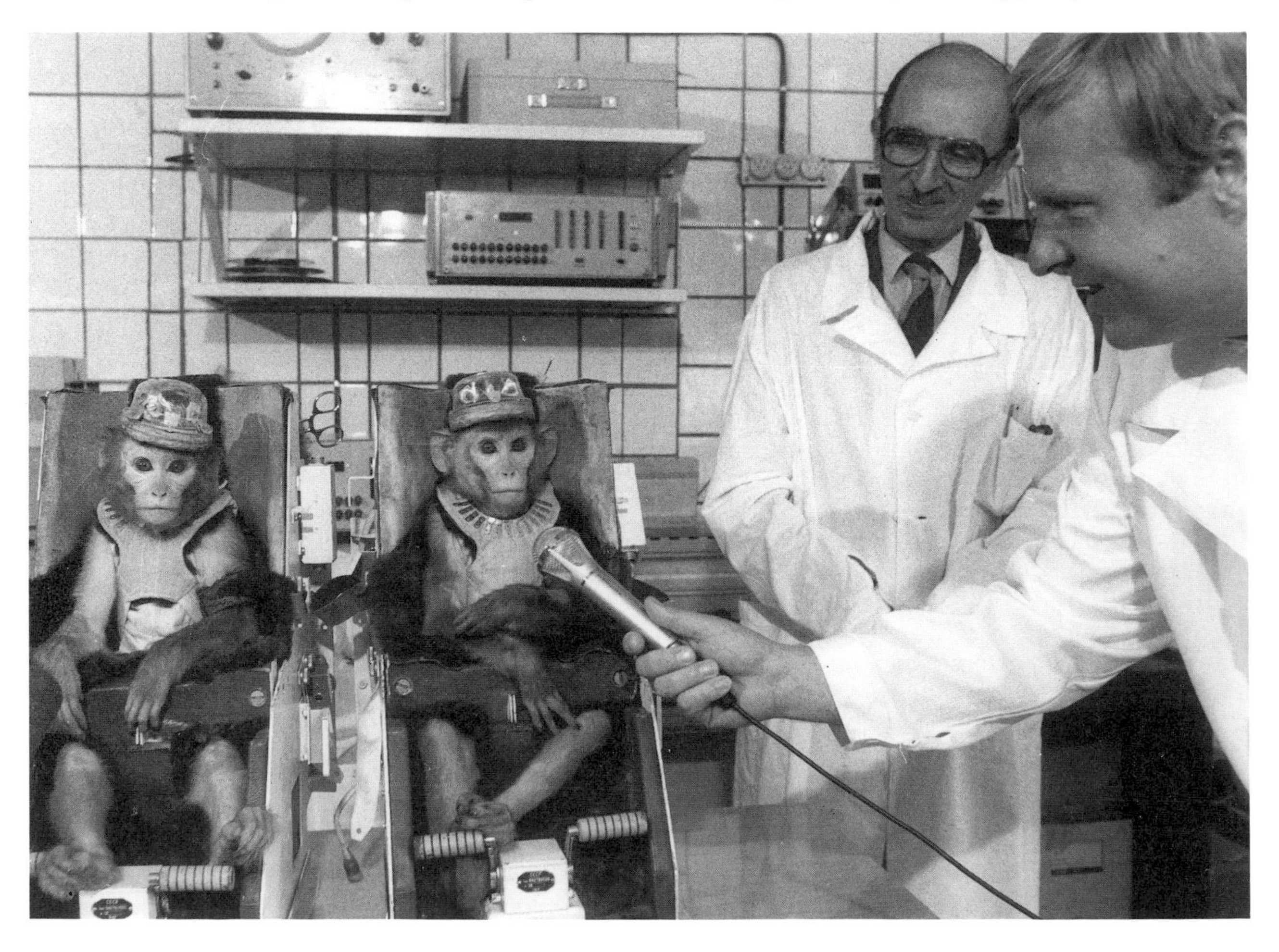

Right: At Moscow's Institute for Biomedical Problems, two more monkeys patiently await their flight aboard a 'Biosatellite'. Dr. Oleg Gazenko, seen next to the monkeys, shows paternal pride while a colleague tries some impromptu reporting. The monkey's reactions are not recorded.

Many have been used for remote sensing which has become an important part of the Soviet economy. In late July 1987, for example, Cosmos 1870 was launched into a polar orbit and, with a weight approaching 21 tonnes, was as large as the polar platform free-fliers ESA and NASA will use in the 1990s as part of the International Space Station. It has been used to look at water resources, observe the weather and produce radar profiles of, and accurately map, land within Soviet borders. Since 1967, 9 Soviet bloc countries have taken part in the Intercosmos programme where the Soviets have launched satellites with instruments provided by foreign countries. The first was Cosmos 261 in December 1968 which investigated the upper atmosphere. The programme has been expanded to include French and Swedish cooperation.

Biological experiments have been an important part of the Cosmos series: by the end of 1988, eight had been launched, five of which had involved U.S. cooperation. In late September 1987 Cosmos 1887 carried two rhesus monkeys as well as 30 other experiments involving rats, frogs and insects into orbit. The monkeys were called Dryoma (the slow one) and Yerosha (little mischief) – who certainly lived up to his name. Strapped into the cabin, Yerosha managed to loosen his left paw on the fifth day and began pressing buttoms within his reach. Soviet scientists decided to let the mission run its 13-day course, as they realized he could do no harm! Ironically, the vehicle landed 3,000 km (1,875 miles) off target, and the monkeys were rescues from the cold of Siberian mid-winter.

EUROPEAN COMMUNICATIONS SATELLITE (ECS)

Europe's own telecommunications organization, known as EUTELSAT, inherited a highly successful series of comsats known as European Communications Satellites (ECS) on its formation in 1985. ECS have provided 26 European countries with telephone and data links as well as transponders for European TV networks. The ECS satellite series has been so successful that the technology used in the series has become part of other communications satellites. Ironically, the programme got off to a false start in September 1977 when the prototype Orbital Test Satellite, OTS-1, crashed into the Atlantic when its Delta booster exploded just under a minute after launch. The back-up craft, however, known as OTS-2, was successfully sent aloft on 11 May 1978, and placed into a geostationary orbit at 10°E where it could cover Western Europe.

The OTS series was designed as a test vehicle for the later operational ECS system, and was stabilized about three axes (i.e. it didn't spin) and consisted of a service module with a communications module attached. The satellite was just over 2 m (6 ft 6 in) high and stretched over 9 m (29 ft 6 in) across when its solar panels were deployed. OTS-2 provided Europe with its first taste of regional communications and proved itself finacially viable with its test TV and telephone transmissions. In May 1988, it celebrated its tenth year in space and was still successfully operating.

In August 1984, ECS-2 was launched by an Ariane-3. Intended as an in-orbit 'spare', ECS-2 is still in operation today. The spacecraft is 2m high, and with solar panels deployed, 14m across. In geostationary orbit, the craft has a mass of 650 kg.

The ECS satellites were based on the OTS design, and, though twice as large, utilized technology developed for the earlier satellites. ECS-1 was launched by Ariane in June 1983 and could carry 12,000 telephone calls. ECS-2 was launched the following August and had more powerful antenna which could transmit and receive signals across a wider range of frequencies. ECS-3 crashed when the 15th Ariane was lost in September 1985, but two years later its replacement ECS-4 successfully reached geostationary orbit. ECS-5 was launched in July 1988 aboard an Ariane 4.

The common technology developed for both the OTS and ECS series was utilized in the building of the MARECS satellites – 'Marine ECS'. They provide direct-dial, teletype, facsimile and computer modem links for international shipping by two satellites. Operated by the International Maritime Satellite Organization (INMARSAT), one is stationed over the Pacific, the other over the Atlantic.

HIGH ENERGY ASTRONOMICAL OBSERVATORIES (HEAO)

The three High Energy Astronomical Observatories (HEAO) launched by NASA in the late 1970s gave astronomers their first insight into the violent processes that occur in the universe which are visible at only X- and Gamma-Ray wavelengths. Whereas earlier X-ray satellites could only resolve down to the scale of a few minutes of arc (1 arcminute is one 60th of a degree), the HEAOs had a resolution capability well into the arcsecond range. So for the first time, stellar objects other than the Sun could be observed in detail at X-ray wavelengths.

Each of the three HEAO satellites consisted of a large 'telescope' known as a 'grazing incidence telescope'. Because X-rays are easily absorbed by most materials, it is not possible to use lenses or mirrors made of glass to focus them. Metals such as gold or nickel have to be used and are placed in concentric rings rather than at right angles as with lenses. Described as 'mirrors', they allow X-rays to be captured and focussed onto detectors. The mirror surfaces run parallel with the telescope body, because only X-rays that hit the surface within about 2 degrees will be reflected – hence the term 'grazing' incidence. At greater angles, the incoming X-rays bombard the mirror surface and are absorbed. These mirrors have to be polished very accurately – to within one millionth of a metre – or else the X-rays will be absorbed by the surface. At the focus of the telescope were four instruments, two of which were capable of imaging, the others being spectrometers. The mirror had a focal length of 3 m (10 ft) and the whole spacecraft weighed 2,720 kg (5,996 lb).

HEAO 1 was launched on 12 August 1977 from Cape Canaveral by an Atlas Centaur. Within six months, it had mapped the whole of the sky at X-ray wavelengths and more than quadrupled the number of known X-ray objects to a total of over 1,400. But by far the most successful was the second in the series, renamed Einstein in honour of the hundredth anniversary of the famous scientist's birth in November 1978. For the next $2\frac{1}{2}$ years, it returned a wealth of new data, despite a four-month period in 1980 when its attitude control system failed; this soon corrected itself. Einstein gave astronomers the chance to investigate the structures of quasars, supernova remnants and hearts of galaxies previously seen only at radio wavelengths. It also revealed that most types of stars emit X-rays, and the total number of X-ray objects was taken up to many thousands. HEAO-3 was launched in September 1979 and in May 1981 ran out of attitude control fuel, thereby causing the end of its operational life. Its instruments were devoted to more energetic gamma rays.

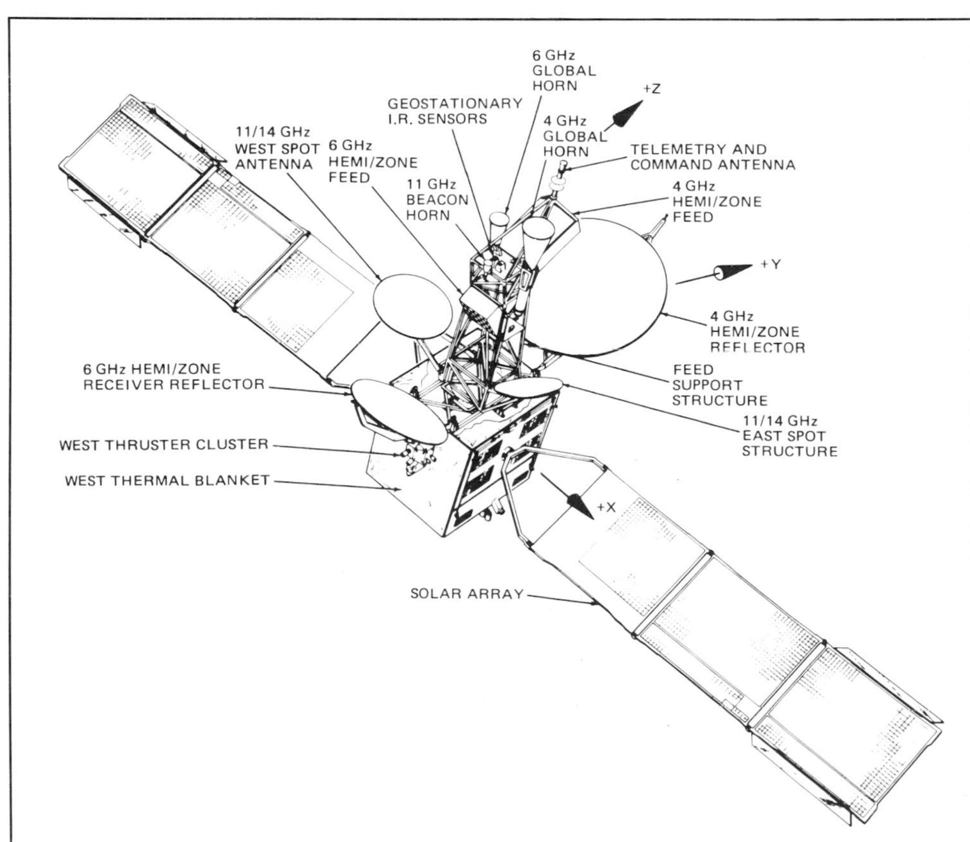

The three-axis INTELSAT 5 is shown here in outline, with its main components. A team of contractors from many countries was responsible for its design, reflecting the international nature of INTELSAT. The many 'spot' beams are shown, as testament to their individual requirements.

INTELSAT

The International Telecommunications Satellite Consortium – known as INTELSAT for short – now operates 16 geostationary communications satellites which provide 165 countries across the globe with most of their TV and telephone transmissions. The United States has the largest market share, and in the early days operated the facilities for the other countries. And though INTELSAT's Headquarters remain in Washington DC, the organization now operates its satellites for itself. The INTELSAT satellites have so many uses that it would not be possible to list them here, but for example, the Live Aid concert in 1985 was seen live in 88 countries thanks to the service the organization offers. Since 1983, the INTELSAT Business Service has provided instant information to the world's business community, and since 1976, a permanent 'hotline' is open between Moscow and Washington in the event of international hostilities.

The first INTELSAT was known as Early Bird and showed that a communications satellite system was a viable proposition. More recently, problems with the Ariane and Shuttle programme have been the cause of delays with the deployment of the 6th generation INTELSAT series. As of time of writing, three of them will be launched on Ariane and two with the Titan Centaur. There had been suggestions that the Soviet Proton launcher might be used, but this no longer seems likely. A further generation of INTELSATS – smaller than the 6th – is expected to be launched in 1992 to provide regional communications with four communications satellites.

As can be seen from the table below, the INTELSATs have grown in size and capacity with each new series.

INTELSAT	First Launch	Launcher	Width	Height	Phone Circuits	TV Channels
1	1965	Delta	0.7 m (2 ft 3 in)	0.6 m (1 ft 11 in)	240	–
2	1967	Delta	1.4 m (4 ft 7 in)	0.7 m (2 ft 3 in)	240	–
3	1968	Delta	1.4 m (4 ft 7 in)	1.0 m (3 ft 3 in)	1,500	–
4	1971	Atlas Centaur	2.4 m (7 ft 10 in)	5.3 m (17 ft 4 in)	4,000	2
4A	1975	Atlas Centaur	2.4 m (7 ft 10 in)	6.8 m (22 ft 3 in)	6,000	2
5	1980	Atlas Centaur or Ariane	2.0 m (6 ft 6 in)	6.4 m (20 ft 11 in)	12,000	2
5A	1985	Atlas Centaur or Ariane	2.0 m (6 ft 6 in)	6.4 m (20 ft 11 in)	15,000	2
6	1989	Ariane 4 or Titan 3	3.6 m (11 ft 9 in)	6.4 m (20 ft 11 in)	120,000	3

INFRA-RED ASTRONOMICAL SATELLITE

On 26 January 1983, the Infra-Red Astronomical Satellite (IRAS) was launched to perform the first all-sky survey at infrared wavelengths. IRAS was an international project with NASA's Jet Propulsion Laboratory providing the telescope, Holland providing the spacecraft, and with mission operations controlled from the United Kingdom. Because the telescope was essentially measuring 'heat', the telescope had to be cooled by liquid helium to within two degrees of absolute zero, the lowest possible temperature, or else the spacecraft would interfere with the measurements. The telescope was essentially a 60 cm (23 in) reflector which focussed infra-red radiation onto an array of 64 very sensitive detectors. The liquid helium which cooled the IRAS optics was to last for 300 days, maximized by thermal insulation of the satellite and a sunshield to minimize external heat.

IRAS was launched by a Delta launcher from the Vandenburg Air Force Base in California into an orbit inclined at 99° to the equator and 900 km

Technicians in the Netherlands prepare the IRAS satellite for launch. Here the main body of the spacecraft is being tested for structural weaknesses. The 'lid' was designed to minimize excess heating from stray sunlight.

Right: Despite recent problems, the LANDSAT series proved the viability of remote sensing. Built by the General Electric Company, the first LANDSAT was based on the Nimbus series of satellites. Its data collection system returned information to Earth from two independent multispectral scanners.

(562 miles) above the Earth's surface. By the time the helium ran out on 21 November, 95 per cent of the sky had been mapped a total of four times. IRAS observed a total of nearly 250,000 objects which has allowed the world's astronomers to reveal a remarkable view of the infra-red sky. In late 1987, it was announced that IRAS had shown that the planet Pluto had methane ice caps. A dust shell surrounding a star called Vega only visible in the infra-red was believed to signal evidence for the formation of a planetary system. Star formation within dark nebulae was also seen, as was the discovery of faint infra-red emission from dust surrounding our galaxy. Beyond our galaxy, IRAS noted that many galaxies emit more infrared radiation than visible light.

INTERNATIONAL ULTRAVIOLET EXPLORER (IUE)

The first astronomical satellite to be launched into a geosynchronous orbit holds the distinction of being perhaps the first satellite to be given a presidential award a decade after its launch! The International Ultraviolet Explorer (IUE) was a collaborative venture between NASA, the United Kingdom and ESA, proposed by Professor Bob Wilson of University College London (UCL), as a European-only project in the late 1960s. However, the European Space Research Organization – a forerunner of ESA – at first declined but eventually joined in after NASA agreed to build the body of the spacecraft. It is essentially octagonal-shaped out of which a 45 cm (18 in) diameter telescope protrudes. A pair of angled-solar panels were provided by ESA and the spacecraft's two spectrometers were built by Professor Wilson's team at University College.

After launch on 26 January 1978 by a NASA Delta at Cape Canaveral, IUE soon held the distinction of being the first 'dual'-operated satellite, with control centres on both sides of the Atlantic. The European ground station is at Villafranca near Madrid, while the Goddard Space Flight Center was NASA's control site. Observing time on the satellite is shared between NASA and a joint ESA-UK scientific committee, with NASA taking two-thirds of the time. UK astronomers are entitled to one-sixth the observing time; because IUE is operating 24 hours a day, this means that eight hours every other day are available to them.

IUE is still operating a decade after launch, and has been described as the longest-lived astronomical satellite ever. All the data taken from the satellite is available in an archive which is open to scientists from around the world, so that discoveries from IUE data have been legion. The explo-

sion of Supernova 1987A – thc nearest supernova explosion to our galaxy for three centuries – allowed IUE to pinpoint the star which exploded in February 1987 – a hot supergiant star, 20 times more massive than the Sun. Another important IUE discovery was the observation that our own galaxy is surrounded by an extensive 'halo' of hot gas which has a temperature of 200,000 degrees Kelvin.

On 10 November 1988, at a ceremony at the White House, Ronald Reagan awarded a prestigious Presidential Design Award For Excellence to the IUE satellite. A total of ten awards out of 500 entries were given, including projects as diverse as a bridge, a transit line and a national memorial. Solar cell degradation at a lower rate than expected means that IUE is expected to operate well until 1990.

LANDSAT

What was planned as a series of satellites to investigate the possibilities of remote sensing from space has been one of the most successful enterprises in space technolgy. LANDSAT 1 – originally called the Earth Resources Technology Satellite 1 (ERTS-1) – was launched into a near polar orbit, such that it circled the Earth 14 times a day. The orbit chosen was Sun-synchronous, which meant that the Sun appeared at the same angle behind the spacecraft when it crossed the equator. This also meant that LANDSAT 1

LANDSAT LOG

	Launch	Orbit	Inclination
1	23 July 1972	901 × 920 km (563 × 575 miles)	99°
2	22 January 1975	907 × 918 km (567 × 573 miles)	99°
3	5 March 1978	900 × 918 km (562 × 573 miles)	99°
4	16 July 1982	689 × 696 km (430 × 435 miles)	98°
5	1 March 1984	705 km (circular) (440 miles)	98°

passed the same point on the globe at the same local time every 18 days. Data from the satellite would show the effects of localized changes on the surface. By May 1973, LANDSAT 1's data storage system had broken down, so that only real-time images could be returned. By January 1978, the satellite finished operations after returning a total of 300,000 pictures. The satellites which followed on from LANDSAT 1 carried more sophisticated scanning instruments.

Each LANDSAT was based on the design of the Nimbus weather satellite (see page 175) with its solar panels running parallel to the main body of the spacecraft which was 3 m (9 ft 10 in) with a 1.5 m (4 ft 11 in) diameter instrument section underneath. Each LANDSAT had a number of multiwavelength scanners, and for the most recent, LANDSAT 5, these consisted of the Multispectral Scanner (MSS) and the Thematic Mapper (TM). The MSS continually scans the ground with a resolution of 80 × 80 m (262 × 262 ft) on the surface across four wavebands covering the visible to the near infrared regions of the spectrum. The Thematic Mapper senses seven wavebands, four similar to the MSS bands, the other three in the infrared. Its maximum resolution is 30 × 30 m (98 × 98 ft) on the surface. The dramatic value of the TM was shown on 29 April 1986, when LANDSAT 5 passed over Chernobyl and returned the first detailed images of the stricken nuclear plant.

The result of the LANDSAT programme has been the return of over two million high resolution views of the Earth, all of which are stored on digital tape at the EROS Data Center in Sioux Falls. The raw data or published images can be obtained there, and scientists in over a hundred countries have done so. Since 1986, the LANDSAT programme has been operated by a company called EOSAT, which will be responsible for a further LANDSAT, to be launched in 1991. In early 1989, the U.S. government threatened to remove its subsidies for EOSAT, but thankfully this was averted.

METEOR

The Soviet Union's weather satellite needs are covered by the Meteor series, now into its third generation of spacecraft. They are launched into polar orbits to provide a daily weather service for Soviet meteorologists with at least three in operation at any time. Each Meteor satellite is cylindrical, carrying meteorological instruments on a scanning platform. Two angled solar panels stretch from either side of the main satellite body, and continually 'lock onto' the Sun to provide power. Two TV cameras and a handful of radiometers provide data that is automatically transmitted to three main data reception stations in the Soviet Union.

The Russians experimented with weather satellites in the mid-1960s, and it is believed that Cosmos 122 was the prototype for the Meteor series – its launch from Baikonur was witnessed by General De Gaulle on 25 June 1966. The first in the Meteor series proper was launched on 26 March 1969, and its orbit inclined at 81° measuring 644 by 713 km (402 × 445 miles) set a pattern for later spacecraft. A second generation Meteor series was ushered in during 1975, equipped with more advanced scanners. In July 1981, a 'Meteor-Priroda' was launched which carried a multi-spectral series of scanners to become Russia's first remote sensing satellite. A third generation Meteor series was introduced in October 1985, and launched into a much higher orbit, to allow greater coverage of the globe.

This artwork shows a Meteor satellite in operation above the Earth. A network of these satellites provides the Soviet Union with data for its weather forecasting services.

METEOSAT

The European Space Agency's Meteosat programme has allowed European weather forecasters to improve their long-range weather forecasts to such an extent that they are the envy of the world. As part of the World Weather Watch, Meteosat was the European contribution to a network of five weather satellites allowing total global coverage. The first was launched in 1977, the second in 1981, and though both were termed 'pre-operational' by ESA, they have been highly successful and more important than prototypes. Each Meteosat is spin-stabilized, making 100 revolutions per minute, and essentially cylindrical in shape. Its scientific instruments included a high-resolution radiometer and a data transmission system. The radiometer is essentially a 40 cm (15 in) aperture telescope which focusses the incoming light onto detectors sensitive to both visable and infra-red light. Because the satellite is spinning, it is able to build up a 'picture' of the Earth every 25 minutes, and returns a new one back to Earth every half-hour.

The detectors allow three spectral bands or 'channels' to be investigated – each reveals different information about the Earth's atmosphere. Pictures at visible wavelengths show the sunlight reflected from the Earth's surface, allowing meteorologists to estimate wind speeds from the motions of clouds. Infra-red pictures show the amount of heat given off from the land and sea, so that their temperature can be calculated. A third channel looks at infra-red radiation from the upper atmosphere, mainly absorbed by water vapour – this allows the humidity of the atmos-

The 'raw' data from Meteosat is familiar to European TV viewers. Here, a picture at 12.25 GMT on 27 June 1986, shows a depression over the Atlantic, whose rainy tendrils have covered the western half of the UK. Water vapour and infra-red data complement the picture returned by Meteosat.

phere to be determined. Each Meteosat also collects local weather information via the data transmission system from automatic weather stations, buoys and aircraft. When combined with the pictures, they allow meteorologists to check the remotely-sensed data with that obtained on the ground.

Meteosat 1 was launched on 23 November 1977, from Cape Canaveral, and entered geostationary orbit above the equator over the Gulf of Guinea. Within a month, it was returning detailed pictures back to Earth and it returned some 40,000 images over the next two years. A day after it had spent two years in space, the radiometer failed, so no more images were returned. When Meteosat 2 was launched in 1981, its radiometer was improved to avoid the earlier failure, but sadly its data transmission system did not work! But in tandem, the first two European weather satellites provided full coverage until the end of 1984. Meteosat 2's images were much sharper and became a regular fixture on TV weather forecasts.

In June 1988, Meteosat P2 was launched on an Ariane 4, and, though still termed pre-operational, returned much valuable information. The launch of Meteosat MOP-1 in March 1989 saw the programme through its operational phase, and is now administered by an organization called EUMETSAT, which numbers 17 member countries, three of which are not ESA member countries.

MOLNIYA

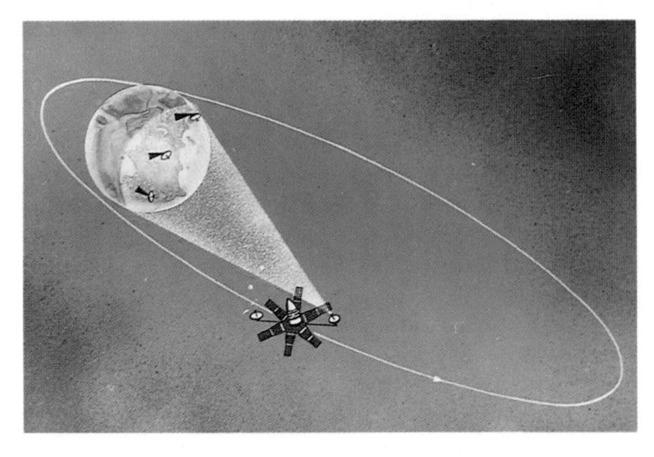

The Soviet Union's telecommunications needs are dictated by its northern location. So the Molniya satellites on highly elliptical orbits provide coverage, each one following on as relays from the next.

The Soviet Union's communications needs are complicated by the northerly location of many of its sparsely-populated regions which would benefit from communications satellites. A 24-hour period geostationary orbit would not allow very strong signals to be transmitted above 65° latitude, so a new orbit is therefore necessary. The answer turned out to be a system of satellites which followed each other in highly elliptical orbits so that one or more would always be over the horizon. The orbits range from 40,000 km (25,000 miles) down to 600 km (375 miles), with the apogee over Siberia and the orbital inclination at 65°. Two orbital passes per day allow a total of 16 hours of communications to be made, relaying Moscow TV to remote regions.

The first Molniya ('Lightning') was launched on 23 April 1965, and by the end of 1988 over 120 had been launched. A prototype seems to have been Cosmos 24 in September 1964, successful enough to allow an average of three per year to be launched until 1987. Each Molniya has a cylindrical body with conical ends to which 0.9 m (3 ft) dish aerials are mounted 180° apart on articulated arms. With six large solar panels, they vaguely resemble a starfish! In 1971 an improved Molniya version was introduced which had greater capacity than the earlier version. This second version was phased out within a decade, only to be replaced with Molniya 3s which had an even greater capacity and the ability to relay colour TV pictures. At present, there are eight Molniya 1s and eight Molniya 3s in operation, separated by 45° with the former concentrating on domestic communications links and the latter on international.

The U.S.S.R. operates a network of geostationary communications satellites under the general heading of 'Statsionar'. A total of 52 TV and communications satellites had been placed into geostationary orbit by the end of 1987. Raduga and Gorizont satellites handle all telephone communications within the Soviet Union and relay international TV coverage (such as 1980's Olympics from Moscow). Another series is called Ekran which relays domestic TV to about 40 per cent of the Soviet Union. The Statsionar network provides telecommunications for over 40 countries as part of the INTERSPUTNIK organization, a Soviet version of INTELSAT, set up in 1971. As well as Eastern bloc countries, Vietnam, Afghanistan and Algeria are now members.

NIMBUS

The Nimbus series of satellites was launched from the Vandenburg Air Force Base into polar orbits from which they monitored the Earth's surface with mainly infra-red sensors. Originally, the series was planned as a weather observation series (Nimbus is the Latin word for cloud). However, as technology improved, the series developed during the 1970s to cover a variety of Earth sciences such as oceanography, geology and cartography.

The basic design of the satellites was the same: a central instrument section was attached to a pair of near vertical solar panels giving them something of a butterfly shape. The central section consisted of Earth-pointing scanners at the bottom and control systems at the top. The satellite's orientation in orbit was maintained by

No.	Launch Date	Inclination	Weight	Orbit
1	28 August 1964	98°	376 kg	422 × 932 km (264 × 583 miles)
2	15 May 1966	100°	413 kg	1,100 × 1,181 km (688 × 738 miles)
3	14 April 1969	99°	575 kg	1,070 × 1,131 km (669 × 707 miles)
4	8 April 1970	107°	675 kg	1,093 × 1,107 km (683 × 692 miles)
5	11 December 1972	99°	768 kg	1,089 × 1,102 km (683 × 689 miles)
6	12 June 1975	99°	827 kg	1,092 × 1,104 km (682 × 690 miles)
7	24 October 1978	99°	907 kg	943 × 953 km (589 × 596 miles)

NIMBUS LOG

gas jets. Each successive satellite was heavier than the previous one, indicating the greater range of experiments carried. The greater power requirements were also met by two thermonuclear generators.

NIMBUS 1

The first in the Nimbus series proved beyond doubt that weather data was easily obtained on the ground with simple portable stations as the satellite passed into range. Over the month of its operation, the satellite returned 27,000 high resolution TV pictures at optical and infra-red wavelengths. For the first time it tracked hurricanes, including Hurricane Cleo.

NIMBUS 2

After the success of the first Nimbus, an infra-red radiometer was added to the satellite's inventory of experiments. This observed the Earth at five different wavelengths stretching from the visible into the infra-red to enable studies of water vapour and carbon dioxide across the globe. The satellite's TV pictures were received by over 300 ground stations.

Nimbus 5 contained 40,000 components which included a camera system and automatic shutter. A typical exposure was 40 milliseconds, enough to produce an 800-line TV picture on its vidicon sensor.

NIMBUS 3

A third Nimbus satellite had been launched on 18 May 1968, but there was a launch failure. Just under a year later, its replacement successfully reached orbit and carried an infra-red spectrometer which could measure temperatures with an accuracy of 2°C at sea level.

NIMBUS 4

Nimbus 4 carried a total of nine experiments, most of which were improved instruments developed from earlier flights. However, it also carried an automatic location monitor which 'homed into' signals from weather balloons, buoys and even a bear in Montana. As it passed overhead, the satellite 'called up' electronic devices attached to them and could therefore pinpoint their movement. This was useful for measuring atmospheric and ocean currents and – in the case of the bear – animal behaviour!

NIMBUS 5

Six new instruments comprised the experiments flown on Nimbus 5. Improved infra-red spectrometers meant that it could measure atmospheric water vapour and temperature profiles through cloud. It also monitored the Gulf Stream off the coast of South America for the benefit of shipping in the area.

NIMBUS 6

The main experiment for the sixth in the series was monitoring iceberg movements and seasonal changes across the North Pole in order to site oilrigs and pipelines there. In 1978 it was used to track the progress of an Arctic explorer, and two years later, that of a lone yachtsman across the Pacific.

NIMBUS 7

The last in the series of Nimbus satellites dramatically highlighted mankind's growing pollution of his environment. Its eight scientific instruments were designed to monitor ozone concentrations in the upper atmosphere, pollution of the seas and chlorophyll distribution in the coastal regions.

NOAA

The U.S. National Oceanic and Atmospheric Administration (NOAA) has launched a number of weather satellites which provide the United States Weather Bureau the data needed to produce accurate forecasts. There are essentially two types of satellite operated by NOAA: the polar orbiting NOAA series which bear the

Copernicus was launched in 1972, and operated for longer than planned. Like most satellites, its lifetime depended on the attitude control fuel which kept its orientation. Degradation of solar panels is another manifestation of wear and tear, due to atmospheric oxygen particles.

administration's name; and geostationary GOES satellites which form the American contribution to the World Weather Watch.

The first American weather satellites were named TIROS, or the Television and Infrared Observation Satellite, the first of which was launched on 1 April 1960. It operated for 78 days and returned nearly 23,000 pictures of the Earth's weather systems. A total of ten TIROS satellites were launched until TIROS 10 in July 1965, which operated for two years. A further six satellites were launched from 1966 to 1969, known as the TIROS Operational Satellites (TOS), and were managed by the forerunner of NOAA, the Environmental Science Services Administration (ESSA). Around 400 ground stations around the world used the data from the TOS satellites along with those from the Nimbus series. In the early 1970s, a further series of satellites were launched, and though originally termed the Improved Tiros Operational System, the satellites were designated as NOAA after launch. NOAA 1 was launched on 11 December 1970 into an orbit inclined at 101° to the equator. With the launch of NOAA 6 in June 1979, a further generation of polar orbiting satellites was introduced, along with the launch of a further series, the TIROS N satellites.

During the early 1970s, NASA launched two prototype geostationary weather satellites known as the Synchronous Meteorological Satellites (SMS) which were then taken over by NOAA. They were re-named the Geostationary Operational Environment Satellites (GOES), the first of which was launched in May 1974, and stationed over the East Atlantic to begin with. The seventh in the series was launched in February 1987, after the loss of its predecessor the previous year when its Delta launcher failed. Each of the GOES satellites carried a radiometer which measures visible and infra-red radiation across the globe, similar to those used on Meteosat.

The next generation of NOAA satellites will be more advanced polar orbiters, which will measure radiation from the oceans and land with greater accuracy. NOAA has also plans for developing a polar platform as part of the International Space Station, *Freedom*.

ORBITING ASTRONOMICAL OBSERVATORIES (OAO)

Though only two of the four Orbiting Astronomical Observatories were successful, the experience NASA gained with the series paved the way for further astronomy satellites. The OAO satellites were designed to make observations at mainly ultraviolet wavelengths of the stars and the matter between them. OAO 1 failed to operate beyond its third day in orbit when its battery failed: OAO 3 failed to reach orbit. However, OAO 2 operated successfully from 1968 to 1973, and OAO 4 from 1972 to 1980. In 1973, the latter was renamed Copernicus because of the 500th anniversary of the great astronomer. The re-naming was apt, for the satellite prompted modern astronomers to change their perception of the Universe.

Ultraviolet radiation is that which exists beyond the range of the human eye, and has a shorter wavelength than violet light, hence its name. Most of the UV radiation which reaches the Earth's surface is absorbed by atmospheric oxygen and ozone. So it is necessary to venture beyond the atmosphere to observe UV radiation. The main result from Copernicus was that there were fewer 'new' sources in ultraviolet light than at X-ray or Gamma wavelengths. But spectroscopy at ultraviolet wavelengths is very valuable because many abundant elements have their strongest spectral lines in this region. Thus Copernicus was able to measure concentrations of hydrogen in dust clouds between the stars. Copernicus also carried a British-built X-ray telescope which monitored the 200 or so known sources at the time. In particular, it was used to monitor the Cygnus X-1 star system in which a black hole is believed to exist.

SOLAR MAX

The first satellite to be repaired in orbit and redeployed was launched on Valentine's Day 1980 and known as the Solar Maximum Mission, or Solar Max for short. The Sun undergoes a definite cycle of activity every 11 years, and the satellite's launch was timed to coincide with a period of greater activity. Solar Max carried seven instruments which concentrated on solar flares, sudden flare-ups from the Sun's uppermost layers which release vast amounts of energy equivalent to several megaton H-bomb explosions. Their purpose was to find where the bursts occur and what triggers the release of energy. Over 1,500 flares were observed by the Solar Max, and the seventh instrument which monitored the Sun's energy output found that the rate of energy given off by our star was constant. The spacecraft was launched into an orbit inclined at 28.5° from Cape Canaveral on a Delta, but a failure in its attitude control system in December 1980 effectively rendered it useless.

In April 1984, astronauts George Nelson and Jim van Hoften performed a successful search and rescue operation on the stricken Solar Max satellite. Ironically, NASA decided to shut the satellite down five years later to save money!

The 11th Shuttle mission (*Challenger* on 41-C) was devoted to the rescue of Solar Max, and was launched into an orbit slightly below the 569 km (355 miles) mark of the satellite. On 7 April 1984, using the MMU which had been test flown on the previous flight, George Nelson (himself an astronomer) attempted to reach Solar Max which was slowly tumbling in space, but failed to latch a trunnion pin docking device onto a 'grappler' built into the satellite for just such a rescue. Try as he might to latch onto the satellite, he failed. Two days later, Commander Bob Crippen nudged *Challenger* underneath SMM and by use of the remote manipulator arm, the Solar Max was slowly lowered into the payload bay. Nelson, along with James van Hoften, donned spacesuits and within an hour had replaced the failed electronics of the attitude control box. An experiment which had failed was also replaced by the astronauts. On 12 April 1984, Solar Max was nudged back into orbit, and after a month of checking out its systems, NASA controllers declared the satellite to be fully operational. It was estimated that replacing the satellite would have cost around $250 million, while the Shuttle rescue had cost a mere $50 million. Solar Max continues to return much more information on the Sun (it also observed Halley's Comet in 1986) though, sadly, NASA has now decided not to bring the craft back to Earth as originally planned.

SPOT SATELLITE PROBATOIRE d'OBSERVATION DE TERRE

On 22 February 1986, the French space agency, CNES, launched the first commercial remote sensing satellite outside of the superpowers. Known as SPOT, it soon lived up to its name as 'the satellite for Earth observation'. Within months it was in the news: its high resolution images of the Chernobyl reactor in the Soviet Union revealed smoke and fire at the reactor site. By the end of 1987, it had returned 500,000 images of the Earth's surface which are marketed through a company called SPOT Image. The first SPOT is in a Sun-synchronous, almost polar (98.7°) orbit at a height of 840 km (525 miles) much lower than the LANDSAT satellites. Its instruments have a resolution of 20 m (65 ft) on the ground, compared with 30 m (98 ft) for LANDSATs 4 and 5. Its orbit is designed so that it will repeat its cycle of observations over the same area every 26 days. Looking directly below, SPOT can cover an area of 60 × 60 km (37 × 37 miles) in three spectral bands. Its cameras can also view 'sideways' to produce three dimensional maps of regions.

The French government gave the go-ahead for SPOT in 1980, to try to corner the European market in remote sensing. The development programme was funded by CNES, which actually owns the satellite, though SPOT Image was set up to market its images with centres in France, Sweden and the U.S. Very soon it was competing with LANDSAT for a share in the market, but the demand in the U.S. has been less than expected. Nevertheless, CNES will launch two more SPOT satellites in the early 1990s. A second generation series of SPOTs is planned for the mid-1990s, to be equipped with more accurate sensors to monitor vegetation for crop forecasts and environmental studies.

TRACKING AND DATA RELAY SATELLITE SYSTEM (TDRSS)

The successful deployment of the second TDRSS satellite by the crew of *Discovery* in October 1988, was a piece of good luck in a programme that has been dogged by problems from the very start. The programme was designed to allow NASA's spacecraft in Low Earth Orbit to relay their telecommunications without the need for extensive ground stations. For many years, NASA has operated a vast, worldwide Space Tracking and Data Network of ground stations which relay data to and from Mission Control. But each station provides communications support for only a small fraction (at most 20 per cent) of a user satellite's lifetime. The TDRSS network will cover nearly the entire orbital period of a satellite's operations.

The TDRSS satellites are the largest privately-owned comsats ever built. Each weighs around 2,267 kg (5,000 lb) and spans more than 17 m (57 ft) from solar panel to solar panel. Each uses two large radio dishes for telecommunications, each of which is 4.8 m(16 ft) in diameter and covers 12.8 m (42 ft) when fully deployed. Along with smaller dishes, TDRSS has seven antennae in all. TDRS-1 was launched on *Challenger's* maiden flight when its booster rocket failed to fire properly, but after two months of delicate manoeuvring, the satellite was coaxed into a geostationary orbit over Brazil. The second TDRS satellite was lost in the *Challenger* accident in January 1986.

Technicians at the Cape prepare TDRS-1 for launch by attaching it to its IUS launcher (at bottom). The gold-coloured antenna is stored at the top of this view. Unfortunately the IUS misfired, putting the first element of TDRSS into a useless orbit.

With only one TDRS in orbit, and no spacecraft to communicate with, TDRS-1 was used during July 1986 by an international team of radio astronomers. TDRS-1 was electronically linked with radio telescopes in Japan and Australia to locate extragalactic objects with unprecedented accuracy – a technique known as Very Long Baseline Interferometry. Sadly, on 28 November 1986 an antenna was lost on TDRS-1.

It was not until 30 September 1988, that the crew of *Discovery* successfully deployed TDRS-3 over the Pacific Ocean just south of Hawaii. The next TDRS satellite was launched from the Shuttle in March 1989 to replace TDRS-1 over the Atlantic, though it will actually be used as a 'spare'. Both spacecraft will allow 23 satellites to communicate with the Earth simultaneously, such as the Hubble Space Telescope. Two further launches are planned for 1990 and 1992 to prepare for the increased traffic due to the International Space Station.

FUTURE SOLAR EXPLORATION

Perhaps the most worrying aspect of the *Challenger* accident has been the 'knock-on' effect in delaying a number of Solar System exploration missions. The missions mentioned in this section have suffered as a result of the delays but will be launched within the next few years – all the spacecraft mentioned have been built and await their launch. Though both ESA and NASA have other spacecraft in the pipeline, experience shows that they are often victims of budgetary and launch vehicle problems.

GALILEO

The next phase of mankind's exploration of the giant planet Jupiter and its moons will be undertaken by NASA's Galileo mission, due for launch in October 1989. By the time it arrives there, it will have been nearly two decades since the project was first proposed. The mission has been repeatedly altered because of cost overruns with the Space Shuttle and the hiatus which followed the loss of *Challenger* in 1986. The choice of booster used to launch the spacecraft has also caused delays, so much so that Galileo will take six years to reach its target because it is using a booster for which it was not originally designed. After arriving at Jupiter in 1995, Galileo will spend 21 months in orbit around the giant planet, and will answer many of the puzzles which Voyager has posed.

The project was first proposed in the early 1970s as a follow-on to the Voyagers and was known as the Jupiter Orbiter and Probe. The spacecraft was planned as a Voyager-based orbiter plus a probe to be deployed into the planet's atmosphere based on those used in the Pioneer Venus missions. The project was approved by Congress in 1977, and later renamed after the discoverer of Jupiter's four largest moons. Though it appears similar to the Voyager spacecraft, the Galileo orbiter contains a revolutionary new feature known as the 'spin-despin' mechanism. Some instruments (such as the imaging cameras) will remain on a three-axis stabilized platform, while others – the fields and particles instruments – will be 'spun' in circular motion around the main hub of the spacecraft. To maintain radio contact with Earth, the orbiter uses a 4.87 m (16 ft) diameter dish and radioisotope thermonuclear generators for power. The instruments carried on the orbiter are similar to those carried on Voyager, although the TV experiment is called the Solid State Imaging Experiment. A Charge Couple Device (CCD) will be used instead of a vidicon tube – it has a greater spectral response and photometric accuracy, allowing more detailed TV pictures to be taken.

After arriving at Jupiter, the orbiter will make 11 different orbits of the planet, with each altered by the Galilean moons to allow at least one close pass of each moon in turn. The Galileo orbiter will also fly through Jupiter's magnetotail, 'downwind' of the planet's interaction with the solar wind. 150 days before the orbiter arrives, the Galileo Probe will be despatched into the planet's atmosphere, aimed to enter the atmosphere at 6°N, returning the first direct measurements. Travelling at many kilometres per second, the spacecraft will decelerate as it reaches the upper atmosphere and then a parachute will unfurl to slow it down so that it will make further measurements for at least an hour. The atmospheric pressure and temperature will be measured as the probe descends, so the probe will reveal the exact structure of the atmosphere. A nephelometer ('cloud-meter') will be used to measure the size and chemical composition of cloud particles. An infra-red detector will measure the temperatures of the clouds and observe the amount of sunlight within each cloud deck.

Above: In October 1995, the Galileo atmospheric probe will be dropped into the seething atmosphere of a giant planet. Here technicians check the centre of gravity of the probe. Its 'nose-cone' is a heatshield designed as protection against the fiery heat of entry into the Jovian atmosphere.

Opposite: This artwork shows Ulysses as it would have appeared had it been launched in 1986. The gold-laminated body of the spacecraft is seen beneath its communications dish. Because of the Challenger accident, Ulysses will now be launched by a solid rocket booster, the IUS. Astronaut crews were worried about flying a potential 'liquid bomb'.

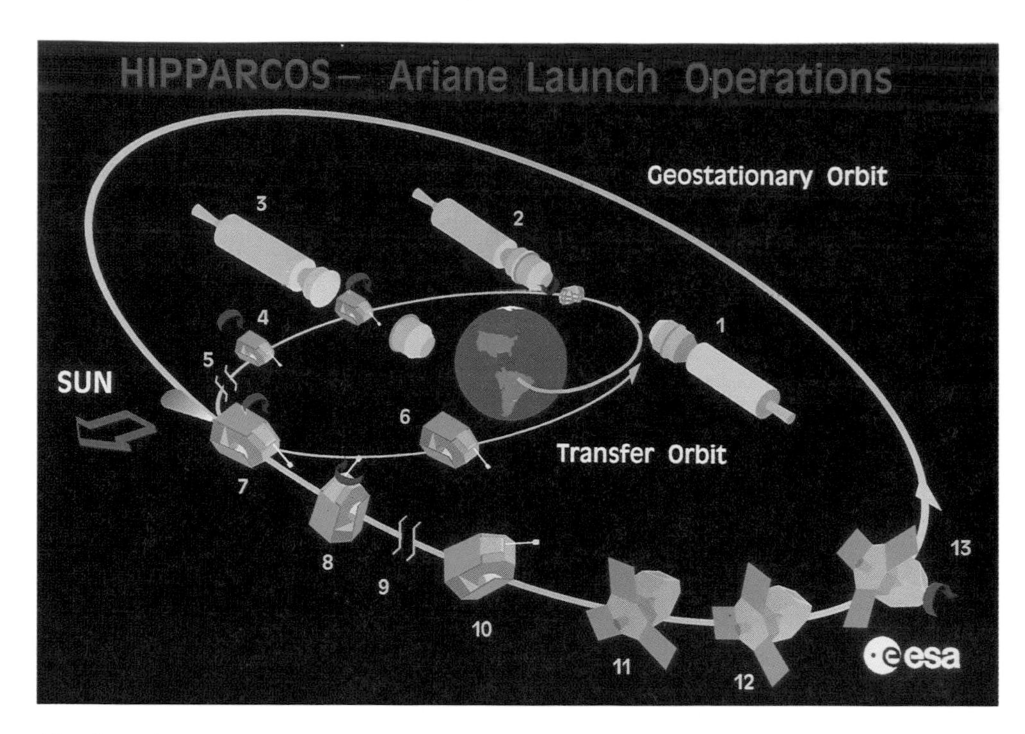

After launch by an Ariane-4, HIPPARCOS will enter a transfer orbit before heading up to geostationary orbit. The transfer orbit phase lasts for at least three orbits, during which time its systems will be tested and checked out.

Another instrument will search for lightning, measuring the static and light associated with such violent discharges of electricity.

When originally planned in the late 1970s, Galileo was to have been launched from the Shuttle in 1982. However, delays meant that the launch was repeatedly altered, a state of affairs compounded by arguments over which booster to use. Originally, a three-stage Inertial Upper Stage (IUS) had been chosen, but this was later changed to the more powerful liquid-fuelled Centaur booster. It was then assigned to a two-stage IUS and then the Centaur again, for a planned launch in August 1986 on Shuttle mission 61-G. The *Challenger* accident raised fears about the wisdom of carrying a liquid-fuelled booster such as the Centaur, which some astronauts regarded as little more than a bomb. So now, Galileo will be launched by the IUS, which is much less powerful than the Centaur. So Galileo will take a circuitous route to reach Jupiter known as Venus-Earth-Earth-Gravity Assist (VEEGA). The only way that Jet Propulsion Laboratory engineers can get Galileo to Jupiter is to launch the spacecraft inwards towards the Sun, so that it picks up speed by making two passes of the Earth and then heading outwards towards Jupiter.

Launch from Earth will take place on Shuttle mission STS-34 in October 1989, with the launch window running from 9 October to 24 November. Galileo will head inwards towards Venus, which will pass in February 1990. The increase in solar radiation means that a 'solar shield' has had to be built to protect the spacecraft. However, the Galileo orbiter's instruments are ideal for observing Venus and its atmosphere, which is a small measure of comfort to planetary scientists! Galileo will then head past the Earth in December 1990, heading out towards the asteroid belt and making a close fly-by of asteroid Gaspra in October 1991. The instruments on board the orbiter will hopefully return the first detailed information on an asteroid. Galileo will then pass the Earth for a final time in December 1992, before passing through the asteroid belt (and a close encounter with asteroid Ida on 27 August 1993) before arriving in Jupiter orbit on 7 December 1995. The delay in launching Galileo has caused so many problems that the spacecraft has had to be stored at the Jet Propulsion Laboratory in Pasadena until mid-1989. The power supplied by the RTGs is already decaying, so much so that by the time the spacecraft reaches Jupiter, there may be not enough power to operate all the instruments simultaneously.

HIPPARCOS

The European Space Agency's HIPPARCOS satellite, launched in August 1989 by an Ariane 4 rocket into a geostationary orbit, has ushered a new era in space-based astronomy. HIPPARCOS will allow the world's astronomers to build up a better picture of the stars and their distances to allow a highly accurate 'stellar population' census to be determined. A greater knowledge of a star's distance means that its luminosity can be determined, from which its mass, temperature and radius may be deduced.

HIPPARCOS is designed as an astrometric satellite: astrometry is the science of accurately measuring the positions of astronomical objects. To do this, the satellite uses an ingenious optical system consisting of a reflecting telescope some 29 cm (11 in) in diameter which receives light from two fields of view, separated by 58 degrees, and then focusses onto a highly sensitive photoelectric detector. The satellite slowly spins on its axis once every two hours to stabilize itself, so a particular star seen in one field of view will enter the next some time later, and by this means the optical detector can accurately determine its position. All stars seen in the two fields of view (measuring only 0.9×0.9 degrees) will have their relative separations in the direction of scan measured with a high degree of precision. The direction of the satellite's axis of rotation can also be changed so it effectively 'precesses' like a spinning top, and will eventually build up a map of the whole of the sky.

The satellite has been named after the acronym High Precision Parallax Collecting Satellite. Parallax is the apparent movement in an object's position when compared to more distant objects as the Earth moves in its orbit around the Sun. This same technique can be used to measure the distances of the nearer stars, but the parallax angles involved are very small. It wasn't until the 1830s that optical equipment was sufficiently accurate for astronomers to perform such measurements. An additional consideration is that the stars themselves are actually moving; they possess their own 'proper motions', a fact that was obvious from the first telescopic observations. Certain stars appear to move from their 'expected' positions quite appreciably over the space of a few tens of years. The introduction of photographic techniques in the late 19th century allowed even greater accuracy in astrometric measurements, but ultimately, the Earth's atmosphere limits their accuracy. At present, the best accuracy is 0.01 seconds of arc (a degree is made up of 60 minutes of arc, which themselves contain 60 seconds of arc). HIPPARCOS will increase that accuracy to 0.002 arc seconds, equivalent to the apparent angular size that a golf-ball held up in London would appear to an observer in New York!

An 'input catalogue' of 120,000 stars will form the basis of HIPPARCOS operations: these are the stars which will be investigated by the satellite. The catalogue has been compiled from existing astrometric measurements obtained from the ground. The satellite has an expected lifetime of about $2\frac{1}{2}$ years, but control systems and electronics are duplicated so it may well last twice this time. As HIPPARCOS will take six months to complete one survey of the whole sky, it will help refine knowledge of the exact positions and motions of the 120,000 candidate stars. A secondary experiment on HIPPARCOS known as TYCHO, will make photometric measurements of nearly 400,000 stars to allow astronomers to infer many of their physical properties mentioned above. Because of the complicated way in which the satellite takes its measurements, it will take two years before the first data catalogues can be produced: the final catalogue will not be ready before 1995.

HUBBLE SPACE TELESCOPE

Aptly described as the most sophisticated scientific satellite ever built, the joint NASA/ESA Hubble Space Telescope will be finally launched in 1990. The project began in earnest in the early 1970s, and was officially started in 1977 when its funding began. In October 1983 it was christened after the pioneering American astronomer, Edwin P. Hubble, who showed in the 1920s that the Universe extended much further than the edges of our Milky Way. The Hubble Space Telescope should have been launched in October 1986, but was delayed in the aftermath of the *Challenger* accident.

The spacecraft is essentially a large cylinder 4.7 m (15 ft 5 in) in diameter and 14.3 m (47 ft) in length with two large solar panels running parallel to the main body of the spacecraft. The heart of the telescope is its 2.4 m (7 ft 10 in) mirror – the largest ever sent into space – which reflects light onto a variety of detectors, described below. The telescope's control and computing facilities are located behind the main mirror, surrounding the detectors. Above the Earth's atmosphere, the telescope will be able to 'see' much further,

British Aerospace in Filton, Bristol, were responsible for the construction of Hubble's solar panels. They will unfurl in the same way that kitchen blinds do! Each 'wing' covers 33 m² (356 sq ft) and contains nearly 50,000 solar cells. Because of an initial error, the cells have had to be replaced because Hubble's power consumption will be greater than originally calculated.

viewing galaxies and objects up to 14 billion light years distant. As the light has taken 14 billion years to reach us, Hubble will allow astronomers to look 'back in time' at how the Universe began and its early stages of formation and evolution. Hubble Space Telescope will also be able to distinguish fine details with ten times the clarity of the best views obtained from the ground. Views of the outer planets as good as those returned from the Voyager spacecraft are expected.

The five instruments are located behind the main mirror and can be brought into focus by the use of another mirror. Even a summary of their uses shows how they will revolutionize astronomy.

HIGH RESOLUTION SPECTOGRAPH

Spectrographs are used by astronomers to analyse light emitted from astronomical objects so that their temperature, chemical composition, density and velocities can be accurately measured. The High Resolution Spectrograph carried on Hubble looks at the ultraviolet regions of the spectrum, light which is absorbed by the Earth's atmosphere. It will detect objects a thousand times dimmer than those seen to date by spacecraft carrying ultraviolet telescopes. Its fine resolution means that the individual stars within clusters can be seen for the first time, as well as separating the stars within binary systems.

WIDE FIELD/PLANETARY CAMERA

Unlike the other HST instruments, the WF/PC can operate in two modes. The 'wide field' capability allows large areas to be observed, allowing the accurate plotting of the location and structure within galaxies and quasars. It also has a high resolution or 'planetary' imaging capability to allow the outer planets to be studied in detail. This mode will enable astronomers to make detailed observations of the structure of galaxies, nebulae and star clusters.

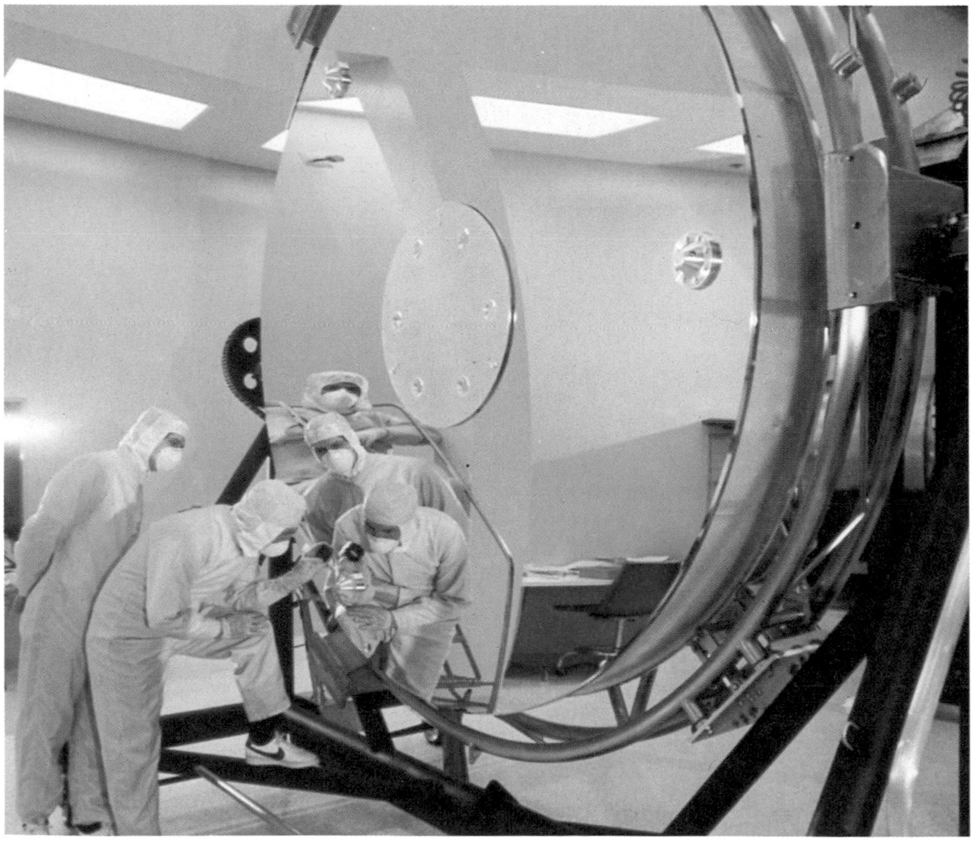

Hubble's Eyes – a 2.4-m (7.9-ft) diameter mirror – is seen after it was silvered and polished in 1981. Launch was then set for 1986, but the *Challenger* accident has delayed this until at least 1990.

FAINT OBJECT SPECTROGRAPH

This instrument complements the High Resolution Spectrograph, providing highly detailed spectroscopic data on very faint objects, but with much less resolution. The FOS will be used to obtain spectra of very faint objects at ultraviolet and visible wavelengths. This will allow, for example, comets to be seen before they are affected by the solar wind and observe quasars which are believed to be the active nuclei of certain galaxies.

HIGH SPEED PHOTOMETER

This instrument is designed to observe the total light emitted from astronomical objects and the variation of their light output over time. The HSP will be able to observe fluctuations in brightness within 10 millionths of a second. As such, it will be used to 'calibrate' the light from very faint stellar objects, allowing them to make precise observations of pulsars, supernovae and binary stars. It will also be used to look at stars in the last stages of their evolution – such as white dwarf stars, neutron stars as well as matter being drawn into black holes.

FAINT OBJECT CAMERA

This instrument, developed by the European Space Agency, can capture images too faint to be seen by the other detectors on Hubble. It will be used to measure the light intensities of extremely faint galaxies and observe closer objects with much greater detail across the wavelength range from ultraviolet to the red part of the spectrum. Astronomers will be able to investigate the possibility of planets around other stars and massive black holes within globular star clusters.

Because it is making such detailed measurements, without the ability to 'lock onto' stellar objects Hubble would be useless. So three fine guidance sensors are used to point the spacecraft towards a desired target within a hundredth of an arcsecond – a level of unprecedented accuracy. Once established at that position, the telescope can be maintained there for over 24 hours to allow all five instruments in turn to look at the object in question.

The delay forced by the *Challenger* accident has been a blessing in disguise, mainly because the spacecraft has been updated on the ground. Less than half the computer programmes needed to operate Hubble's instruments were written by

After arriving in orbit around Venus, Magellan will use its powerful radar to map the whole of the planet's surface. One mission scientist has described it as 'wrapping a noodle around Venus'. The dish antenna (a Voyager spare) will alternatively be used to transmit tape-recorded data back to Earth and then acquire further data.

mid-1986, so the delay has allowed the software to be refined and upgraded. It was also realized that the solar panels would not provide sufficient energy for all the telescope systems, so they have been upgraded by British Aerospace, the prime contractor. Hubble has been stored at Lockheed's facility in Sunnyvale, California, with a launch now scheduled for STS-31 in March 1990. Astronauts Bruce McCandless and Kathy Sullivan will deploy the 12-ton telescope from the Shuttle payload bay, while astronomer-astronaut Steve Hawley will oversee operations from inside the Orbiter *Atlantis*. The Space Telescope will be released in a far higher orbit than planned, nearly 600 km (370 miles) above the Earth. Because its deployment has been delayed, the effects of increased solar activity are coming into play. The Earth's atmosphere has effectively 'ballooned', so the telescope's orbit has been increased to minimize problems with atmospheric drag.

It is hoped that by early 1990, Hubble will be in full operation. NASA has created the Space Telescope Science Institute in Baltimore, Maryland, as the control centre for the telescope. After a three-month 'checkout' period in early 1990, Hubble will be ready to begin its work. The telescope has been designed to be continually upgraded, allowing any failing instruments to be replaced at later dates by astronauts. In December 1988, NASA announced that another instrument will be added to the telescope in 1994 and replace one of the five instruments described above. The Near Infrared Camera and Multi-Object Spectrometer (NICMOS) will thus become the first infra-red instrument on Hubble.

MAGELLAN

NASA's first Shuttle-launched planetary mission, the Magellan Venus radar mapper, was safely despatched in May 1989 after launch from *Atlantis* on STS-30. Another mission to have been beset by delays, it started out as the Venus Orbiting Imaging Radar mission (VOIR – an apt acronym, the French verb 'to see'). It was later modified to a much smaller spacecraft, and renamed Magellan, after the first explorer to circumnavigate the Earth. The spacecraft was built 'on the cheap' using a spare Voyager radio dish and back-up equipment from the Galileo spacecraft by Martin Marietta. The 3.7 m (12 ft) Voyager radio dish will be used for radar imaging as well as telecommunications with the Earth. An equipment bay known as the Forward Equipment Module is located underneath the main dish and contains the electronics for the radar system among other components. Underneath the module is a ten-sided main spacecraft bus which contains the main computers and other subsystems. At the base of the spacecraft is a Star-48 solid rocket motor which will be used to allow the spacecraft to enter orbit around Venus. Power is provided by solar panels attached to the main spacecraft body.

Magellan was delivered to Cape Canaveral in October 1988, after which a curious mishap took place during routine ground tests of the spacecraft's sub-systems. Technicians were checking power systems before the scientific instruments were added. Unfortunately, a design fault in the batteries led to a short circuit when two electrical connectors were joined. According to the official NASA report, the technician responsible 'suddenly observed sparks which caused him to jump backward. Sparks, flames, and smoke were immediately observed.' Luckily, the fire was contained and the Magellan electrical mishap was rectified, with a resultant design change to the batteries themselves. The spacecraft was mated with *Atlantis* in early March 1989, preparing for the launch window which was only 20 days long and opened on 25 April. Magellan had to be launched during that period and in the small hours of 5 May, it was safely on its way.

Magellan was boosted away from *Atlantis* with the relatively unpowerful Inertial Upper Stage. It will not arrive at Venus until August 1990, when it will enter a near-polar orbit after which the spacecraft's systems will be checked out. Its Synthetic Aperture Radar is one of the most advanced radar mapping instruments ever built, and transmits a radar pulse across a 'strip' of the planet's surface at right angles to the spacecraft's direction of flight. The strength and nature of the returned radar signal depends on the roughness of the surface and its composition. So radar data can be used to piece together a map of the surface strip by strip. Magellan will orbit the planet every 3.15 hours from 250 km to 1,900 km (156 to 1,187 miles) and its SAR will operate for about 40 minutes on each orbit. The radar instrument will map a swath 24 km (15 miles) wide and 16,000 km (10,000 miles) long during that time.

The mission will last for a nominal Venusian year – 243 days – and it is hoped that a total of 1,852 swaths will be made covering 90 per cent of the total surface area. However, as experience with most space missions has shown, the technology lasts for much longer than planned, so it may be possible to repeat up to nine 'mapping cycles'. The limiting factor is the attitude control fuel which points the spacecraft's antenna in the correct direction. The SAR will hopefully return images of the surface down to 300 m (984 ft) across and 100 m (328 ft) vertically at the surface. The more mapping cycles that are made, the more complete the coverage of the planet.

In the early 1990's NASA's Mars Observer will put into place the final pieces of the planetary jigsaw that is Mars. The large booms beneath the spacecraft are part of the magnetometer which will return badly-needed data about the Martian core.

MARS OBSERVER

In the 1990s, a new generation of Mars-bound spacecraft will be launched, the first of which is NASA's Mars Observer. Designed as the first in a new series of planetary observer spacecraft, the basic variant will be used for later exploration of the Moon, Venus, and the asteroids. Though both NASA and the Soviets are planning more advanced sample return missions and rovers for the late 1990s, plans for these missions are still sufficiently fluid that they are liable to change. Certainly, both NASA and the Soviets will use the data returned from Mars Observer in planning these later missions.

The spacecraft was originally proposed as the Mars Geoscience/Climatology Orbiter, designed to investigate the question of water and climate on Mars. It had been intended to launch the craft in 1990. The spacecraft is based on RCA's Satcom series, and shares many design features which have minimized development costs. The instruments flown on the spacecraft point downwards to the planet, so a costly scanning platform has been avoided. The Mars Observer is scheduled to be launched in September 1992, and will use a Transfer Orbit Stage (TOS) to boost it towards the red planet. The TOS is a new upper stage which shares similar technology to the IUS and uses solid propellants. Because of delays with the Shuttle, the Mars Oberver will be launched to Mars by a Titan 3 – the TOS is compatible with both.

After a journey of 353 days, the spacecraft will enter a highly elliptical orbit around Mars. After a few months, once its onboard systems have been checked, the orbit will be changed to a low – 883 km (224 miles) – polar orbit that is Sun-synchronous. As seen from the spacecraft, the Sun will be at the same angle whenever it crosses the equator – this means that changes in atmospheric conditions and on the surface can be monitored, as they will become apparent by comparison to earlier observations. The instruments carried by the Mars Observer will survey the whole of the planet in about 50 days, and because the mission is scheduled to last for a Martian year (687 days) over 12 observation cycles may be made. The spacecraft will be able to monitor seasonal changes of the weather in the atmosphere and on the surface.

The Mars Observer carries eight instruments, including a line-scanning TV camera which will monitor the planet's weather, and return detailed views of the surface down to 1.5 m (4 ft 11 in) on the surface. Two instruments are of particular note, which should clear up some of the more pressing problems about Mars. The first is a magnetometer, located on a 6 m (20 ft) boom well away from the spacecraft which will measure the strength of the planet's magnetic field, which is not well understood. A better knowledge of the magnetic field will reveal information about the planet's interior and core. A gamma ray spectrometer will return data about the minerology of the planet's surface, poorly understood at present.

ULYSSES

Mankind's first look at the poles of the Sun will be possible due to a joint NASA/ESA project called Ulysses, due for launch in October 1990. Yet another victim of the delays caused by the *Challenger* accident, Ulysses also fell foul of budgetary problems in the early 1980s with the American space programme. At that time, the plan was for two spacecraft called the International Solar Polar Mission (ISPM), which would fly over the north and south poles of the Sun respectively. One craft would be provided by ESA, the other NASA. Sadly, NASA cancelled its spacecraft in 1982, so ESA had to 'go it alone' in building a single spacecraft. However, NASA did not back out completely, as NASA provided the launch facilities, tracking and radioisotope thermonuclear generator for the spacecraft. The project was renamed Ulysses after the reference in Dante's *Inferno* to a world beyond the Sun which the hero wanted to explore.

The vagaries of international cooperation ensured further setbacks with the loss of *Challenger* in 1986, as Ulysses was scheduled for a launch in May 1986, in the same launch window that would have seen Galileo despatched to Jupiter. This is because Ulysses will make use of the powerful gravitational field of Jupiter to swing it underneath the south pole of the Sun. With the return to flight operations of the Shuttle, it has been decided that Galileo will use the first launch window to reach Jupiter in October 1989, which means Ulysses will not be launched until a year later. It will reach Jupiter in February 1992, passing within half a million km (312,500 miles) of the planet's north pole and head downwards, at right angles to the orbital plane of the planets. Ulysses will observe the south polar regions of the Sun (below 70°S latitude) from May to September 1994 before passing over the north pole (again, above 70°N) between May and September of the following year. The mission is planned to last a total of five years.

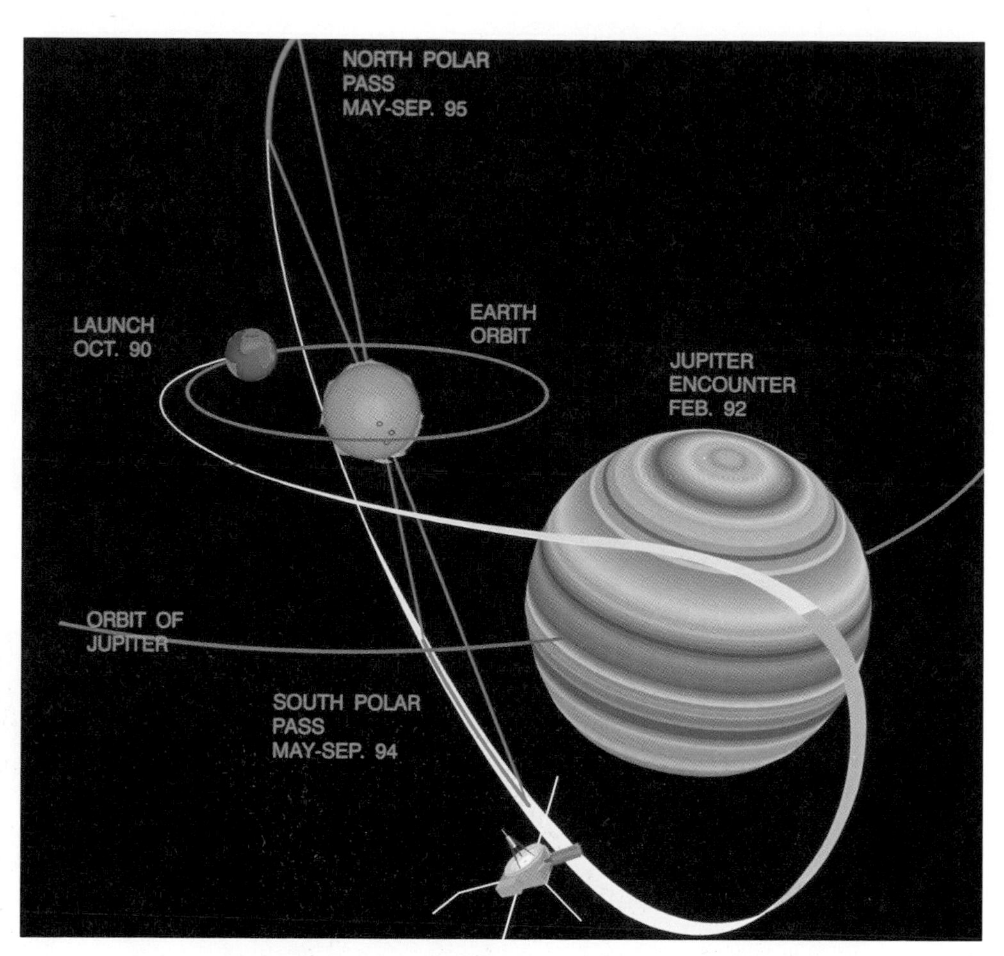

The route the ESA spacecraft Ulysses will take on its way to the south pole and the north pole of the Sun.

The Ulysses spacecraft is spin-stabilized, and carries a total of nine experiments. They are designed to look at the solar wind and how it is generated in the upper reaches of the Sun's atmosphere, the solar magnetic field and its interaction with the interplanetary field, as well as the solar plasma environment and solar and galactic cosmic rays. Ulysses will allow the first three-dimensional picture of solar activity to be made and will allow scientists to test their theories that the solar wind is generated from holes in the Sun's upper atmosphere. Because the craft is heading within Jupiter's dangerous radiation fields, its components have been 'radiation hardened' to ensure they are not damaged by Jupiter's radiation belts.

MILESTONES

Even the briefest look into the history of space travel will show that the rate of progress has increased remarkably over the past few years. As an easy guide to the history of space travel, the following chapter provides a summary of the events that have taken place. It is not an exhaustive list, merely intended as a quick and easy reference.

850 The Chinese use gunpowder to make fireworks to celebrate religious festivals.

1686–1687 Sir Isaac Newton's *Principia Mathematica* published in which he states his laws of motion.

1799 The battle of Seringapatam in which British forces are routed by the rockets of the forces of Tippu Sultan.

1807 Admiral Sir William Congreve, after adapting Indian rockets at Woolwich Arsenal, uses them in the Napoleonic wars at Boulogne.

1814 British use Congreve rockets to attack Fort McHenry in Baltimore during the War of Independence.

1857 Konstantin Tsiolkovskii born in Russia.

1865 Jules Verne's novel *From The Earth To The Moon* published.

1882 Robert Goddard born in Massachusetts.

1883 Tsiolkovskii's *Free Space* published, in which he shows that a rocket will work in vacuum as a result of 'action-reaction' propulsion.

1894 Hermann Oberth born in Germany.

1901 Author H.G. Wells uses a substance with 'anti-gravity' properties to launch men to the Moon in *The First Men in The Moon*.

1903 Tsiolkovskii produces *Exploring Space With Devices* in which he discusses liquid propellants.

1909 Goddard's studies lead him to conclude that liquid hydrogen and oxygen would make for efficient propulsion.

1919 Goddard submits *A Method Of Attaining Extreme Altitudes* for publication by the Smithsonian Institution.

1923 Hermann Oberth's *The Rocket Into Interplanetary Space* published.

1924 Tsiolkovskii's *Cosmic Rocket Trains* discusses the concept of multi-stage rockets for the very first time.

1925 Walter Hohmann describes the principles on which interplanetary flight is based in *The Attainability of Celestial Bodies*.

1926 Goddard successfully tests the world's first liquid-fuelled rocket at Auburn, Massachusetts.

1927 German rocket enthusiasts form the Society for Space Travel (VfR) which holds regular meetings at the Golden Sceptre alehouse.

The Gas Dynamics Laboratory (GDL) is established in Leningrad.

1928 Oberth becomes consultant to Fritz Lang's *Frau Im Mond* to build a rocket for premiere publicity. A prototype is built which later explodes on the launchpad.

1931 On 14 May VfR successfully launch a liquid-fuelled rocket named Repulsor 1 which reaches the height of 60 m (200 ft).

1932 Von Braun and others demonstrate a liquid-fuelled Mirak rocket to the German Army at Kummersdorf, south of Berlin. Though it crashed before its parachute could open, von Braun later becomes a civilian employee charged with developing liquid-fuelled rockets under Captain Walter Dornberger, himself in charge of solid-fuelled rockets within the Army.

1933 Soviets launch first hybrid solid/liquid fuelled rocket which reaches a height of 400 m (1,312 ft) after launch near Moscow.

1936 Scientists at the California Institute of Technology begin rocket experiments in an arroyo north of Pasadena. This is the birth of the Jet Propulsion Laboratory.

1937 Von Braun's group move to specially-built rocket test establishment at Peenemunde on the Baltic Coast of Germany.

1938 Goddard begins to develop high-speed pumps for liquid-fuelled rocket engines.

1942 The first successful launch of the A-4

The dramatic moment of launch is captured in this remarkable view of *Challenger* on the Spacelab 2 mission on 30 October 1985.

rocket from Peenemunde reaches a height of 85 km (53 miles), 190 km (120 miles) downrange.

1944 In September, the first operational A-4, renamed the V2, is launched against London, the first of over a thousand.

1945 Arthur C. Clarke proposes the concept of the geostationary orbit for global telecommunications in an article in the October issue of *Wireless World*.

On 5 May, Peenemunde is captured by the Red Army but most facilities have already been destroyed by the personnel. Von Braun and Dornberger have already left, to be captured by the U.S. Army and relocated at White Sands Proving Ground in New Mexico under 'Operation Paperclip'.

1946 German rocket engineers arrive in Moscow to begin work with Soviet research groups. Sergei Korolev builds rocket using V2 technology.

The first V2s shipped from Germany arrive at White Sands to be test-fired under U.S. Army supervision.

1947 Russians begin launch tests of V2 missiles at Kapustin Yar.

1949 On 24 February a two-stage rocket is launched from White Sands. It consists of a V2 to which an American WAC-Corporal upper stage is added.

1951 The first German technicians to help the Russians are repatriated.

1952 Von Braun proposes a wheel-shaped space station which becomes the inspiration for countless science fiction authors and artists.

1955 Construction of the Baikonur Cosmodrome begins in earnest.

President Eisenhower announces that a small scientific satellite will be launched as part of the International Geophysical Year (IGY). The Navy Vanguard Project is chosen after it had been previously announced that the Army Jupiter C would be the launcher.

1957 Soviet Union launches the world's first artificial satellite, Sputnik 1, on 4 October.

Sputnik 2 is launched on 3 November with the dog, Laika, the first animal in space.

America's first attempt to launch a satellite fails when the Vanguard launcher explodes on the launch pad on 6 December.

1958 Explorer 1 is successfully launched on 31 January and enters the history books as America's first satellite. It detects the van Allen belts of radiation around the Earth.

The National Aeronautics and Space Administration (NASA) is formed on 1 October to oversee the U.S. space programme. Its first major project is the manned Mercury Project.

1959 Soviets launch the first probe to hit the Moon, Luna 2, in September.

In October, Luna 3 circumnavigates the Moon and returns the first pictures of the far side.

1960 First applications satellites launched by NASA, Tiros 1, weather satellite, in April, Echo 1, communications satellite, landed in August.

Two dogs (Strelka and Belka) in Sputnik 5 become the first animals to be successfully returned to Earth in August.

1961 Yuri Gagarin becomes the first human being to orbit the Earth in his Vostok 1 capsule on 12 April.

Alan Shepard becomes the first American to enter space on a short sub-orbital flight on 5 May. Three weeks later, President Kennedy declares that America will send a man to the Moon and return him safely before the end of the decade.

Gherman Titov spends more than a day in space (25 hours) in the second Vostok manned flight in August.

1962 John Glenn becomes the first American to orbit the Earth in his Friendship 7 Mercury capsule on 20 February.

In July, NASA launches Telstar 1, the world's first privately financed communications satellite.

Andrian Nikolayev in Vostok 3 and Pavel Popovich in Vostok 4 enter the history books as the first dual flight in August.

NASA's Mariner 2 becomes the first human artefact to reach another planet when it flies past Venus in December.

1963 The first woman to fly in space is Valentina Tereshkova in June in Vostok 7, another dual flight with Valeri Bykovsky in Vostok 6.

The Soviets launch their first meteorological satellite, Cosmos 23, in December.

1964 At the end of July, NASA's Ranger 7 returns the first close-up TV pictures of the surface of the Moon.

Syncom 3 becomes the first successful communications satellite to operate from geosynchronous orbit.

The first three-man crew (Komarov, Feoktistov and Yegorov) orbits the Earth in Voskhod 1.

1965 Alexei Leonov becomes the first human being to walk in space after leaving the Voskhod 2 capsule on 18 March.

NASA's first two-man Gemini mission (Gemini 3 – Grissom and Young) becomes the first manned spacecraft to manoeuvre and change orbit.

Mariner 4 becomes the first successful probe to fly past Mars in July, returning 11 pictures of the surface.

In December, Schirra and Stafford in Gemini 6 rendezvous to within a few metres of Borman and Lovell in Gemini 7.

1966 Luna 9 becomes the first probe to make a soft-landing on the surface of the Moon on 31 January.

Armstrong and Scott in Gemini 8 perform the first docking with an unmanned target vehicle.

On 3 April, Luna 10 becomes the first successful probe to enter orbit around the Moon.

1967 Astronauts Grissom, White and Chaffee are killed when a fire engulfs their Apollo capsule in a launch pad rehearsal at Cape Canaveral.

Vladimir Komarov is killed when his Soyuz 1 spacecraft crashes to the ground after its parachute fails to deploy properly.

Cosmos 186 and 188 become the first unmanned probes to automatically dock in October.

1968 Peter Glaser suggests building solar power satellites in Earth orbit to beam energy to the ground by microwaves.

Soviets successfully recover unmanned Zond 5 capsule in September after it circumnavigates the Moon and lands in the Indian Ocean.

Apollo 8 (Borman, Lovell and Anders) becomes the first manned spacecraft to orbit the Moon on Christmas Eve.

1969 In mid-January, Soyuz 4 (Shatalov) and Soyuz 5 (Yeliseyev, Khrunov and Volynov) achieve the first successful docking by manned spacecraft. Yeliseyev and Khrunov transfer to Soyuz 4 for return to Earth.

In March, Apollo 9 crew (McDivitt, Scott and Schweickart) test the Lunar Module in Earth orbit for the first time.

On 20 July, Neil Armstrong and Buzz Aldrin successfully land the Apollo 11 Lunar Module on the surfacc of the Moon. Armstrong is the first human being to walk on the surface of another planet.

In early August, NASA's Mariner 6 and 7 fly past Mars, returning pictures of its southern hemisphere.

1970 In April, the crew of Apollo 13 (Lovell, Haise and Swigert) survive a fuel tank explosion en route to the Moon and are safely returned to Earth.

In September, the Soviets launch Luna 16 which becomes the first unmanned probe to return soil samples from the Moon. Two months later, Luna 17 despatches the first remote-controlled roving vehicle onto the Moon's surface (Lunokhod 1).

1971 The crew of Soyuz 11 (Dobrovolsky, Volkov and Patsayev) perish when a pressure valve malfunctions and the cabin

Academician S. Koroloyov with Yuri Gagarin prior to the launching in September 1961.

Right: Neil Armstrong, Michael Colins and Ed Aldrin awaiting pickup after splashdown of the Apollo 11 Spacecraft Command Module on 24 July 1969. The fourth man in the life raft is a U.S. Navy underwater demolition swimmer.

atmosphere is lost on re-entry. They had been the first crew to work in a space station – Salyut 1.

NASA's Mariner 9 becomes the first probe to enter orbit around Mars in late November. At the same time, the Soviet Mars 2 lander crash lands: and at the start of December, the Mars 3 lander ceases transmission after it lands, due to a fierce dust storm.

1972 On 5 January, President Nixon announces his approval of the Space Transportation System – better known as the Space Shuttle.

On 3 March, NASA launches Pioneer 10, destined to become the first object to leave the Solar System.

In July, the first Earth Resources Technology Satellite (ERTS) is launched. Later renamed LANDSAT, it is the first in a series of highly successful remote sensing satellites.

On 19 December, the crew of Apollo 17 (Cernan, Schmitt and Evans) returns to Earth, signalling the end of the Apollo project.

1973 NASA's Pioneer 11 is launched on 6 April, following in the trail of Pioneer 10 which becomes the first probe to reach Jupiter in December.

On 14 May, NASA launches its first space station, Skylab. Despite damage to the craft during launch, the station is successfully repaired by astronauts and is home to three different crews over the next year.

Vasili Lazarev and Oleg Makarov successfully test the modified Soyuz 12 capsule, later used to ferry crews to and from Salyut space stations.

1974 Mariner 10 is the first probe to fly past Mercury in March. Its orbit around the Sun is designed so that it encounters Mercury again in September 1974 and March 1975.

On 30 May, ATS-6 is launched which beams educational TV programmes to the Indian sub-continent as an experiment.

In December, Pioneer 11 reaches Jupiter, and then is re-directed towards Saturn for an encounter scheduled for five years later.

1975 On 5 April, Lazarev and Makarov survive the first ever manned launch malfunction when the upper stage of their booster fails to separate.

In July, the Apollo–Soyuz Test Project takes place with the first joint Superpower docking of manned spacecraft.

In October, the Soviet Union's Venera 9 and 10 landers return the first pictures of the surface of Venus.

1976 On 20 July, NASA's Viking 1 lander successfully touches down on the surface of Mars, returning a wealth of data. Viking 2 repeats the success with a landing in early September.

Below: Pioneers 10 and 11 carried this famous plaque in the event they were apprehended by a alien civilization.

In August, Indonesia becomes the first south-east Asian country to operate its own telecommunications satellite, named Palapa.

1977 Starting in August, NASA conducts five airborne tests of the Space Shuttle *Enterprise* after release from a 747 over the Mojave desert in California.

NASA launches Voyagers 1 and 2 on their more advanced journeys through the outer Solar System.

1978 In January, the Soviet Union launches the first Progress craft carrying supplies and other cargo for crews onboard Salyut 6.

In March, Vladimir Remek (Czech) becomes the first Soviet bloc cosmonaut to fly in space on Soyuz 28 and spends a week onboard the Salyut 6 station with his Soviet hosts.

In December a 'flotilla' of spacecraft arrive at Venus. NASA's Pioneer Venus 1 and 2 enter orbit and each drop four probes onto the planet. Russia's Venera 11 and 12 successfully despatch probes to the surface.

1979 Between February and August, Vladimir Lyakhov and Valeri Ryumin spend a total of 175 days on Salyut 6, setting a new record for long-duration space flight.

In March, Voyager 1 reaches Jupiter, producing a wealth of discoveries. Four months later, Voyager 2 follows in its wake.

In July, the empty Skylab space station re-enters the atmosphere over the barren outback of Australia.

On September 1, Pioneer 11 becomes the first human artefact to reach Saturn and continues on its way out of the Solar System.

On Christmas Eve, the European Space Agency successfully launches its Ariane booster from a launch site in French Guiana.

1980 Between April and October, Leonid Popov and Valeri Ryumin extend the long-duration record to 184 days. During their stay on Salyut 6 they are visited by a number of crews, including the first Hungarian, Vietnamese and Cuban cosmonauts.

In November, Voyager 1 reaches Saturn, revolutionizing our view of the multi-ringed planet and its retinue of moons. Voyager 2 reaches Saturn the following August.

1981 On 12 April, John Young and Bob Crippen are launched in the Space Shuttle *Columbia* on its maiden test flight. They return two days later to a perfect landing in California.

The second flight of a reusable spacecraft takes place in November when Dick Truly and Joe Engle fly *Columbia* into space again.

1982 In March, two further Venera craft (13 and 14) land on the surface of Venus and return colour TV pictures.

Salyut 7 is launched in April, and from May to December a new long-duration record is established of 211 days by Lebedev and Berezevoi. At the end of June, they are briefly joined by the first Western European in space, French *spationaute*, Jean-Loup Chrétien. In August, they are joined by the second woman to venture into space, Svetlana Savitskaya.

After the fourth successful flight of *Columbia* (27 June–4 July) NASA declares the Space Transportation System 'fully operational'. In November, *Columbia* takes to the skies for a fifth time with a four-man crew.

1983 On 25 January, the Infra-Red Astronomical Satellite (IRAS) is launched from Vandenburg in California. It operates for 300 days in Earth orbit and results in a plethora of new discoveries.

In June 1983, Pioneer 10 passes the orbit of Neptune, so has left the Solar System altogether – the first probe to do so.

The Space Shuttle *Challenger* comes into operation with the sixth mission in the series in April. On its next flight in June, one of the five crew members is Dr. Sally Ride, America's first woman in space.

On 27 September, cosmonauts Titov and Strekalov survive the first known explosion on a launch pad with a crew onboard. One of the main engines of their Soyuz T-7 booster catches fire but they fire the craft's emergency rocket system which successfully pulls them away from the explosion.

In October, the next in the Venera series (15 and 16) enter orbit around Venus and begin their radar mapping mission.

At the end of November, the first European Spacelab is flown onboard *Columbia* with ESA astronaut Ulf Merbold as one of the six crew members.

1984 On 25 January, President Reagan announces that he has directed NASA to begin development of a space station, with international partners.

On 7 February, Bruce McCandless becomes the first human satellite when he flies the manned Maneuvering Unit, a self-propelled backpack out of *Challenger*'s payload bay on the 10th Shuttle flight.

On 8 February, cosmonauts Kizim, Solovyev and Atkov are launched in their Soyuz T-10 craft to dock with Salyut 7. They spend some 236 days in orbit, establishing a new long-duration record.

1985 En route to Halley's Comet, the Soviet VeGa probes deploy French-built balloons into the atmosphere of Venus in June.

During the summer months, Vladimir Dzhanibekov and Viktor Savinykh bring back the Salyut 7 space station 'to life' after its power systems fail.

In September, NASA's International Cometary Explorer (ICE) becomes the first man-made spacecraft to fly through the dusty tail of a comet.

1986 On 24 January, Voyager 2 becomes the first human artefact to reach Uranus, discovering new rings and new moons around the planet.

On 28 January, the Shuttle *Challenger* explodes 73 seconds after lift-off, killing all seven crew members. Further Shuttle flights are indefinitely postponed.

In February, the Soviet Mir space station is launched and is later occupied by Kizim and Solovyov in March. They use their Soyuz T-15 craft to fly to Salyut 7 and return to Mir before landing back on Earth.

In early March a 'flotilla' of probes reach Halley's Comet, including two Japanese (Sagikake and Suisei) and the Soviet VeGas 1 and 2. ESA's Giotto probe becomes the first to journey within 500 km (312 miles) of the icy nucleus at the heart of the comet.

1987 In February, Yuri Romanenko and Alexander Laveikin are launched in the latest Soyuz variant, TM-2, to Mir. Romanenko spends a record 326 days in orbit, returning on 29 December.

On 15 May, the Soviet Energia is test launched, the most powerful booster ever built which will allow the Soviet Union to advance its plans for the industrialization of space in the next century.

1988 In September, the Space Shuttle successfully returns to flight operations with the launch of *Discovery* and its crew of five.

In November the Soviet shuttle Buran makes its unmanned maiden flight attached to the Energia booster and returns safely to Earth.

On 21 December, Musa Manarov and Vladimir Titov return to Earth after spending a year in space.

1989 On May 5, Magellan is safely despatched to Venus – the first in a 'new wave' of planetary probes for the 1990s.

On 24 August, Voyager 2 reaches Neptune, the eighth planet from the Sun – and its final planetary encounter, before heading towards the stars.

Right: The long-familiar sight of a Soyuz launch has become a regular occurrence.

SUMMARY

Year	Event	Spacecraft	Country
1957	Launch of first satellite	Sputnik 1	U.S.S.R.
1959	First probe to reach the Moon	Luna 2	U.S.S.R.
1960	First applications satellites	Tiros 1 Echo 1	U.S.A.
1961	First man in space (Yuri Gagarin)	Vostok 1	U.S.S.R.
1962	First probe to reach Venus	Mariner 2	U.S.A.
1963	First woman in space (Valentina Tereshkova)	Vostok 6	U.S.S.R.
1964	First close-up pictures of the Moon's surface	Ranger 7	U.S.A.
1965	First probe to reach Mars	Mariner 4	U.S.A.
1966	First probe to land on the Moon	Luna 9	U.S.S.R.
1967	First death in space (Vladimir Komarov)	Soyuz 1	U.S.S.R.
1968	First men to fly around the Moon (Borman, Lovell and Anders)	Apollo 8	U.S.A.
1969	First men to land on the Moon (Armstrong and Aldrin)	Apollo 11	U.S.A.
1971	First operational space station	Salyut 1	U.S.S.R.
1972	First probe to orbit Mars	Mariner 9	U.S.A
1973	First probe to reach Jupiter	Pioneer 10	U.S.A.
1974	First probe to reach Mercury	Mariner 10	U.S.A.
1975	First international manned mission	Apollo–Soyuz Test Project	U.S.A./ U.S.S.R.
1976	First probe to land on Mars	Viking 1	U.S.A.
1978	First Soviet-bloc cosmonaut (Vladimir Remek)	Soyuz 28/ Salyut 6	U.S.S.R.
1979	First probe to reach Saturn	Pioneer 11	U.S.A.
1981	First flight of re-usable spacecraft	*Columbia* STS-1	U.S.A.
1983	First probe to leave the Solar System	Pioneer 10	U.S.A.
1984	First Human Satellite (Bruce McCandless)	*Challenger*	U.S.A.
1985	First probe to reach a comet	ISEE-3/ICE	U.S.A.
1986	First probe to reach Uranus	Voyager 2	U.S.A.
1987	First permanently-occupied space station	Mir	U.S.S.R.
1988	Longest manned spaceflight – 365 days (Titov and Manarov)	Mir	U.S.S.R.
1989	First probe to reach Neptune	Voyager 2	U.S.A.

General Index

Figures in *italics* refer to illustrations

Index of Astronauts

Index of Vehicles and Equipment